LINEAR PROGRAMMING

Active Set Analysis
and Computer Programs

LINEAR PROGRAMMING
Active Set Analysis and Computer Programs

MICHAEL J. BEST

Department of Combinatorics and Optimization
University of Waterloo
Waterloo, Ontario N2L 3G1
Canada

KLAUS RITTER

Institut für Angewandte Mathematik und Statistik
Technische Universität München
8000 München 2
West Germany

Prentice-Hall Inc., Englewood Cliffs New Jersey 07632

Library of Congress Cataloging in Publication Data

Best, Michael J.
   Linear programming.

   Bibliography: p.
   Includes index.
   1. Linear programming.   I. Ritter, K. (Klaus).
II. Title.
T57.74.B47   1985   519.7'2   84-26595
ISBN 0-13-536996-7

Editorial/production supervision: *Mary Carnis*
Cover design: *20/20 Service, Inc.*
Manufacturing buyer: *Anthony Caruso*
Page layout: *Diane Heckler Koromhas*

Although each computer program has been tested by the authors, no warranty, express or implied, is made by the authors or by the publisher, Prentice-Hall, Inc., as to the accuracy and functioning of the program and related program material, nor shall the act of the distribution constitute any such warranty, and no responsibility is assumed by the authors in connection therewith.

ISBN 0-13-536996-7 01

Prentice-Hall International, Inc., *London*
Prentice-Hall of Australia Pty. Limited, *Sydney*
Editora Prentice-Hall do Brasil, Ltda., *Rio de Janeiro*
Prentice-Hall Canada Inc., *Toronto*
Prentice-Hall Hispanoamericana, S.A., *Mexico*
Prentice-Hall of India Private Limited, *New Delhi*
Prentice-Hall of Japan, Inc., *Tokyo*
Prentice-Hall of Southeast Asia Pte. Ltd., *Singapore*
Whitehall Books Limited, *Wellington, New Zealand*

To   *Laurie, Gillian, David, Kerri*

and

Liselotte, Catharina, Christiane

CONTENTS

PREFACE

The essential idea of a solution method for a linear programming problem is to move from one extreme point of the feasible region to an adjacent one in such a way as to improve the objective function value. In virtually any presentation of the subject, this is illustrated by graphical examples. Our method of presentation is new in that it is based on a *direct* generalization of this geometric notion. The traditional approach of the simplex method forces the use of a higher-dimensional problem. In our experience, this makes it more difficult to appreciate the simplicity of underlying geometrical principles.

It is our belief that the formulation of any algorithm should be unambiguous and sufficiently detailed so that a reader with basic computer programming experience should be able to write a program to implement the algorithm in a short period of time. For this reason, each of our algorithm statements includes a formal statement of the problem to be solved, the initial data requirements, and a precise statement of the general iteration.

Linear programming is a special case of a more general optimization problem having a nonlinear objective function and nonlinear constraints. For these nonlinear problems, there is a well-developed theory concerned with optimality conditions and duality. Our presentation of the theory of linear programming provides a natural introduction for analogous topics in nonlinear programming.

Prerequisites and Scope

The prerequisite for this text is elementary linear algebra: matrix and vector manipulation, and linear independence.

We have endeavored to provide the reader with a complete presentation of the essential aspects of linear programming. We also include a thorough presentation of some of the more advanced topics, such as parametric methods, dual methods, upper bounding, and generalized upper bounding methods. We have included a sampling of the many applications of linear programming in a variety of areas, such as production planning and graduation of actuarial data. Detailed model development is given in two cases taken from civil engineering and economic analysis. This gives the reader examples of the importance of the modeling process.

We include an appendix giving computer programs for all of the algorithms. The appendix also gives the output from these programs corresponding to example problems from the main text. Following the presentation of an algorithm and an example in the main text, a reference is made to the relevant computer program in the appendix.

Linear programming is a broad area. Necessarily, there are some areas that could not be included: network flows, transportation problems, and decomposition. We have chosen to express the solution of linear equations in terms of inverse matrices. It is beyond the scope of this book to discuss how numerical analysis may be used to perform computations more efficiently and to reduce round-off errors.

Figure P.1   Chapter interdependence.

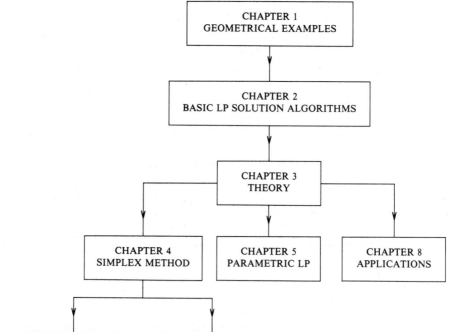

Course Presentation

The interdependence of chapters is shown as in Figure P.1.

A basic course would consist of Chapters 1 through 4 with selected applications from Chapter 8. As time permits, material from Chapters 5, 6, and 7 can be included.

Acknowledgments

We take pleasure in acknowledging the patient and helpful assistance of R. J. Caron and R. A. Cornale.

Michael J. Best

Klaus Ritter

January 1985

GEOMETRICAL EXAMPLES

The geometry of linear programming plays a major role throughout this text. In this chapter, we present two graphical examples to illustrate properties of an optimal solution and an algorithm for obtaining an optimal solution.

1.1 PROPERTIES OF AN OPTIMAL SOLUTION: EXAMPLE

We begin our analysis of a linear programming problem with a simple two-dimensional example.

Example 1.1

$$
\begin{array}{lrcll}
\text{minimize:} & -\ x_1\ -\ x_2 & & & \\
\text{subject to:} & x_1\ +\ 3x_2 & \le & 9, & (1) \\
& 2x_1\ +\ x_2 & \le & 8, & (2) \\
& -\ x_1 & \le & -1, & (3) \\
& -\ x_2 & \le & -1. & (4)
\end{array}
$$

The *objective function*[1] for this problem is $f(x) = -x_1 - x_2$. Letting

$$
x = \begin{bmatrix} x_1 \\ x_2 \end{bmatrix},
$$

---

[1] The *italicized* terms are introduced informally in this chapter. Precise definitions are given in subsequent chapters.

the *feasible region* for this example is

$$R = \{ x \mid x \text{ satisfies constraints (1) to (4)} \}.$$

A point is *feasible* for the problem if it is in R and is *infeasible* otherwise. Each constraint divides 2-space into two half-spaces: one being the set of points for which the left-hand side is "less than or equal to" the right-hand side and the other corresponding to "greater than or equal to." The line for which equality holds is the common boundary of the half-spaces. Figure 1.1 shows a graph of R. An optimal solution for the example is a point in R for which the objective function value is smallest.

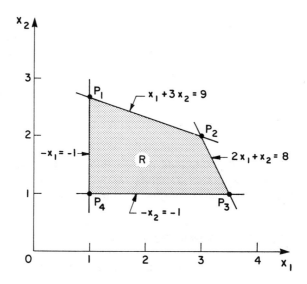

Figure 1.1   Feasible region for Example 1.1.

Because this example has only two variables, we can obtain the optimal solution graphically as follows. Figure 1.2 shows R and the line $[f(x) =] -x_1 - x_2 = -3$. The shaded subregion of R consists of points for which $f(x) < -3$. The heavily shaded subregion shows points of R for which $f(x) < -4$. The objective function can be decreased by moving the line $-x_1 - x_2 = $ a constant up and to the right until the shaded region is reduced to a single point. The resulting point is the optimal solution $P_2 = (3, 2)'^{2}$ and the minimum objective function value is -5.

Two critical observations can be made from Example 1.1. These observations will be generalized in subsequent sections.

---

[2] Prime ($'$) denotes transposition.

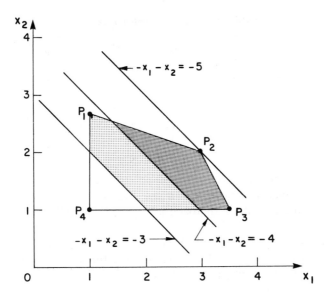

Figure 1.2   Optimal solution for Example 1.1.

Observation 1.

The optimal solution occurs at an *extreme point* (geometrically a "corner" point) of the feasible region.

R for Example 1.1 has four extreme points labeled P_1 through P_4 in Figure 1.1. P_1 is the solution of the two simultaneous linear equations

$$x_1 + 3x_2 = 9,$$
$$-x_1 \qquad = -1,$$

and P_2 is the solution of

$$x_1 + 3x_2 = 9,$$
$$2x_1 + x_2 = 8.$$

A constraint is said to be *active* at a point if the left-hand side is equal to the right-hand side, *inactive* if strictly less than, and *violated* if strictly greater than. Table 1.1 shows the active constraints and their associated objective function values for each of the extreme points in Example 1.1.

Based on Observation 1, one can immediately formulate an algorithm for the solution of a linear programming problem. Compute all extreme points and their corresponding objective function values. The one having the smallest objective function value is optimal. This algorithm is not unreasonable for two variable problems having a small number of constraints. Each extreme point requires the solution of two linear equations in two unknowns. However, the solution of some of the linear equations may have to be discarded because it violates one or more of the remaining inequality constraints.

TABLE 1.1   Extreme Points for Example 1.1

Extreme Point	Active Constraints	Objective Function Value
$P_1 = (1, 8/3)'$	1, 3	$-11/3$
$P_2 = (3, 2)'$	1, 2	-5
$P_3 = (7/2, 1)'$	2, 4	$-9/2$
$P_4 = (1, 1)'$	3, 4	-2

Furthermore, enumeration of extreme points becomes exponentially tedious (and computationally expensive) as the number of problem variables increases. A more organized approach is required and will be developed subsequently.

Before making the second observation, we require the geometrical notions of gradient and cone. The *gradient* of a function of one or more variables is the vector of first partial derivatives of the function. For example, the gradient of the linear function $3x_1 + 5x_2$ is the 2-vector

$$\begin{bmatrix} 3 \\ 5 \end{bmatrix}.$$

The inequality constraint $3x_1 + 5x_2 \leq 15$ divides the plane into two half-spaces. The gradient of this constraint function can be drawn at their common boundary by observing that it is orthogonal to the line $3x_1 + 5x_2 = 15$. Furthermore, since the gradient points in the direction of maximum local increase of the function, the gradient for our example points outside the feasible region. This is illustrated in Figure 1.3.

Figure 1.3   Gradient example.

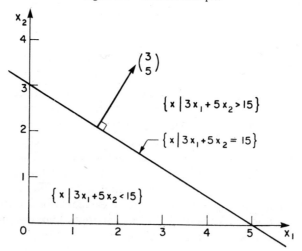

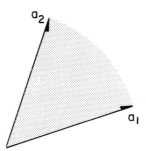

Figure 1.4   Cone spanned by a_1 and a_2.

The *cone* spanned by a set of vectors is the set of all nonnegative linear combinations of the vectors. The coefficients of the linear combination are called the *multipliers* of the cone. In the plane, a cone has the appearance its name suggests, as shown in Figure 1.4.

With these two concepts, we return to the geometry of Example 1.1. Let

$$c = \begin{bmatrix} -1 \\ -1 \end{bmatrix}$$

denote the gradient of the objective function. Then $-c$ points in the direction of maximum local decrease of the objective function. Indeed, this is the direction we followed in obtaining a graphical solution in Figure 1.2. Figure 1.5 shows $-c$ and the cones spanned by the gradients of the active constraints for each of the extreme points $P_1, \ldots, P_4$. Observe that at the optimal solution P_2, $-c$ lies in the cone spanned by the gradients of the active

Figure 1.5   Optimality conditions for Example 1.1.

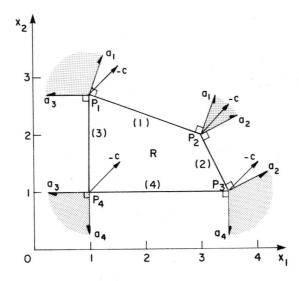

constraints. Furthermore, the optimal solution is the only extreme point for which this is the case.

Observation 2.

A point is optimal if and only if it is feasible and the negative gradient of the objective function lies in the cone spanned by the gradients of the active constraints.

We have given a geometrical verification that P_2 satisfies the foregoing optimality conditions. Next, we do the same thing algebraically. Constraints (1) and (2) are active at P_2. Their gradients are

$$\begin{bmatrix} 1 \\ 3 \end{bmatrix} \quad \text{and} \quad \begin{bmatrix} 2 \\ 1 \end{bmatrix},$$

respectively. The cone part of the optimality conditions requires that there are nonnegative numbers u_1 and u_2 such that

$$u_1 \begin{bmatrix} 1 \\ 3 \end{bmatrix} + u_2 \begin{bmatrix} 2 \\ 1 \end{bmatrix} = \begin{bmatrix} 1 \\ 1 \end{bmatrix}.$$

These may be rewritten as two linear equations

$$u_1 + 2u_2 = 1,$$
$$3u_1 + u_2 = 1,$$

which have solution $u_1 = 1/5$ and $u_2 = 2/5$. Since this solution is nonnegative, we have algebraically verified that the optimality conditions are satisfied at P_2.

1.2 LP SOLUTION ALGORITHM: EXAMPLE

We illustrate the basic algorithm for the solution of a linear programming problem (LP) with the following example. In the next chapter we develop an algorithm which will solve an LP having an arbitrary number of variables and constraints. It will be a straightforward generalization of the geometrical notions used in

Example 1.2

$$
\begin{aligned}
\text{minimize:} \quad & -5x_1 + 2x_2 \\
\text{subject to:} \quad & -2x_1 + x_2 \leq 2, && (1) \\
& x_1 + 2x_2 \leq 14, && (2) \\
& 4x_1 + 3x_2 \leq 36, && (3) \\
& -x_1 \leq 0, && (4) \\
& -x_2 \leq 0. && (5)
\end{aligned}
$$

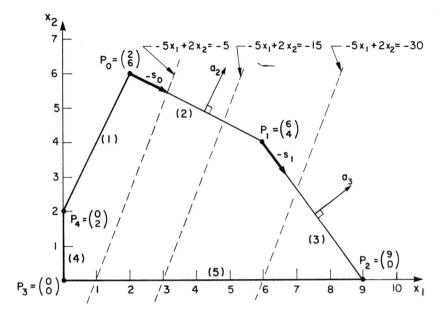

Figure 1.6 Geometry of Example 1.2.

The feasible region, extreme points, and optimal solution P_2 are shown in Figure 1.6.

Applying the same analysis as in the preceding section, it is easy to see that the optimal solution is P_2. However, here we are not particularly interested in the optimal solution per se, but rather in the solution procedure with which we will locate it.

Suppose that we are given the initial extreme point $x_0 = P_0$. By moving along the *edge* joining it to P_1, we can reduce the objective function until the adjacent extreme point $x_1 = P_1$ is obtained. This gives a new extreme point with a smaller objective function value. The process can be repeated at x_1 by moving from x_1 along the edge joining it to P_2. This gives a second extreme point $x_2 = P_2$ with a smaller objective function value. Since x_2 is optimal, the objective function can no longer be reduced by moving along an edge emanating from it.

We give a slightly more formalized version of this algorithm in Figure 1.7.

Beginning with the extreme point $x_0 = P_0 = (2, 6)'$, we apply the prototype algorithm to Example 1.2. In doing so, it is useful to employ the following notation. Let

$$a_1 = \begin{bmatrix} -2 \\ 1 \end{bmatrix}, \quad a_2 = \begin{bmatrix} 1 \\ 2 \end{bmatrix}, \quad a_3 = \begin{bmatrix} 4 \\ 3 \end{bmatrix},$$

$$a_4 = \begin{bmatrix} -1 \\ 0 \end{bmatrix}, \quad a_5 = \begin{bmatrix} 0 \\ -1 \end{bmatrix}, \quad \text{and} \quad c = \begin{bmatrix} -5 \\ 2 \end{bmatrix}$$

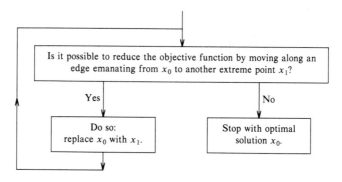

Figure 1.7   Prototype algorithm.

denote the gradients of constraints (1) to (5) and the objective function, respectively. We begin with the extreme point $x_0 = P_0 = (2, 6)'$ at which constraints (1) and (2) are active. Points along the edge joining P_0 and P_1 can be written as

$$x_0 - \sigma s_0,$$

where σ is a scalar and s_0 is a 2-vector orthogonal to a_2. Our immediate problem is to find s_0 and the value of σ (call it σ_0) for which $x_0 - \sigma_0 s_0 = P_1$. We call s_0, σ, and σ_0 the *search direction, step size*, and *maximum feasible step size*, respectively. Now if we did not already know the components of x_0, we would solve the linear equations

$$-2x_1 + x_2 = 2, \qquad (1)$$
$$x_1 + 2x_2 = 14. \qquad (2)$$

Letting

$$D_0 = \begin{bmatrix} -2 & 1 \\ 1 & 2 \end{bmatrix} \quad \text{and} \quad b_0 = \begin{bmatrix} 2 \\ 14 \end{bmatrix},$$

we solve $D_0 x_0 = b_0$ to obtain $x_0 = D_0^{-1} b_0$. From Exercise 1.9, we obtain

$$D_0^{-1} = \begin{bmatrix} -2/5 & 1/5 \\ 1/5 & 2/5 \end{bmatrix}.$$

Now we want s_0 such that $a_2' s_0 = 0$. By definition of the inverse matrix, column 1 of D_0^{-1} is orthogonal to a_2, so that s_0 is immediately available as

$$s_0 = \begin{bmatrix} -2/5 \\ 1/5 \end{bmatrix}.$$

Now consider points of the form

$$x_0 - \sigma s_0 = \begin{bmatrix} 2 \\ 6 \end{bmatrix} - \sigma \begin{bmatrix} -2/5 \\ 1/5 \end{bmatrix}.$$

As σ varies from $-\infty$ to $+\infty$, $x_0 - \sigma s_0$ is the set of points on the line $x_1 + 2x_2 = 14$. Furthermore, observe that

$$c'(x_0 - \sigma s_0) = 2 - \frac{12}{5}\sigma \quad \text{and} \quad a_1'(x_0 - \sigma s_0) = 2 - \sigma,$$

so that increasing σ from 0 decreases the objective function and renders constraint (1) inactive. Substituting $x_0 - \sigma s_0$ into each of the five problem constraints imposes restrictions on σ:

$$\sigma \geq 0, \qquad (1)$$
$$0 \leq 0, \qquad (2)$$
$$\sigma \leq 10, \qquad (3)$$
$$\tfrac{2}{5}\sigma \geq -2, \qquad (4)$$
$$\tfrac{1}{5}\sigma \leq 6. \qquad (5)$$

Therefore, $x_0 - \sigma s_0$ is feasible for all σ with $0 \leq \sigma \leq 10$. Indeed, this set of points is precisely the edge joining P_0 and P_1. Thus the largest value of σ for which $x_0 - \sigma s_0$ is feasible is $\sigma_0 = 10$. The maximum feasible step size is $\sigma_0 = 10$.

We continue by setting

$$x_1 = x_0 - \sigma_0 s_0 = \begin{bmatrix} 2 \\ 6 \end{bmatrix} - 10 \begin{bmatrix} -2/5 \\ 1/5 \end{bmatrix} = \begin{bmatrix} 6 \\ 4 \end{bmatrix},$$

and observe that $x_1 = P_1$, as required.

This completes one iteration of the prototype algorithm. We continue by repeating with different data. Since constraints (1) and (3) are inactive and active at x_1, respectively, we obtain D_1' from D_0' by replacing column 1 ($= a_1$) with a_3. This gives

$$D_1 = \begin{bmatrix} 4 & 3 \\ 1 & 2 \end{bmatrix} \quad \text{and} \quad D_1^{-1} = \begin{bmatrix} 2/5 & -3/5 \\ -1/5 & 4/5 \end{bmatrix}.$$

We want constraint (2) to become inactive and constraint (3) to remain active (i.e., $a_3's_1 = 0$). By definition of the inverse matrix, column 2 of D_1^{-1} is orthogonal to a_3. Accordingly, we set

$$s_1 = \begin{bmatrix} -3/5 \\ 4/5 \end{bmatrix}.$$

Now consider points of the form

$$x_1 - \sigma s_1 = \begin{bmatrix} 6 \\ 4 \end{bmatrix} - \sigma \begin{bmatrix} -3/5 \\ 4/5 \end{bmatrix}.$$

As σ varies from $-\infty$ to $+\infty$, $x_1 - \sigma s_1$ is the set of points on the line $4x_1 + 3x_2 = 36$. Furthermore, observe that

$$c'(x_1 - \sigma s_1) = -22 - \frac{23}{5}\sigma \quad \text{and} \quad a_2'(x_1 - \sigma s_1) = 14 - \sigma,$$

so that increasing σ from 0 decreases the objective function and renders constraint (2) inactive. Substituting $x_1 - \sigma s_1$ into each of the five problem constraints imposes restrictions on σ:

$$2\sigma \geq -10, \qquad (1)$$
$$\sigma \geq 0, \qquad (2)$$
$$0 \leq 0, \qquad (3)$$
$$\frac{3}{5}\sigma \geq -6, \qquad (4)$$
$$\frac{4}{5}\sigma \leq 4. \qquad (5)$$

Consequently, $x_1 - \sigma s_1$ is feasible for all σ with $0 \leq \sigma \leq 5$, and this set of points is the edge joining P_1 and P_2. The maximum feasible step size is $\sigma_1 = 5$.

We continue by setting

$$x_2 = x_1 - \sigma_1 s_1 = \begin{bmatrix} 6 \\ 4 \end{bmatrix} - 5 \begin{bmatrix} -3/5 \\ 4/5 \end{bmatrix} = \begin{bmatrix} 9 \\ 0 \end{bmatrix},$$

$$D_2' = \begin{bmatrix} a_3, a_5 \end{bmatrix} = \begin{bmatrix} 4 & 0 \\ 3 & -1 \end{bmatrix} \quad \text{and} \quad D_2^{-1} \doteq \begin{bmatrix} 1/4 & 3/4 \\ 0 & -1 \end{bmatrix}.$$

We observe from Figure 1.6 that there is no edge emanating from x_2 along which the objective function can be reduced, so the prototype algorithm terminates with optimal solution x_2.

EXERCISES

1.1. Draw the equivalent of Figure 1.5 for Example 1.2.

1.2. Sketch the feasible region R defined by

$$4x_1 - 3x_2 \geq -15, \qquad (1)$$
$$x_1 \geq -3, \qquad (2)$$
$$x_1 + x_2 \geq -4, \qquad (3)$$
$$-x_1 - 3x_2 \leq 9, \qquad (4)$$
$$x_1 - 3x_2 \leq 6. \qquad (5)$$

Determine graphically the minimum of each of the following objective functions on R.

(a) $x_1 + 3x_2$,

(b) $x_1 + x_2$,

(c) $4x_1 - 3x_2$,

(d) $-x_1 + x_2$.

1.3. Give a graphical solution for each of the following linear programming problems. Sketch the feasible region and indicate the behavior of the objective function. Show graphically and algebraically that the optimality conditions are satisfied at the optimal solution. Show graphically that these conditions are <u>not</u> satisfied at every other extreme point.

(a)   minimize: $2x_1 - 5x_2$

      subject to: $x_1 + 3x_2 \le 10,$ (1)

                    $2x_1 - 3x_2 \le 0,$ (2)

                  $-x_1 + x_2 \le 3,$ (3)

                  $-x_1 + 2x_2 \le 1.$ (4)

(b)   minimize: $-4x_1 - 5x_2$

      subject to: $x_1 + 2x_2 \le 3,$ (1)

                    $x_1 - x_2 \le 2,$ (2)

                  $2x_1 + x_2 \le 3,$ (3)

                  $3x_1 + 4x_2 \le 8,$ (4)

                  $-x_1 \le 0.$ (5)

(c)   minimize: $23x_1 - 7x_2$

      subject to: $-4x_1 + x_2 \le -2,$ (1)

                    $x_1 + x_2 \le 5,$ (2)

                 $-x_1 - x_2 \le -1,$ (3)

               $-3x_1 + 2x_2 \le 1,$ (4)

                $-x_1 \le 0,$ (5)

                  $-x_2 \le 0.$ (6)

(d)   minimize: $-9x_1 - x_2$

      subject to: $6x_1 + 5x_2 \le 10,$ (1)

                    $3x_1 - 2x_2 \le 8,$ (2)

                  $2x_1 + x_2 \le 3,$ (3)

                $-x_1 + 4x_2 \le -4.$ (4)

(e)   minimize: $-x_1 + x_2$

subject to: $-x_1 - 2x_2 \leq 6,$ (1)

$x_1 - 2x_2 \leq 4,$ (2)

$-x_1 + x_2 \leq 1,$ (3)

$x_1 \qquad \leq 0,$ (4)

$x_2 \leq 0.$ (5)

(f)   maximize: $x_1 - x_2$

subject to: $3x_1 + x_2 \geq 3,$ (1)

$x_1 + 2x_2 \geq 4,$ (2)

$x_1 - x_2 \leq 1,$ (3)

$x_1 \qquad \leq 5,$ (4)

$x_2 \leq 5.$ (5)

(Write this problem in standard form $\min\{c'x \mid Ax \leq b\}$ first.)

1.4. Starting with $x_0 = (1,1)'$, use the prototype algorithm to solve the problem in Example 1.1.

1.5. In Example 1.2, show algebraically that x_2 satisfies the optimality conditions as formulated in Observation 2. Can D_2^{-1} be used to simplify the calculations? *Hint:* $(D_2')^{-1} = (D_2^{-1})'$.

1.6. Consider the problem

minimize: $-6x_1 - 4x_2 - 8x_3$

subject to: $x_1 + x_2 + x_3 \leq 4,$ (1)

$x_1 + 2x_2 - x_3 \leq -1,$ (2)

$2x_1 \qquad + 3x_3 \leq 8,$ (3)

$2x_2 + 4x_3 \leq 10,$ (4)

$x_1 + 2x_2 + 3x_3 \leq 7.$ (5)

Let $x_0 = (1,0,2)'$.
(a) Is x_0 feasible? What constraints are active at x_0? Is x_0 an extreme point?
(b) Show that at x_0, $-c$ is in the cone spanned by the gradients of those constraints active at x_0. Determine the multipliers.

1.7. Use the prototype algorithm with the indicated initial point x_0 to solve each of the problems of Exercise 1.3.
(a) $(-9, -6)'$,
(b) $(0, -2)'$,
(c) $(5, 0)'$,
(d) $(16/9, -5/9)'$,
(e) $(-8/3, -5/3)'$,
(f) $(-2/3, 5)'$.

1.8. Let $x_0 = (0, 2, 0)'$ and $s_0 = (-1, -1, 0)'$. For what values of σ does the point $x_1 = x_0 - \sigma s_0$ satisfy the constraint $x_1 + x_2 + 4x_3 \leq 6$? For what value of σ does the constraint become active?

1.9. Show that if $ad \neq bc$, then

$$\begin{bmatrix} a & b \\ c & d \end{bmatrix}^{-1} = \frac{1}{ad - bc} \begin{bmatrix} d & -b \\ -c & a \end{bmatrix}.$$

1.10. Let x_0 be a feasible point for the model LP

$$\min \{ c'x \mid a_i'x \leq b_i, \ i = 1, \ldots, m \}$$

and let s_0 be a given search direction. What conditions on (nonnegative) σ must be satisfied in order that $x_0 - \sigma s_0$ is also feasible? Derive a formula for the largest such σ.

BASIC LP SOLUTION ALGORITHMS

In Chapter 1 an algorithm has been developed for solving linear programming problems with two variables. The algorithm is based on the observation that if a given extreme point is not optimal, it is possible to move along an edge of the feasible region to an adjacent extreme point with a smaller value of the objective function. The purpose of this chapter is to extend this algorithm to general linear programming problems with n variables.

2.1 MODEL PROBLEM AND DEFINITIONS

We consider the model problem

$$\min\{c'x \mid Ax \le b\}, \qquad (2.1)$$

where c and b are given n- and m-vectors, respectively, and A is a given (m, n) matrix. Thus n denotes the number of problem variables, m denotes the number of inequality constraints, and x is an n-vector whose optimal value is to be determined. We adopt the convention that all vectors are column vectors and transposition is denoted by a prime (e.g., x' and A'). Sometimes it is helpful to write (2.1) in the equivalent form:

$$\min\{c'x \mid a_i'x \le b_i, \ i = 1, \ldots, m\}. \qquad (2.2)$$

Thus

$$
b = \begin{bmatrix} b_1 \\ b_2 \\ \vdots \\ b_m \end{bmatrix} \quad \text{and} \quad A = \begin{bmatrix} a_1' \\ a_2' \\ \vdots \\ a_m' \end{bmatrix}, \quad \text{or} \quad A' = \begin{bmatrix} a_1, a_2, \ldots, a_m \end{bmatrix},
$$

where $a_1, a_2, \ldots, a_m$ are n-vectors.

The ith constraint is $a_i'x \leq b_i$ and a_i is the gradient[1] of the ith constraint.

Let

$$
R = \{ x \mid Ax \leq b \}.
$$

Here R is called the feasible region for (2.1) and $c'x$ is called the objective function. A point x_0 is feasible for (2.1) if $x_0 \in R$ and infeasible otherwise. Constraint i is inactive, active, or violated at x_0 according to whether $a_i'x_0 < b_i$, $a_i'x_0 = b_i$, or $a_i'x_0 > b_i$, respectively. x_0 is an optimal solution (or simply optimal) if $x_0 \in R$ and $c'x_0 \leq c'x$ for all $x \in R$.

The algorithm discussed in Chapter 1 is based on the geometric concept of extreme points and edges of the feasible region. Therefore, our first task is to define extreme points and edges for the model problem (2.1). Returning to Figure 1.6, we observe that each of the five extreme points $P_0, P_1, \ldots, P_4$ is determined by two active constraints and that it is the only point at which precisely these two constraints are active.

Now let $x_0 \in R$ and assume that $a_i'x_0 = b_i$, $i = 1, \ldots, k$; $a_i'x_0 < b_i$, $i = k + 1, \ldots, m$ (i.e., the first k constraints are active at x_0). Clearly, x_0 is the only feasible solution at which these constraints are active if and only if x_0 is the unique solution of the linear equations

$$
a_i'x = b_i, \quad i = 1, \ldots, k. \tag{2.3}
$$

It is not difficult to verify that the solution of (2.3) is unique if and only if the set $\{ a_1, \ldots, a_k \}$ contains n linearly independent vectors (see Exercise 2.14).

By analogy to the two-dimensional case we therefore call x_0 an extreme point of R if $x_0 \in R$ and there are n constraints having linearly independent gradients active at x_0. This definition does not necessarily imply that there are exactly n constraints active at an extreme point. There could be more. x_0 is a nondegenerate extreme point if it is an extreme point at which exactly n constraints are active. x_0 is a degenerate extreme point if it is an extreme point at which more than n constraints are active.

---

[1] Terms which are underlined are formal definitions.

In Figure 2.1, $P_1, P_2, \ldots, P_5$ are nondegenerate extreme points. In Figure 2.2, there are three active constraints at P_1. Thus P_1 is a degenerate extreme point. Similarly, P_4 is also a degenerate extreme point since there are five active constraints at it. Note, however, that the feasible regions are identical in both cases. In this chapter we are concerned primarily with non-degenerate extreme points. Degenerate extreme points cause some technical problems which we will discuss in detail in Appendix A.

Referring to Figure 1.6 once more, we recall that the edges of the feasible region that emanate from P_0 are the line segments which connect P_0 to the extreme points P_1 and P_4, respectively. Each edge is determined by $n - 1$ of the constraints that are active at P_0.

Let x_0 be a nondegenerate extreme point of R. Renumbering constraints, if necessary, we may assume that the first n constraints are active at x_0. An <u>edge</u> emanating from x_0 is then a line segment of the form

$$x_0 - \sigma s_k, \quad 0 \le \sigma \le \hat{\sigma}_k,$$

where $\hat{\sigma}_k$ is the largest scalar such that $x_0 - \sigma s_k \in R$, and s_k is an n-vector with the property

$$a_i' s_k = 0, \quad i = 1, \ldots, n, \quad i \ne k,$$

$$a_k' s_k = 1.$$

From this definition we see immediately that there are n different edges emanating from the extreme point x_0. Two extreme points are said to be <u>adjacent</u> if they are connected by an edge.

Figure 2.1 Nondegenerate extreme points.

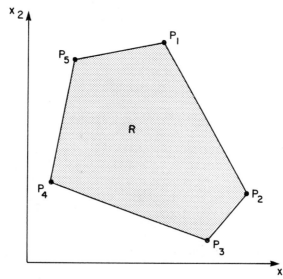

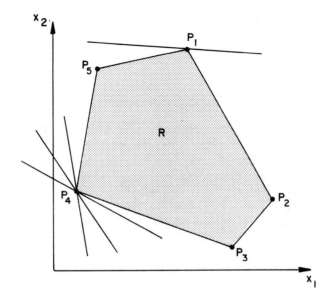

Figure 2.2 Degenerate extreme points P_1 and P_4.

Finally, it has been observed in Example 1.1 that a feasible point is optimal if the negative gradient of the objective function lies in the cone spanned by the gradients of the active constraints. The following theorem shows that this is true in general.

Theorem 2.1.

Let x_0 be a feasible solution for the model problem (2.1) with

$$a_i' x_0 = b_i, \quad i = 1, \ldots, k; \quad a_i' x_0 < b_i, \quad i = k + 1, \ldots, m.$$

Then x_0 is an optimal solution if $-c$ lies in the cone spanned by a_1, $\ldots, a_k$; that is, if there are nonnegative multipliers $u_1, \ldots, u_k$ such that

$$-c = \sum_{i=1}^{k} u_i a_i. \tag{2.4}$$

Proof:

Let x be any feasible solution. We have to show that $c'x \geq c'x_0$. For $i = 1, \ldots, k$ we have $a_i' x_0 = b_i$ and $a_i' x \leq b_i$. Thus $a_i' x \leq a_i' x_0$ or $a_i'(x - x_0) \leq 0, i = 1, \ldots, k$. Scalar multiplication of (2.4) with $x - x_0$ gives

$$-c'(x - x_0) = \sum_{i=1}^{k} u_i a_i'(x - x_0) \leq 0,$$

where the inequality follows from $u_i \geq 0$ and $a_i'(x - x_0) \leq 0, i = 1$ $\ldots, k$. Therefore, we have $c'(x - x_0) \geq 0$ or $c'x \geq c'x_0$. ∎

2.2 THE BASIC ALGORITHM

We next proceed to refine the algorithm of Chapter 1 as it applies to our model problem

$$\min \{ c'x \mid a_i'x \le b_i, \ i = 1, \ldots, m \}. \tag{2.5}$$

The basic idea is as simple as in Example 1.2. If $x_0 \in R$ is any extreme point, we use the optimality conditions of Theorem 2.1 to decide whether x_0 is an optimal solution. If this is not the case, we have to determine a new extreme point of R for which the objective function has a strictly smaller value by moving along an edge emanating from x_0. If this second extreme point is not optimal, we construct a third one, which again gives a strictly smaller value of the objective function, and so on. Because the objective function decreases strictly, no extreme point occurs twice. Therefore, the method terminates after a finite number of steps with an optimal extreme point or the information that the given problem has no optimal solution.

In order to focus on the essential points, we temporarily assume the following:

1. A nondegenerate extreme point x_0 for (2.5) is known.
2. The first n constraints are active at x_0; that is,

$$a_i'x_0 = b_i, \ i = 1, \ldots, n; \quad a_i'x_0 < b_i, \ i = n + 1, \ldots, m.$$

We define

$$D_0' = \left[a_1, \ldots, a_n \right] \quad \text{and} \quad b_0 = (b_1, \ldots, b_n)'.$$

Thus the columns of the (n, n) matrix D_0' are the gradients of the constraints active at x_0. Since x_0 is a nondegenerate extreme point $a_1, \ldots, a_n$ are linearly independent and D_0 is nonsingular. Let

$$D_0^{-1} = \left[c_1, \ldots, c_n \right]$$

(i.e., c_i is the ith column of D_0^{-1}). Let k be any integer satisfying $1 \le k \le n$. By definition of the inverse matrix

$$a_i'c_k = 0, \ i = 1, \ldots, n, \ i \ne k,$$

$$a_k'c_k = 1.$$

Therefore, for every $\sigma > 0$,

$$a_i'(x_0 - \sigma c_k) = a_i'x_0 = b_i, \ i = 1, \ldots, n, \ i \ne k \tag{2.6}$$

and

$$a_k'(x_0 - \sigma c_k) = a_k'x_0 - \sigma = b_k - \sigma < b_k. \tag{2.7}$$

Equations (2.6) and (2.7) show that the n edges of R emanating from x_0 can be represented in the form

$$x_0 - \sigma c_k, \quad \sigma > 0, \quad k = 1, \ldots, n.$$

Now choose any integer k satisfying $1 \leq k \leq n$ and consider the effect of defining the search direction s_0 and the new point x_1 according to

$$s_0 = c_k \quad \text{and} \quad x_1 = x_0 - \sigma s_0, \quad \text{with } \sigma > 0. \tag{2.8}$$

The objective function value at x_1 is

$$c'x_1 = c'(x_0 - \sigma s_0) = c'x_0 - \sigma c's_0,$$

and will be strictly less than $c'x_0$ provided that

$$c'c_k > 0 \quad \text{and} \quad \sigma > 0. \tag{2.9}$$

Furthermore, it follows from (2.6) that constraints $1, \ldots, k - 1, k + 1 , \ldots, n$ remain active at $x_0 - \sigma s_0$ for all σ, whereas (2.7) implies that constraint k becomes inactive at $x_0 - \sigma s_0$ as σ is increased from 0. Thus $x_0 - \sigma s_0$ is feasible for all $\sigma \geq 0$ satisfying

$$a_i'(x_0 - \sigma s_0) \leq b_i, \quad i = n + 1, \ldots, m. \tag{2.10}$$

For those i for which $a_i's_0 \geq 0$, (2.10) is satisfied for all $\sigma \geq 0$ and our feasibility requirement reduces to

$$\sigma \leq \frac{a_i'x_0 - b_i}{a_i's_0}, \quad \text{all } i = n + 1, \ldots, m \quad \text{with } a_i's_0 < 0.$$

The largest value of σ satisfying these inequalities is called the <u>maximum feasible step size</u> σ_0 and is obtained from

$$\sigma_0 = \min\left\{ \frac{a_i'x_0 - b_i}{a_i's_0} \;\middle|\; \text{all } i = n + 1, \ldots, m \text{ with } a_i's_0 < 0 \right\}. \tag{2.11}$$

It is useful to record the index of the constraint for which the minimum occurs in (2.11). There may be ties so that this index is not, in general, uniquely determined. We adopt the convention of recording the smallest such index. At present, this choice is arbitrary. However, this choice turns out to be appropriate in resolving problems due to degeneracy, as will be shown in Appendix A. We thus define l as the smallest integer such that

$$\sigma_0 = \frac{a_l'x_0 - b_l}{a_l's_0}.$$

We now set

$$x_1 = x_0 - \sigma_0 s_0,$$

and observe that constraints $1, 2, \ldots, k - 1, k + 1, \ldots, n$ and constraint l are active at x_1. These are n in number and it is straightforward to show that $a_1, a_2, \ldots, a_{k-1}, a_{k+1}, \ldots, a_n$ together with a_l are linearly independent (Exercise 2.15). Consequently, x_1 is an extreme point. Furthermore, x_1 is a nondegenerate extreme point provided that the index l associated with σ_0 is uniquely determined.

Choosing s_0 according to (2.8) ensures that $x_1 = x_0 - \sigma_0 s_0$ is feasible and an extreme point adjacent to x_0. Choosing k to satisfy $c' c_k > 0$ ensures that the objective function decreases when moving from x_0 to x_1. Indeed, any k with $c' c_k > 0$ will do. However, the greatest rate of decrease is obtained by choosing k such that

$$c' c_k = \max\{ c' c_i \mid 1 \leq i \leq n \}. \tag{2.12}$$

We note that this choice of k will not in general guarantee that x_1 has the smallest objective function value among all extreme points adjacent to x_0. This is because a rapid rate of decrease may be limited by a smaller step size, whereas a slower rate of decrease may be associated with a very large step size, thus producing a larger total reduction in the objective function (see Exercise 2.16). However, it is traditional to choose the "leaving" constraint according to (2.12).

Choosing the leaving constraint in this manner results in an additional benefit. Suppose that $c' c_k \leq 0$. Then the definition of k implies that all $c' c_i \leq 0$. This implies that x_0 is optimal. Indeed, writing $-c$ as

$$-c = u_1 a_1 + \cdots + u_n a_n$$

and multiplying with c_i we obtain the multipliers

$$u_i = -c' c_i \geq 0, \quad i = 1, \ldots, n.$$

Thus it follows from Theorem 2.1 that x_0 is optimal if all $c' c_i \leq 0$.

Therefore, the extreme point x_0 is either optimal or there is at least one edge emanating from x_0 along which the objective function decreases. This edge leads to the adjacent extreme point x_1 provided that the maximum feasible step size σ_0 is defined. This is the case if $a_i' s_0 < 0$ for at least one i. If $a_i' s_0 \geq 0$ for all i, our model problem (2.5) has no optimal solution. Indeed, if all $a_i' s_0 \geq 0$, then

$$a_i'(x_0 - \sigma s_0) = a_i' x_0 - \sigma a_i' s_0 \leq b_i$$

for all $\sigma \geq 0$ and $i = 1, \ldots, m$. Thus $x_0 - \sigma s_0$ is feasible for all $\sigma \geq 0$. Since $c'(x_0 - \sigma s_0) = c' x_0 - \sigma c' s_0$ and $c' s_0 > 0$ imply that $c'(x_0 - \sigma s_0)$ becomes arbitrarily small as σ increases indefinitely, the objective function is not bounded from below on R. Thus no optimal solution exists. This situation is illustrated in Figure 2.3.

Suppose that σ_0 is defined and consider the new extreme point x_1. Since $a_1, \ldots, a_{k-1}, a_{k+1}, \ldots, a_n, a_l$ are n linearly independent gradients

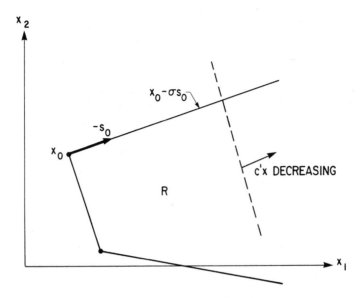

Figure 2.3 Objective function unbounded from below on R.

(Exercise 2.15) of constraints which are active at x_1, the matrix D_1' that we want to associate with x_1 is given by

$$D_1' = \left[a_1, \ldots, a_{k-1}, a_l, a_{k+1}, \ldots, a_n \right].$$

This matrix differs from D_0' only in the kth column, where we have replaced the gradient a_k of the kth constraint, not active at x_1, with the gradient a_l of the newly active constraint.

Now the steps above can be repeated with x_0 and D_0^{-1} replaced by x_1 and D_1^{-1}, respectively. Thus the main features of the algorithm may be summarized as in Figure 2.4.

Before we can give a detailed description of the algorithm we have to study more carefully the matrices D_j' and D_j^{-1} associated with the extreme point x_j. Returning to the extreme point x_0, we observe that the matrix D_0' is not used explicitly. It only serves the purpose of illustrating the definition of the matrix D_0^{-1}. However, as we have seen, moving from x_0 along the edge $x_0 - \sigma c_k$ to the new extreme point x_1 changes the status of the constraint with gradient equal to the kth column of D_0' from active to inactive. For x_0 and every subsequently encountered extreme point x_j it is, therefore, important to know which gradient (i.e., which vector a_i), is in a given column of the matrix D_j' associated with x_j. In order to achieve this we define an ordered index set

$$J_j = \left\{ \alpha_{1j}, \alpha_{2j}, \ldots, \alpha_{nj} \right\}.$$

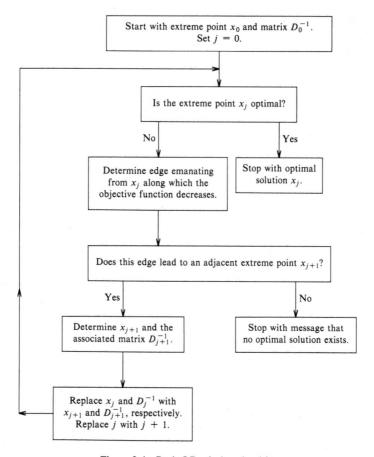

Figure 2.4   Basic LP solution algorithm.

Each of the numbers α_{ij} is an element of the set $\{1, 2, \ldots, m\}$ where $\alpha_{ij} = k$ if and only if the ith column of D_j' is equal to a_k. Thus for $i = 1, \ldots, n$, the number α_{ij} is equal to the index of the constraint with gradient given by the ith column of D_j', and D_j' may be expressed as

$$D_j' = \left[a_{\alpha_{1j}}, a_{\alpha_{2j}}, \ldots, a_{\alpha_{nj}} \right].$$

For instance, for the extreme point x_0 defined at the beginning of this section, we have $D_0' = \left[a_1, \ldots, a_n \right]$. Therefore,

$$J_0 = \{ \alpha_{10}, \ldots, \alpha_{n0} \} = \{ 1, \ldots, n \}.$$

Similarly, for the extreme point x_1 with

$$D_1' = \left[a_1, \ldots, a_{k-1}, a_l, a_{k+1}, \ldots, a_n \right]$$

the index set J_1 has the form

$$J_1 = \{\alpha_{11}, \ldots, \alpha_{n1}\}$$

with $\alpha_{k1} = l$ and $\alpha_{i1} = i, i = 1, \ldots, n, i \neq k$.

In addition to x_1 and J_1 we need the matrix D_1^{-1}. Of course, D_1^{-1} could be obtained by inverting the matrix D_1. However, to invert an (n, n) matrix completely requires approximately n^3 additions and multiplications. When n is other than very small, the arithmetic work for the matrix inversion is quite time consuming even on a modern computer. Now D_0^{-1} is known and since D_1' differs from D_0' in just one column, it seems reasonable to expect that there is a computationally more efficient method of computing D_1^{-1} from D_0^{-1} and the new column. Indeed there is, and we formulate one such method as follows.

Procedure Φ.

Let $D' = \begin{bmatrix} d_1, \ldots, d_n \end{bmatrix}$ be an (n, n) nonsingular matrix and let $D^{-1} = \begin{bmatrix} c_1, \ldots, c_n \end{bmatrix}$. Let d and k be a given n-vector and an index, respectively, such that $d'c_k \neq 0$. Define

$$\hat{D}' = \begin{bmatrix} d_1, \ldots, d_{k-1}, d, d_{k+1}, \ldots, d_n \end{bmatrix};$$

that is, $\hat{D}'$ is obtained from D' by replacing column k with d. We use the notation $\Phi(D^{-1}, d, k)$ to denote $\hat{D}^{-1}$. One way of computing $\hat{D}^{-1} = \begin{bmatrix} \hat{c}_1, \ldots, \hat{c}_n \end{bmatrix}$ is

$$\hat{c}_i = c_i - \frac{d'c_i}{d'c_k} c_k, \quad \text{for all } i = 1, \ldots, n, \ i \neq k,$$

and

$$\hat{c}_k = \frac{1}{d'c_k} c_k.$$

The procedure for modifying the inverse of a matrix due to a change in a row or column was first formulated by Sherman and Morrison [1949] and Woodbury [1950].

The reader will be asked to show (Exercise 2.23) that Procedure Φ requires approximately n^2 arithmetic operations rather than the n^3 required for a complete matrix inversion. We illustrate Procedure Φ as follows.

Example 2.1

Let

$$D_0' = \begin{bmatrix} 1 & 2 \\ 3 & 4 \end{bmatrix}, \quad \text{then} \quad D_0^{-1} = \begin{bmatrix} -2 & 3/2 \\ 1 & -1/2 \end{bmatrix}.$$

Suppose that we obtain D_1' from D_0' by replacing column 1 with $(-2, 1)'$. We thus compute

$$D_1^{-1} = \Phi\left(D_0^{-1}, \begin{bmatrix} -2 \\ 1 \end{bmatrix}, 1\right) = \begin{bmatrix} c_{11}, c_{21} \end{bmatrix},$$

where

$$c_{11} = \frac{1}{5}\begin{bmatrix} -2 \\ 1 \end{bmatrix} = \begin{bmatrix} -2/5 \\ 1/5 \end{bmatrix}$$

and

$$c_{21} = \begin{bmatrix} 3/2 \\ -1/2 \end{bmatrix} - \frac{(-7/2)}{5}\begin{bmatrix} -2 \\ 1 \end{bmatrix} = \begin{bmatrix} 1/10 \\ 1/5 \end{bmatrix}.$$

Thus

$$D_1^{-1} = \begin{bmatrix} -2/5 & 1/10 \\ 1/5 & 1/5 \end{bmatrix}.$$

We remark that the calculations required for Procedure Φ are rather tedious for hand calculation. However, for implementation on a digital computer, Procedure Φ is quite appropriate.

Next we give a detailed formulation of the basic algorithm.

ALGORITHM 1

Model Problem: $\min\{c'x \mid a_i'x \le b_i, \ i = 1, \ldots, m\}$

Initialization:
Start with extreme point x_0, $J_0 = \{\alpha_{10}, \ldots, \alpha_{n0}\}$, and D_0^{-1}, where $D_0' = \begin{bmatrix} d_1, \ldots, d_n \end{bmatrix}$ is nonsingular, and, $d_i = a_{\alpha_{i0}}$ for all $i = 1, \ldots, n$. Compute $c'x_0$ and set $j = 0$.

Step 1:   Computation of Search Direction s_j.
Let $D_j^{-1} = \begin{bmatrix} c_{1j}, \ldots, c_{nj} \end{bmatrix}$ and $J_j = \{\alpha_{1j}, \ldots, \alpha_{nj}\}$. Compute the multipliers

$$v_i = c'c_{ij}, \ i = 1, \ldots, n$$

and determine the smallest index k such that

$$v_k = \max\{v_i \mid i = 1, \ldots, n\}.$$

If $v_k \le 0$, stop with optimal solution x_j. Otherwise, set $s_j = c_{kj}$ and go to Step 2.

Step 2:   Computation of Maximum Feasible Step Size σ_j.
If $a_i's_j \geq 0$ for $i = 1, \ldots, m$, print the message "problem is unbounded from below" and stop. Otherwise, compute the smallest index l and σ_j such that

$$\sigma_j = \frac{a_l'x_j - b_l}{a_l's_j} = \min\left\{\frac{a_i'x_j - b_i}{a_i's_j} \mid \text{ all } i \notin J_j \text{ with } a_i's_j < 0\right\},$$

and go to Step 3.

Step 3:   Update.
Set $x_{j+1} = x_j - \sigma_j s_j$, compute $c'x_{j+1}$, and obtain D_{j+1}' from D_j' by replacing column k with a_l. Set $D_{j+1}^{-1} = \Phi(D_j^{-1}, a_l, k)$ and $J_{j+1} = \{\alpha_{1,j+1}, \ldots, \alpha_{n,j+1}\}$, where

$$\alpha_{i,j+1} = \alpha_{ij}, \quad \text{for all } i \text{ with } i \neq k, \quad \text{and} \quad \alpha_{k,j+1} = l.$$

Replace j with $j + 1$ and go to Step 1.

We note that in Algorithm 1 it is not necessary to form or store D_j' explicitly. It is used implicitly to understand the algorithm, but only D_j^{-1} need be stored and updated. We remark that the most basic concept of Algorithm 1 is the active set. Put in its simplest terms, Algorithm 1 is a method which iteratively changes one element of the active set. Steps 1 and 2 determine the elements to be deleted and added, respectively. This is sometimes referred to as "dropping" and "adding" an active constraint, respectively. Step 3 performs the actual update of the active set.

We next illustrate Algorithm 1 by applying it to the problem of Example 1.2.

Example 2.2

$$\begin{array}{rl}
\text{minimize:} & -5x_1 + 2x_2 \\
\text{subject to:} & -2x_1 + x_2 \leq 2, \qquad (1) \\
& x_1 + 2x_2 \leq 14, \qquad (2) \\
& 4x_1 + 3x_2 \leq 36, \qquad (3) \\
& -x_1 \qquad\quad \leq 0, \qquad (4) \\
& -x_2 \leq 0. \qquad (5)
\end{array}$$

Initialization:

$$x_0 = \begin{bmatrix} 2 \\ 6 \end{bmatrix}, \quad D_0^{-1} = \begin{bmatrix} -2/5 & 1/5 \\ 1/5 & 2/5 \end{bmatrix}, \quad J_0 = \{1, 2\}, \quad c'x_0 = 2, \quad j = 0.$$

Iteration 0

Step 1: $v_1 = \max\left\{\dfrac{12}{5}, \dfrac{-1}{5}\right\} = \dfrac{12}{5}, \quad k = 1, \quad s_0 = \begin{bmatrix} -2/5 \\ 1/5 \end{bmatrix}.$

Step 2:[2] $\sigma_0 = \min\left\{-, -, \dfrac{-10}{-1}, -, \dfrac{-6}{-1/5}\right\} = 10, \quad l = 3.$

Step 3: $x_1 = \begin{bmatrix} 2 \\ 6 \end{bmatrix} - 10\begin{bmatrix} -2/5 \\ 1/5 \end{bmatrix} = \begin{bmatrix} 6 \\ 4 \end{bmatrix}, \quad c'x_1 = -22,$

$D_1^{-1} = \begin{bmatrix} 2/5 & -3/5 \\ -1/5 & 4/5 \end{bmatrix}, \quad J_1 = \{3, 2\}, \quad j = 1.$

Iteration 1

Step 1: $v_2 = \max\left\{\dfrac{-12}{5}, \dfrac{23}{5}\right\} = \dfrac{23}{5}, \quad k = 2, \quad s_1 = \begin{bmatrix} -3/5 \\ 4/5 \end{bmatrix}.$

Step 2: $\sigma_1 = \min\left\{-, -, -, -, \dfrac{-4}{-4/5}\right\} = 5, \quad l = 5.$

Step 3: $x_2 = \begin{bmatrix} 6 \\ 4 \end{bmatrix} - 5\begin{bmatrix} -3/5 \\ 4/5 \end{bmatrix} = \begin{bmatrix} 9 \\ 0 \end{bmatrix}, \quad c'x_2 = -45,$

$D_2^{-1} = \begin{bmatrix} 1/4 & 3/4 \\ 0 & -1 \end{bmatrix}, \quad J_2 = \{3, 5\}, \quad j = 2.$

Iteration 2

Step 1: $v_1 = \max\left\{\dfrac{-5}{4}, \dfrac{-23}{4}\right\} = \dfrac{-5}{4}, \quad k = 1.$

$v_1 \leq 0$; stop with optimal solution x_2.

A computer program which implements Algorithm 1 is given in Appendix B, Section B.1.1. The output from applying this program to the problem of Example 2.2 is shown in Figure B.10.

---

[2] We use a dash to indicate those constraints for which $a_i's_j \geq 0$.

We further illustrate Algorithm 1 by applying it to a problem that is unbounded from below.

Example 2.3

$$\begin{array}{ll}
\text{minimize:} & -x_1 - 5x_2 \\
\text{subject to:} & 2x_1 - x_2 \le 7, \quad (1) \\
& -x_2 \le -1, \quad (2) \\
& -x_1 - x_2 \le -4, \quad (3) \\
& -x_1 \le -1. \quad (4)
\end{array}$$

Initialization:

$$x_0 = \begin{bmatrix} 3 \\ 1 \end{bmatrix}, \quad D_0^{-1} = \begin{bmatrix} 1 & -1 \\ -1 & 0 \end{bmatrix}, \quad J_0 = \{2, 3\}, \quad c'x_0 = -8, \quad j = 0.$$

Iteration 0

Step 1: $\quad v_1 = \max\{4, 1\} = 4, \quad k = 1, \quad s_0 = \begin{bmatrix} 1 \\ -1 \end{bmatrix}.$

Step 2: $\quad \sigma_0 = \min\left\{-, -, -, \dfrac{-2}{-1}\right\} = 2, \quad l = 4.$

Step 3: $\quad x_1 = \begin{bmatrix} 3 \\ 1 \end{bmatrix} - 2\begin{bmatrix} 1 \\ -1 \end{bmatrix} = \begin{bmatrix} 1 \\ 3 \end{bmatrix}, \quad c'x_1 = -16,$

$$D_1^{-1} = \begin{bmatrix} -1 & 0 \\ 1 & -1 \end{bmatrix}, \quad J_1 = \{4, 3\}, \quad j = 1.$$

Iteration 1

Step 1: $\quad v_2 = \max\{-4, 5\} = 5, \quad k = 2, \quad s_1 = \begin{bmatrix} 0 \\ -1 \end{bmatrix}.$

Step 2: $\quad a_i's_1 \ge 0, i = 1, \ldots, 4$; stop, the problem is unbounded from below.

The feasible region for the problem above, as well as the progress of Algorithm 1, is illustrated in Figure 2.5.

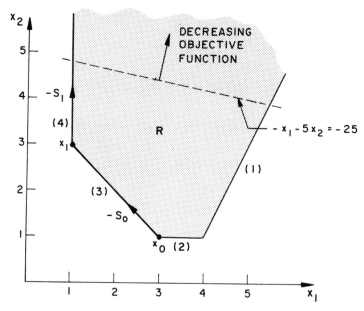

Figure 2.5   Feasible region for Example 2.3.

2.3 PROPERTIES OF ALGORITHM 1

We now establish the properties of Algorithm 1. Some are obvious. By construction, at every iteration j,

$$a_i'x_j = b_i, \quad \text{for all } i \in J_j,$$

and

$$a_i'x_j \le b_i, \quad \text{for all } i \notin J_j.$$

Furthermore, if the algorithm terminates with the message "problem is unbounded from below," then

$$c's_j > 0 \quad \text{and} \quad a_i's_j \ge 0, \ i = 1, \ldots, m.$$

This implies that $x_j - \sigma s_j$ is feasible for all $\sigma \ge 0$ and that $c'(x_j - \sigma s_j)$ is a strictly decreasing function of σ. Thus $c'x$ is indeed unbounded from below on the set of feasible solutions.

If termination occurs with "optimal solution x_j," we have

$$v_i = c'c_{ij} \le 0, \ i = 1, \ldots, n.$$

Let

$$D_j' = \left[a_{\alpha_{1j}}, a_{\alpha_{2j}}, \ldots, a_{\alpha_{nj}} \right], \quad D_j^{-1} = \left[c_{1j}, c_{2j}, \ldots, c_{nj} \right],$$

and write $-c$ as

$$u_1 a_{\alpha_{1j}} + u_2 a_{\alpha_{2j}} + \cdots + u_n a_{\alpha_{nj}} = -c,$$

or equivalently,

$$D_j' u = -c.$$

These equations may be solved in terms of D_j^{-1}:

$$u' = -c' D_j^{-1},$$

or in component form,

$$u_i = -c' c_{ij}, \quad i = 1, \ldots, n.$$

Thus, $u_i = -v_i \geq 0$, $i = 1, \ldots, n$. This implies that $-c$ is in the cone spanned by the gradients of the constraints active at x_j. Applying Theorem 2.1, we conclude that x_j is indeed an optimal solution for our model problem.

It remains to show that Algorithm 1 terminates after a finite number of steps. This is the case if every extreme point, encountered by the algorithm, is nondegenerate. Let x_j be such a point. Then

$$a_i' x_j < b_i, \quad \text{for all } i \notin J_j.$$

Since $a_i' s_j \geq 0$ for all $i \in J_j$, we have $\sigma_j > 0$. From Step 1, $c' s_j > 0$, so that

$$c' x_{j+1} = c' x_j - \sigma_j c' s_j < c' x_j;$$

that is, the sequence of objective function values is strictly decreasing. No extreme point occurs twice. By definition, every extreme point is determined by n active constraints. Thus the number of extreme points cannot exceed the number of different subsets of n constraints. This number is given by

$$\binom{m}{n} = \frac{m!}{(m-n)! n!}.$$

This bound is finite but may be very large; for example, for $m = 100$ and $n = 10$ there are potentially

$$\binom{100}{10} = \frac{100!}{90! 10!} \approx 1.73 \times 10^{13}$$

extreme points. However, in practice the number of iterations is drastically smaller.

Summarizing the results above, we obtain the following theorem.

Theorem 2.2.

Let the extreme points $x_1, x_2, \ldots, x_j, \ldots$ be obtained by applying Algorithm 1 to the model problem

$$\min\{c'x \mid a_i'x \le b_i, \ i = 1, \ldots, m\}.$$

Assume that each x_j is nondegenerate. Then Algorithm 1 terminates in a finite number of steps with either an optimal solution or the information that the problem is unbounded from below.

If x_j is a degenerate extreme point, it is possible to have $\sigma_j = 0$, $x_{j+1} = x_j$, and $c'x_{j+1} = c'x_j$. The key argument used to establish Theorem 2.2 breaks down and finite termination is no longer assured. We illustrate this situation by applying Algorithm 1 to a problem with a degenerate extreme point.

Example 2.4

$$
\begin{array}{lrrl}
\text{minimize:} & x_1 - 11x_2 & & \\
\text{subject to:} & -9x_1 + 2x_2 & \le 10, & (1) \\
& -2x_1 + x_2 & \le 5, & (2) \\
& -3x_1 + 5x_2 & \le 25, & (3) \\
& x_1 + 3x_2 & \le 29, & (4) \\
& 3x_1 + 2x_2 & \le 45, & (5) \\
& -x_1 & \le 0, & (6) \\
& -x_2 & \le 0. & (7)
\end{array}
$$

Initialization:

$$x_0 = \begin{bmatrix} 0 \\ 0 \end{bmatrix}, \quad D_0^{-1} = \begin{bmatrix} -1 & 0 \\ 0 & -1 \end{bmatrix}, \quad J_0 = \{6, 7\}, \quad c'x_0 = 0, \quad j = 0.$$

Iteration 0

Step 1: $v_2 = \max\{-1, 11\} = 11, \quad k = 2, \quad s_0 = \begin{bmatrix} 0 \\ -1 \end{bmatrix}.$

Step 2: $\sigma_0 = \min\left\{\dfrac{-10}{-2}, \dfrac{-5}{-1}, \dfrac{-25}{-5}, \dfrac{-29}{-3}, \dfrac{-45}{-2}, -, -\right\} = 5,$

$l = 1.$

Step 3: $x_1 = \begin{bmatrix} 0 \\ 0 \end{bmatrix} - 5\begin{bmatrix} 0 \\ -1 \end{bmatrix} = \begin{bmatrix} 0 \\ 5 \end{bmatrix}, \quad c'x_1 = -55,$

$D_1^{-1} = \begin{bmatrix} -1 & 0 \\ -9/2 & 1/2 \end{bmatrix}, \quad J_1 = \{6, 1\}, \quad j = 1.$

Iteration 1

Step 1: $v_1 = \max \left\{ \dfrac{97}{2}, \dfrac{-11}{2} \right\} = \dfrac{97}{2}, \quad k = 1, \quad s_1 = \begin{bmatrix} -1 \\ -9/2 \end{bmatrix}.$

Step 2: $\sigma_1 = \min \left\{ -, \dfrac{0}{-5/2}, \dfrac{0}{-39/2}, \dfrac{-14}{-29/2}, \dfrac{-35}{-12}, -, - \right\} = 0,$

$l = 2.$

Step 3: $x_2 = x_1 = \begin{bmatrix} 0 \\ 5 \end{bmatrix}, \quad c'x_2 = -55,$

$D_2^{-1} = \begin{bmatrix} 2/5 & -1/5 \\ 9/5 & -2/5 \end{bmatrix}, \quad J_2 = \{2, 1\}, \quad j = 2.$

Iteration 2

Step 1: $v_2 = \max \left\{ \dfrac{-97}{5}, \dfrac{21}{5} \right\} = \dfrac{21}{5}, \quad k = 2, \quad s_2 = \begin{bmatrix} -1/5 \\ -2/5 \end{bmatrix}.$

Step 2: $\sigma_2 = \min \left\{ -, -, \dfrac{0}{-7/5}, \dfrac{-14}{-7/5}, \dfrac{-35}{-7/5}, -, - \right\} = 0, \quad l = 3.$

Step 3: $x_3 = x_2 = \begin{bmatrix} 0 \\ 5 \end{bmatrix}, \quad c'x_3 = -55,$

$D_3^{-1} = \begin{bmatrix} -5/7 & 1/7 \\ -3/7 & 2/7 \end{bmatrix}, \quad J_3 = \{2, 3\}, \quad j = 3.$

Iteration 3

Step 1: $v_1 = \max \{4, -3\} = 4, \quad k = 1, \quad s_3 = \begin{bmatrix} -5/7 \\ -3/7 \end{bmatrix}.$

Step 2: $\sigma_3 = \min \left\{ -, -, -, \dfrac{-14}{-2}, \dfrac{-35}{-3}, -, - \right\} = 7, \quad l = 4.$

Step 3: $x_4 = \begin{bmatrix} 0 \\ 5 \end{bmatrix} - 7 \begin{bmatrix} -5/7 \\ -3/7 \end{bmatrix} = \begin{bmatrix} 5 \\ 8 \end{bmatrix}, \quad c'x_4 = -83,$

$D_4^{-1} = \begin{bmatrix} 5/14 & -3/14 \\ 3/14 & 1/14 \end{bmatrix}, \quad J_4 = \{4, 3\}, \quad j = 4.$

Iteration **4**

Step 1: $v_2 = \max\{-2, -1\} = -1$, $k = 2$.

$v_2 \leq 0$; stop with optimal solution x_4.

In Example 2.4 the extreme point $P_1 = (0, 5)'$ is degenerate. Constraints (1), (2), (3), and (6) are active at this point (see Figure 2.6). The algorithm starts with the extreme point P_0 and reaches P_1 after the first iteration. The next two iterations bring no progress. We have

$$x_1 = x_2 = x_3 \quad \text{and} \quad J_1 = \{6, 1\}, \quad J_2 = \{2, 1\}, \quad J_3 = \{2, 3\}.$$

At Iteration 3 the algorithm escapes the degenerate extreme point and arrives at the optimal solution.

Because P_1 is a degenerate extreme point it can be represented by several subsets of two active constraints. The algorithm arrives at P_1 with the active set $J_1 = \{6, 1\}$, and then works its way through $J_2 = \{2, 1\}$ to the active set $J_3 = \{2, 3\}$ before it finds a search direction which allows a positive step size.

Example 2.4 shows that finite termination can occur even if the algorithm encounters degenerate extreme points. However, it could happen that the algorithm generates points $x_j = x_{j+1} = \cdots = x_{j+k}$ and index sets $J_j, J_{j+1}, \ldots, J_{j+k}$ with $J_{j+k} = J_j$. In theory this means that after k iterations the algorithm uses the same data, namely $x_{j+k} = x_j$ and $D_{j+k}^{-1} = D_j^{-1}$, again, and will therefore repeat this cycle indefinitely. In practice, however,

Figure 2.6   Feasible region for Example 2.4.

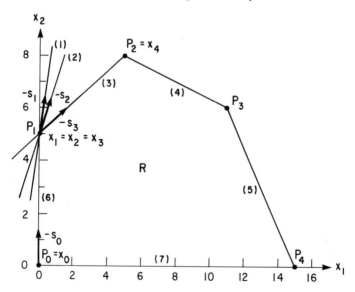

this has hardly ever been observed, because, due to rounding errors, the data used after k iterations, even though theoretically identical with the previous data, will be different.

Therefore, it is common to state algorithms without any provision to avoid the possibility of cycling. For theoretical arguments, however, especially for proving finite termination it is important to deal with this problem. It is somewhat surprising that a simple change in the rules for selecting k in Step 1 of the algorithm is sufficient to assure finite termination in the presence of degenerate extreme points.

In Step 1 of Algorithm 1, k is determined as the smallest integer such that

$$v_k = \max\{v_i \mid i = 1, \ldots, n\}. \tag{2.13}$$

The modified rule specifies k as the smallest integer such that $v_k > 0$. This procedure, together with the "smallest l" rule of Step 2, was first suggested by Bland [1977] for the simplex method (see Chapter 4). We refer to these as Bland's rules.

If k is computed by (2.12), the resulting search direction corresponds to that with greatest rate of decrease in the objective function. Application of Bland's rules, on the other hand, could give an edge along which the objective function barely decreases and thus slow down the progress of the algorithm considerably. It is, therefore, desirable to invoke Bland's rules only when a zero step size indicates the possibility of cycling. For details the reader is referred to Appendix A.

2.4 INITIALIZATION WITH AN ARBITRARY FEASIBLE POINT

Algorithm 1 requires an initial extreme point x_0 and an associated matrix D_0^{-1}. For many problems these data may not be readily available. Therefore, we will now modify Algorithm 1 so that it can be initiated with an arbitrary feasible point x_0 and an arbitrary matrix D_0^{-1}.

Suppose that x_0 is a known feasible point for our model problem

$$\min\{c'x \mid Ax \leq b\}.$$

We associate with x_0 a nonsingular (n, n) matrix D_0' and an index set J_0,

$$D_0' = \left[d_1, \ldots, d_n\right], \quad J_0 = \{\alpha_{10}, \ldots, \alpha_{n0}\}.$$

Some of the columns of D_0' may be gradients of constraints active at x_0. The remaining columns are arbitrary vectors such that D_0 is nonsingular. The two types of columns are differentiated by means of the index set. If d_i is an arbitrary column, we assign the value zero to its associated index α_{i0}. If $d_i = a_k$, we assign α_{i0} the value k. Thus, $d_i = a_{\alpha_{i0}}$ for all i with $1 \leq \alpha_{i0} \leq m$.

We allow the possibility of $\alpha_{i0} = 0$, $i = 1, \ldots, n$. Since we require

$$D_0^{-1} = \left[c_{10}, \ldots, c_{n0} \right],$$

it is convenient in this case to choose D_0' equal to the (n, n) identity matrix. We also allow the possibility of $1 \leq \alpha_{i0} \leq m$, $i = 1, \ldots, n$ so that Algorithm 1 will be a special case.

Consider the effect of setting $s_0 = c_{k0}$, where k is such that $\alpha_{k0} = 0$. We have $c'(x_0 - \sigma s_0) < c'x_0$ provided that $c's_0 > 0$ and $\sigma > 0$. If $c's_0 > 0$, we can proceed as in Algorithm 1. If $c's_0 \leq 0$, we replace s_0 with $-c_{k0}$ to obtain $c's_0 \geq 0$. Since the kth column of D_0' is not the gradient of an active constraint, we can expect that both choices,

$$s_0 = c_{k0} \quad \text{and} \quad s_0 = -c_{k0}$$

lead to a positive step size σ_0 (compare Figure 2.7). Therefore, an appropriate choice of search direction is

$$s_0 = \begin{cases} c_{k0}, & \text{if } c'c_{k0} > 0, \\ -c_{k0}, & \text{if } c'c_{k0} \leq 0. \end{cases} \tag{2.14}$$

Then $c's_0 = |c'c_{k0}|$ and

$$c'(x_0 - \sigma s_0) = c'x_0 - \sigma |c'c_{k0}|.$$

Thus the greatest rate of decrease is obtained by choosing k such that

$$|c'c_{k0}| = \max\{ |c'c_{i0}| \mid \text{all } i \text{ with } \alpha_{i0} = 0 \}. \tag{2.15}$$

Having chosen k according to (2.15) and s_0 from (2.14) we continue by using Steps 2 and 3 of Algorithm 1. This gives $x_1 = x_0 - \sigma_0 s_0$ with $c'x_1 \leq c'x_0$, and constraint l (using the notation of Step 2) which was inactive at x_0 has become active at x_1. The number of active constraints has thus increased by 1. D_1' is obtained from D_0' by replacing the "artificial" column d_k with a_l. Thus $J_1 = \{ \alpha_{11}, \ldots, \alpha_{n1} \}$ with $\alpha_{i1} = \alpha_{i0}$, $i \neq k$, and $\alpha_{k1} = l$. Let p be the number of strictly positive elements in J_0. Repeating this process $n - p$ times will locate an extreme point x_{n-p} provided that the maximum feasible step size is always defined. If this is not the case, there is $j < n - p$ such that $x_j - \sigma s_j$ is feasible for all $\sigma \geq 0$. If $|c'c_{kj}| > 0$, then $c'(x_j - \sigma s_j) = c'x_j - \sigma |c'c_{kj}|$ implies that the problem is unbounded from below. If $|c'c_{kj}| = 0$ for some $j < n - p$, the objective function cannot be decreased if $s_j = \pm c_{ij}$ for any i with $\alpha_{ij} = 0$. In this case, we proceed as in Step 1 of Algorithm 1 by dropping an active constraint.

Incorporating the foregoing procedure into Step 1 of Algorithm 1, we obtain

ALGORITHM 2

Model Problem: $\min\{ c'x \mid a_i'x \leq b_i, \ i = 1, \ldots, m \}$

Initialization:

Start with any feasible point x_0, $J_0 = \{\alpha_{10}, \ldots, \alpha_{n0}\}$, and D_0^{-1}, where $D_0' = \left[d_1, \ldots, d_n\right]$ is nonsingular, and, $d_i = a_{\alpha_{i0}}$ for all i with $1 \leq \alpha_{i0} \leq m$. Compute $c'x_0$ and set $j = 0$.

Step 1:   Computation of Search Direction s_j.

Let $D_j^{-1} = \left[c_{1j}, \ldots, c_{nj}\right]$ and $J_j = \{\alpha_{1j}, \ldots, \alpha_{nj}\}$. If $\alpha_{ij} \geq 1$ for $i = 1, \ldots, n$, go to Step 1.2. Otherwise, go to Step 1.1.

Step 1.1:

Compute

$$v_i = c'c_{ij}, \quad \text{for all } i \text{ with } \alpha_{ij} = 0.$$

Determine the smallest index k such that

$$|v_k| = \max\{|v_i| \mid \text{all } i \text{ with } \alpha_{ij} = 0\}.$$

If $v_k = 0$, go to Step 1.2. Otherwise, set $s_j = c_{kj}$ if $v_k > 0$ and $s_j = -c_{kj}$ if $v_k < 0$. Go to Step 2.

Step 1.2:

Compute

$$v_i = c'c_{ij}, \quad \text{for all } i \text{ with } \alpha_{ij} \geq 1.$$

Determine the smallest index k such that

$$v_k = \max\{v_i \mid \text{all } i \text{ with } \alpha_{ij} \geq 1\}.$$

If $v_k \leq 0$, stop with optimal solution x_j. Otherwise, set $s_j = c_{kj}$ and go to Step 2.

Step 2:   Computation of Maximum Feasible Step Size σ_j.
Same as Step 2 of Algorithm 1.

Step 3:   Update.
Same as Step 3 of Algorithm 1.

We illustrate Algorithm 2 by applying it to the following example.

Example 2.5

$$
\begin{array}{lrcll}
\text{minimize:} & -6x_1 - 8x_2 & & & \\
\text{subject to:} & -3x_1 + 5x_2 & \leq & 25, & (1) \\
& x_1 + 3x_2 & \leq & 29, & (2) \\
& 3x_1 + 2x_2 & \leq & 45, & (3) \\
& -x_1 & \leq & 0, & (4) \\
& -x_2 & \leq & 0. & (5)
\end{array}
$$

Initialization:

$$x_0 = \begin{bmatrix} 2 \\ 3 \end{bmatrix}, \quad D_0' = \begin{bmatrix} 1 & 0 \\ 0 & 1 \end{bmatrix}, \quad J_0 = \{0, 0\}, \quad D_0^{-1} = \begin{bmatrix} 1 & 0 \\ 0 & 1 \end{bmatrix},$$

$$c'x_0 = -36, \quad j = 0.$$

Iteration 0

Step 1:   Transfer to Step 1.1.

Step 1.1: $|v_2| = \max\{|-6|, |-8|\} = 8, \quad k = 2, \quad s_0 = \begin{bmatrix} 0 \\ -1 \end{bmatrix}.$

Step 2: $\sigma_0 = \min\left\{ \dfrac{-16}{-5}, \dfrac{-18}{-3}, \dfrac{-33}{-2}, -, - \right\} = \dfrac{16}{5}, \quad l = 1.$

Step 3: $x_1 = \begin{bmatrix} 2 \\ 3 \end{bmatrix} - \dfrac{16}{5} \begin{bmatrix} 0 \\ -1 \end{bmatrix} = \begin{bmatrix} 2 \\ 31/5 \end{bmatrix}, \quad c'x_1 = \dfrac{-308}{5},$

$$D_1^{-1} = \begin{bmatrix} 1 & 0 \\ 3/5 & 1/5 \end{bmatrix}, \quad J_1 = \{0, 1\}, \quad j = 1.$$

Iteration 1

Step 1:   Transfer to Step 1.1.

Step 1.1:[3] $|v_1| = \max\left\{ \left| \dfrac{-54}{5} \right|, - \right\} = \dfrac{54}{5}, \quad k = 1,$

$$s_1 = \begin{bmatrix} -1 \\ -3/5 \end{bmatrix}.$$

Step 2: $\sigma_1 = \min\left\{ -, \dfrac{-42/5}{-14/5}, \dfrac{-133/5}{-21/5}, -, - \right\} = 3, \quad l = 2.$

Step 3: $x_2 = \begin{bmatrix} 2 \\ 31/5 \end{bmatrix} - 3 \begin{bmatrix} -1 \\ -3/5 \end{bmatrix} = \begin{bmatrix} 5 \\ 8 \end{bmatrix}, \quad c'x_2 = -94,$

$$D_2^{-1} = \begin{bmatrix} 5/14 & -3/14 \\ 3/14 & 1/14 \end{bmatrix}, \quad J_2 = \{2, 1\}, \quad j = 2.$$

---

[3] We use a dash to indicate those columns of D_j^{-1} that are not currently candidates for the search direction.

Iteration 2

Step 1:   Transfer to Step 1.2.

Step 1.2: $v_2 = \max\left\{\dfrac{-27}{7}, \dfrac{5}{7}\right\} = \dfrac{5}{7}, \quad k = 2, \quad s_2 = \begin{bmatrix} -3/14 \\ 1/14 \end{bmatrix}.$

Step 2: $\sigma_2 = \min\left\{-, -, \dfrac{-14}{-1/2}, -, \dfrac{-8}{-1/14}\right\} = 28, \quad l = 3.$

Step 3: $x_3 = \begin{bmatrix} 5 \\ 8 \end{bmatrix} - 28\begin{bmatrix} -3/14 \\ 1/14 \end{bmatrix} = \begin{bmatrix} 11 \\ 6 \end{bmatrix}, \quad c'x_3 = -114,$

$$D_3^{-1} = \begin{bmatrix} -2/7 & 3/7 \\ 3/7 & -1/7 \end{bmatrix}, \quad J_3 = \{2,3\}, \quad j = 3.$$

Iteration 3

Step 1:   Transfer to Step 1.2.

Step 1.2: $v_2 = \max\left\{\dfrac{-12}{7}, \dfrac{-10}{7}\right\} = \dfrac{-10}{7}, \quad k = 2.$

$v_2 \leq 0$; stop with optimal solution x_3.

Figure 2.7 shows the progress of Algorithm 2 for this example.

Figure 2.7   Example 2.5.

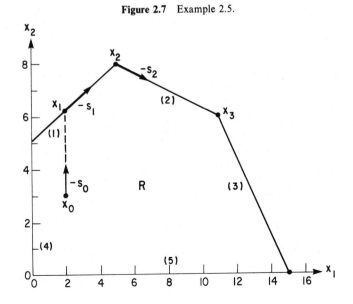

A computer program which implements Algorithm 2 is given in Appendix B, Section B.1.2. The output from applying this program to the problem of Example 2.5 is shown in Figure B.15.

In the preceding example Algorithm 2 terminated with an optimal extreme point. The following example shows that the algorithm may terminate with an optimal solution that is not an extreme point.

Example 2.6

$$
\begin{array}{rlrll}
\text{minimize:} & & x_1 - 2x_2 & & \\
\text{subject to:} & -x_1 + 2x_2 & \leq & 4, & (1) \\
& x_1 & \leq & 8, & (2) \\
& -x_1 & \leq & -2, & (3) \\
& -x_2 & \leq & 0. & (4)
\end{array}
$$

Initialization:

$$
x_0 = \begin{bmatrix} 5 \\ 1 \end{bmatrix}, \quad D_0' = \begin{bmatrix} 1 & 0 \\ 0 & 1 \end{bmatrix}, \quad J_0 = \{0, 0\}, \quad D_0^{-1} = \begin{bmatrix} 1 & 0 \\ 0 & 1 \end{bmatrix},
$$

$c'x_0 = 3, \quad j = 0.$

Iteration 0

Step 1:   Transfer to Step 1.1.

Step 1.1: $|v_2| = \max\{|1|, |-2|\} = 2, \quad k = 2, \quad s_0 = \begin{bmatrix} 0 \\ -1 \end{bmatrix}.$

Step 2: $\sigma_0 = \min\left\{\dfrac{-7}{-2}, -, -, -\right\} = \dfrac{7}{2}, \quad l = 1.$

Step 3: $x_1 = \begin{bmatrix} 5 \\ 1 \end{bmatrix} - \dfrac{7}{2}\begin{bmatrix} 0 \\ -1 \end{bmatrix} = \begin{bmatrix} 5 \\ 9/2 \end{bmatrix}, \quad c'x_1 = -4.$

$$
D_1^{-1} = \begin{bmatrix} 1 & 0 \\ 1/2 & 1/2 \end{bmatrix}, \quad J_1 = \{0, 1\}, \quad j = 1.
$$

Iteration 1

Step 1:   Transfer to Step 1.1.

Step 1.1: $|v_1| = \max\{|0|, -\} = 0, \quad k = 1.$ Transfer to Step 1.2.

Step 1.2: $v_2 = \max\{-, -1\} = -1, \quad k = 2.$

$v_2 \leq 0$; stop with optimal solution x_1.

Figure 2.8 shows the progress of the algorithm for the problem of Example 2.6.

Figure 2.8 shows that x_1 is an optimal solution, but not an extreme point. Furthermore, it is easy to verify that any point on the line segment connecting P_1 and P_2 is an optimal solution. Why did the algorithm not locate one of the optimal extreme points P_1 and P_2?

In Step 1.1 of Iteration 1 we have $c'c_{k1} = 0$. This results in a transfer to Step 1.2. If instead, we set $s_1 = c_{k1}$ and transferred to Step 2, we would have

$$\sigma_1 = \min\left\{-, -, \frac{-3}{-1}, \frac{-9/2}{-1/2}\right\} = 3, \quad l = 3.$$

Thus

$$x_2 = \begin{bmatrix} 5 \\ 9/2 \end{bmatrix} - 3\begin{bmatrix} 1 \\ 1/2 \end{bmatrix} = \begin{bmatrix} 2 \\ 3 \end{bmatrix}.$$

This is precisely the extreme point P_1. But now suppose that we remove constraints (3) and (4) from the problem. If we again set $s_1 = c_{k1}$ and transfer to Step 2, σ_1 is not defined and the algorithm terminates with the message that the problem is unbounded from below. To avoid this error we have chosen to transfer to Step 1.2 whenever $c'c_{kj} = 0$ in Step 1.1 of the algorithm.

Figure 2.8 Example 2.6.

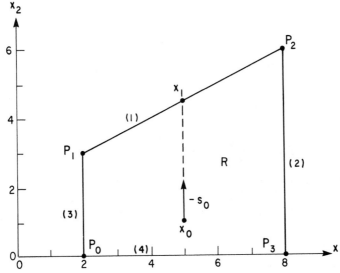

In Examples 2.5 and 2.6 the initial point x_0 is an interior point of R (i.e., no constraints are active at x_0). The following example shows that Algorithm 2 is also applicable if x_0 is on the boundary of R.

Example 2.7

We apply Algorithm 2 to the problem of Example 2.5 with $x_0 = (14, 3/2)'$ as the initial point. Constraint (3) is active at x_0.

Initialization:

$$x_0 = \begin{bmatrix} 14 \\ 3/2 \end{bmatrix}, \quad D_0' = \begin{bmatrix} 1 & 0 \\ 0 & 1 \end{bmatrix}, \quad J_0 = \{0, 0\}, \quad D_0^{-1} = \begin{bmatrix} 1 & 0 \\ 0 & 1 \end{bmatrix},$$

$$c'x_0 = -96, \quad j = 0.$$

Iteration 0

Step 1:   Transfer to Step 1.1.

Step 1.1: $|v_2| = \max \{ |-6|, |-8| \} = 8, \quad k = 2, \quad s_0 = \begin{bmatrix} 0 \\ -1 \end{bmatrix}.$

Step 2: $\sigma_0 = \min \left\{ \dfrac{-119/2}{-5}, \dfrac{-21/2}{-3}, \dfrac{0}{-2}, -, - \right\} = 0, \quad l = 3.$

Step 3: $x_1 = x_0 = \begin{bmatrix} 14 \\ 3/2 \end{bmatrix}, \quad c'x_1 = -96,$

$$D_1^{-1} = \begin{bmatrix} 1 & 0 \\ -3/2 & 1/2 \end{bmatrix}, \quad J_1 = \{0, 3\}, \quad j = 1.$$

Iteration 1

Step 1:   Transfer to Step 1.1.

Step 1.1: $|v_1| = \max \{ |6|, - \} = 6, \quad k = 1, \quad s_1 = \begin{bmatrix} 1 \\ -3/2 \end{bmatrix}.$

Step 2: $\sigma_1 = \min \left\{ \dfrac{-119/2}{-21/2}, \dfrac{-21/2}{-7/2}, -, \dfrac{-14}{-1}, - \right\} = 3, \quad l = 2.$

Step 3: $x_2 = \begin{bmatrix} 14 \\ 3/2 \end{bmatrix} - 3 \begin{bmatrix} 1 \\ -3/2 \end{bmatrix} = \begin{bmatrix} 11 \\ 6 \end{bmatrix}, \quad c'x_2 = -114,$

$$D_2^{-1} = \begin{bmatrix} -2/7 & 3/7 \\ 3/7 & -1/7 \end{bmatrix}, \quad J_2 = \{2, 3\}, \quad j = 2.$$

Iteration 2

Step 1:   Transfer to Step 1.2.

Step 1.2: $v_2 = \max \left\{ \dfrac{-12}{7}, \dfrac{-10}{7} \right\} = \dfrac{-10}{7}, \quad k = 2.$

$v_2 \leq 0$; stop with optimal solution x_2.

At Iteration 0 no progress is made (i.e., $x_1 = x_0$). The algorithm recognizes constraint (3) as active and incorporates its gradient into the matrix D_1'. Iteration 1 yields the optimal solution.

In the following theorem we will establish the main properties of Algorithm 2. In order to prove finite termination we will assume that for every point encountered by the algorithm, the gradients of the active constraints are linearly independent. This is analogous to the assumption of nondegenerate extreme points used in Section 2.3.

Theorem 2.3.

Let Algorithm 2 be applied to the model problem

$$\min \{ c'x \mid Ax \leq b \}$$

beginning with an arbitrary feasible x_0 and let $x_1, x_2, \ldots, x_j, \ldots$ be the sequence of iterates so obtained. If for every j the gradients of those constraints that are active at x_j are linearly independent, then Algorithm 2 terminates after a finite number of steps with either an optimal solution or the information that the problem is unbounded from below.

Proof:

By construction, at every iteration j,

$$a_i' x_j = b_i, \quad \text{for all } i \in J_j,$$

and

$$a_i' x_j \leq b_i, \quad \text{for all } i \notin J_j.$$

If the algorithm terminates with the message "problem is unbounded from below," $c' s_j > 0$ and $a_i' s_j \geq 0$ for $i = 1, \ldots, m$. Thus $x_j - \sigma s_j$ is feasible for all $\sigma \geq 0$ and $c'(x_j - \sigma s_j)$ is a strictly decreasing function of σ. The problem is indeed unbounded from below. If termination occurs with "optimal solution x_j," then we have $v_i = c' c_{ij} = 0$ for all i with $\alpha_{ij} = 0$ and $v_i = c' c_{ij} \leq 0$ for all i with $\alpha_{ij} \geq 1$. Let

$$D_j' = \begin{bmatrix} d_{1j}, \ldots, d_{nj} \end{bmatrix}$$

with $d_{ij} = a_{\alpha_{ij}}$ for all i with $\alpha_{ij} \geq 1$. Then

$$-c = u_1 d_{1j} + u_2 d_{2j} + \cdots + u_n d_{nj}$$

and $u_i = -c' c_{ij}$ for $i = 1, \ldots, n$. Thus $u_i = 0$ for all i with $\alpha_{ij} = 0$ and $u_i \geq 0$ for all i with $\alpha_{ij} \geq 1$. It follows from Theorem 2.1 that x_j is indeed an optimal solution.

To prove finite termination we observe that the number of zero elements in the index set cannot increase. It decreases by one whenever s_j is determined by Step 1.1 of the algorithm. Thus there is j_1 such that, for $j \geq j_1$, control is always transferred to Step 1.2. If all elements of J_{j_1} are positive, x_{j_1} is an extreme point and Algorithm 2 is, for $j \geq j_1$ identical with Algorithm 1. In this case finite termination follows from Theorem 2.2.

Now suppose that J_{j_1} contains some zero elements. For convenience assume that $\alpha_{ij_1} > 0$ for $i = 1, \ldots, p < n$ and $\alpha_{ij_1} = 0$ for $i = p + 1, \ldots, n$. Then we have

$$\alpha_{ij} > 0, \quad i = 1, \ldots, p; \quad \alpha_{ij} = 0, \quad i = p + 1, \ldots, n, \quad \text{for all } j \geq j_1.$$

We first show that $\sigma_j > 0$ for all $j \geq j_1$. The proof of this statement is by contradiction. Suppose that $\sigma_j = 0$ for some $j \geq j_1$. Renumbering constraints if necessary, we may assume that $\alpha_{ij} = i, i = 1, \ldots, p, k = p$, and $l = p + 1$, where k and l are the indices determined in Steps 1.2 and 2 of Algorithm 2, respectively. Then

$$s_j = c_{pj}, \quad a'_{p+1} x_{j+1} = b_{p+1}, \quad a'_{p+1} s_j < 0,$$

$$D'_j = \left[a_1, \ldots, a_p, d_{p+1,j}, \ldots, d_{nj} \right],$$

$$D'_{j+1} = \left[a_1, \ldots, a_{p-1}, a_{p+1}, d_{p+1,j}, \ldots, d_{nj} \right],$$

and

$$c = \sum_{i=1}^{p} (c' c_{ij}) a_i, \quad \text{with } c' c_{pj} = c' s_j > 0. \tag{2.16}$$

Let

$$a_{p+1} = \lambda_1 a_1 + \cdots + \lambda_p a_p + \lambda_{p+1} d_{p+1,j} + \cdots + \lambda_n d_{nj}. \tag{2.17}$$

Multiplying this equality by c_{pj} we obtain $\lambda_p = a'_{p+1} c_{pj} > 0$. Thus solving for a_p we have

$$a_p = \frac{1}{\lambda_p} (a_{p+1} - \lambda_1 a_1 - \cdots - \lambda_{p-1} a_{p-1} - \lambda_{p+1} d_{p+1,j} - \cdots - \lambda_n d_{nj}).$$

Substituting this expression into (2.16) we obtain

$$c = v_1 a_1 + \cdots + v_{p-1} a_{p-1} + v_p a_{p+1} + v_{p+1} d_{p+1,j} + \cdots + v_n d_{nj},$$

where $v_p = c' c_{pj} / \lambda_p$, $v_i = c' c_{ij} - \lambda_i c' c_{pj} / \lambda_p$ for $i = 1, \ldots, p - 1$, and $v_i = -\lambda_i c' c_{pj} / \lambda_p$ for $i = p + 1, \ldots, n$. Because gradients of active constraints are assumed to be linearly independent it follows from (2.17) that at least one of the $\lambda_{\rho+1}, \ldots, \lambda_n$ is different from zero. Thus at least one of the multipliers $v_{\rho+1}, \ldots, v_n$ is different from zero. This implies that in the next iteration s_j is determined by Step 1.1. Since $j \geq j_1$ this is a contradiction. Thus $\sigma_j > 0$ and $c'x_{j+1} \leq c'x_j$ for all $j \geq j_1$.

For any $j \geq j_1$ let x_j, J_j, and D_j' be as in Algorithm 2. Then

$$c = \sum_{i=1}^{\rho} (c'c_{ij})a_{\alpha_{ij}}. \tag{2.18}$$

Let

$$T_j = \{x \mid a_{\alpha_{ij}}'x = b_{\alpha_{ij}}, \ i = 1, \ldots, \rho\}.$$

Then $x_j \in T_j$ and for every $\hat{x} \in T_j$ we have $a_{\alpha_{ij}}'(\hat{x} - x_j) = 0$ for $i = 1, \ldots, \rho$. Therefore, it follows from (2.18) that

$$c'\hat{x} = c'x_j + c'(\hat{x} - x_j) = c'x_j$$

(i.e., $c'x$ is constant on the set T_j). Since $c'x_{j+1} < c'x_j$ for all $j \geq j_1$, there are at most as many iterates x_j with $j \geq j_1$ as there are different sets T_j. Each T_j is determined by a subset of ρ constraints. Thus

$$\binom{m}{\rho} = \frac{m!}{\rho!(m - \rho)!}$$

is an upper bound for the number of sets T_j. This completes the proof of the theorem. ∎

If the algorithm encounters an x_j at which the gradients of active constraints are linearly dependent, cycling is theoretically possible. As in the case of Algorithm 1, this difficulty can be overcome by a change in the choice of the index k in Step 1.2. For details, the reader is referred to Appendix A.

In Example 2.6 we have seen that Algorithm 2 need not terminate with an optimal extreme point. For post-optimality analysis and for parametric programming, however, it is convenient to be able to assume the availability of an optimal solution which is an extreme point. Therefore, we describe briefly how to locate an optimal extreme point.

Suppose that Algorithm 2 terminates with an optimal solution x_j. Let the index set J_j and the matrix D_j^{-1} be associated with x_j. If all elements of J_j are positive, then x_j is an extreme point. Otherwise, we have

$$c'c_{ij} = 0, \quad \text{for all } i \text{ with } \alpha_{ij} = 0.$$

We determine k as in Step 1.1 of Algorithm 2. Assuming that

$$a_i'c_{kj} \neq 0, \quad \text{for at least one } i,$$

we set $s_j = c_{kj}$ if there is at least one i with $a_i' c_{kj} < 0$ and $s_j = -c_{kj}$ if $a_i' c_{kj} \geq 0$ for all i. Then the algorithm continues with Step 2. Observe that σ_j is well-defined. The iteration terminates with an index set J_{j+1} which has one zero less than J_j.

Repeating this process we either arrive at an optimal extreme point or find a c_{kj} such that $a_i' c_{kj} = 0$ for $i = 1, \ldots, m$. In this case the feasible region of our model problem has no extreme points. Indeed, since every a_i is orthogonal to the nonzero vector c_{kj} the set $\{a_1, \ldots, a_m\}$ contains at most $n - 1$ linearly independent vectors. However, an extreme point requires n active constraints with linearly independent gradients.

This shows that if our model problem

$$\min\{c'x \mid Ax \leq b\}$$

has optimal solutions, it has at least one optimal solution which is an extreme point if and only if there is a subset of n constraints with linearly independent gradients. This would be the case, for example, if the constraints included nonnegativity restrictions on all variables.

The problem of the following example has no extreme points.

Example 2.8

$$\begin{aligned}
\text{minimize:} \quad & x_1 - 5x_2 \\
\text{subject to:} \quad & -2x_1 + 10x_2 \leq 50, \quad (1) \\
& 3x_1 - 15x_2 \leq -15. \quad (2)
\end{aligned}$$

Every point on the line $-2x_1 + 10x_2 = 50$ is an optimal solution (see Figure 2.9).

Figure 2.9   Feasible region for Example 2.8.

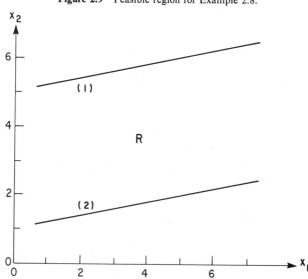

This may seem to be a rather contrived example. However, it can occur in practice due to errors in problem formulation. If large amounts of data are involved, input errors should be expected and may result in an "apparent" problem of the same type as in Example 2.8.

2.5 DETERMINATION OF AN INITIAL FEASIBLE POINT

Algorithm 2 can be applied to solve our model problem

$$\min \{ c'x \mid a_i'x \le b_i, \ i = 1, \ldots, m \} \tag{2.19}$$

provided that an initial feasible point x_0 is known. The purpose of this section is to show how to determine such an x_0. Consider the LP

$$\min \{ \alpha \mid a_i'x - \alpha \le b_i, \ i = 1, \ldots, m, \ -\alpha \le 0 \}. \tag{2.20}$$

For reasons to become apparent, we call (2.20) the initial point or phase 1 problem for (2.19). Observe the following.

1. Equation (2.20) is an LP in $n + 1$ variables (α is a scalar) and $m + 1$ constraints ($-\alpha \le 0$ is constraint number $m + 1$).
2. A feasible point for (2.20) is any x_0 together with any α_0 satisfying

    $$\alpha_0 \ge \max \{ 0, a_i'x_0 - b_i \mid i = 1, \ldots, m \},$$

    $x_0 = 0$ is a convenient choice.
3. Equation (2.20) may be solved by applying Algorithm 2 with initial point as in point 2.

Note that the objective function for (2.20) can be written as

$$(0', 1) \begin{bmatrix} x \\ \alpha \end{bmatrix},$$

so that for (2.20) the "c" of our usual model problem is

$$\begin{bmatrix} 0 \\ 1 \end{bmatrix}.$$

Let x^*, α^* be an optimal solution for the initial point problem. There are two cases to be considered.

Case 1: $\alpha^* > 0$

This implies that (2.19) has no feasible solution, for, suppose to the contrary that there is an $\hat{x}$ feasible for (2.19). But then $\hat{x}$ and $\alpha = 0$ is feasible for (2.20) and $\alpha = 0 < \alpha^*$ contradicts the optimality of α^*. Thus, solving (2.20) to obtain $\alpha^* > 0$ also solves our model problem (2.19) by showing that it has no feasible solution.

Case 2: $\alpha^* = 0$

In this case, x^* is our desired feasible point for (2.19). Algorithm 2 may now be applied to solve (2.19) beginning with $x_0 = x^*$.

The initial point problem and its solution are illustrated in the following example.

Example 2.9

Find a feasible point for the constraints

$$
\begin{array}{rcl}
x_1 + x_2 &\leq& 10, \qquad (1) \\
-x_1 - x_2 &\leq& -8, \qquad (2) \\
x_1 - x_2 &\leq& 6, \qquad (3) \\
-x_1 + x_2 &\leq& -4. \qquad (4)
\end{array}
$$

The initial point problem for these constraints is

$$
\begin{array}{ll}
\text{minimize:} & \alpha \\
\text{subject to:} &
\end{array}
$$

$$
\begin{array}{rcl}
x_1 + x_2 - \alpha &\leq& 10, \qquad (1) \\
-x_1 - x_2 - \alpha &\leq& -8, \qquad (2) \\
x_1 - x_2 - \alpha &\leq& 6, \qquad (3) \\
-x_1 + x_2 - \alpha &\leq& -4, \qquad (4) \\
-\alpha &\leq& 0. \qquad (5)
\end{array}
$$

Using $x_0 = 0$, we solve this using Algorithm 2 as follows.

Initialization:

$$
x_0 = \begin{bmatrix} 0 \\ 0 \end{bmatrix}, \quad \alpha_0 = 8 = \max\{0, -10, 8, -6, 4\}, \quad D_0^{-1} = \begin{bmatrix} 1 & 0 & 0 \\ 0 & 1 & 0 \\ 0 & 0 & 1 \end{bmatrix},
$$

$$
J_0 = \{0, 0, 0\}, \quad j = 0.
$$

Iteration 0

Step 1:   Transfer to Step 1.1.

Step 1.1: $|v_3| = \max\{|0|, |0|, |1|\} = 1, \quad k = 3, \quad s_0 = \begin{bmatrix} 0 \\ 0 \\ 1 \end{bmatrix}.$

Step 2: $\sigma_0 = \min\left\{\dfrac{-18}{-1}, \dfrac{0}{-1}, \dfrac{-14}{-1}, \dfrac{-4}{-1}, \dfrac{-8}{-1}\right\} = 0, \quad l = 2.$

Step 3:
$$\begin{bmatrix} x_1 \\ \alpha_1 \end{bmatrix} = \begin{bmatrix} x_0 \\ \alpha_0 \end{bmatrix} = \begin{bmatrix} 0 \\ 0 \\ 8 \end{bmatrix}, \quad D_1^{-1} = \begin{bmatrix} 1 & 0 & 0 \\ 0 & 1 & 0 \\ -1 & -1 & -1 \end{bmatrix},$$

$$J_1 = \{0, 0, 2\}, \quad j = 1.$$

Iteration 1

Step 1:   Transfer to Step 1.1.

Step 1.1: $|v_1| = \max\{\,|-1|, |-1|, -\,\} = 1, \quad k = 1, \quad s_1 = \begin{bmatrix} -1 \\ 0 \\ 1 \end{bmatrix}.$

Step 2: $\sigma_1 = \min\left\{ \dfrac{-18}{-2}, -, \dfrac{-14}{-2}, -, \dfrac{-8}{-1} \right\} = 7, \quad l = 3.$

Step 3:
$$\begin{bmatrix} x_2 \\ \alpha_2 \end{bmatrix} = \begin{bmatrix} 0 \\ 0 \\ 8 \end{bmatrix} - 7 \begin{bmatrix} -1 \\ 0 \\ 1 \end{bmatrix} = \begin{bmatrix} 7 \\ 0 \\ 1 \end{bmatrix},$$

$$D_2^{-1} = \begin{bmatrix} 1/2 & 0 & -1/2 \\ 0 & 1 & 0 \\ -1/2 & -1 & -1/2 \end{bmatrix}, \quad J_2 = \{3, 0, 2\}, \quad j = 2.$$

Iteration 2

Step 1:   Transfer to Step 1.1.

Step 1.1: $|v_2| = \max\{\,-, |-1|, -\,\} = 1, \quad k = 2, \quad s_2 = \begin{bmatrix} 0 \\ -1 \\ 1 \end{bmatrix}.$

Step 2: $\sigma_2 = \min\left\{ \dfrac{-4}{-2}, -, -, \dfrac{-4}{-2}, \dfrac{-1}{-1} \right\} = 1, \quad l = 5.$

Step 3:
$$\begin{bmatrix} x_3 \\ \alpha_3 \end{bmatrix} = \begin{bmatrix} 7 \\ 0 \\ 1 \end{bmatrix} - \begin{bmatrix} 0 \\ -1 \\ 1 \end{bmatrix} = \begin{bmatrix} 7 \\ 1 \\ 0 \end{bmatrix},$$

$$D_3^{-1} = \begin{bmatrix} 1/2 & 0 & -1/2 \\ -1/2 & 1 & -1/2 \\ 0 & -1 & 0 \end{bmatrix}, \quad J_3 = \{3, 5, 2\}, \quad j = 3.$$

Iteration 3

Step 1:   Transfer to Step 1.2.

Step 1.2: $v_1 = \max\{0, -1, 0\} = 0, \quad k = 1.$

$v_1 \leq 0$; stop with optimal solution x_3 and α_3.

Figure 2.10 shows the feasible region for Example 2.9 and the progress of Algorithm 2 in solving the associated initial point problem. A geometrical interpretation of the initial point problem can be made from Figure 2.10. The

Figure 2.10   Solution of the initial point problem for Example 2.9.

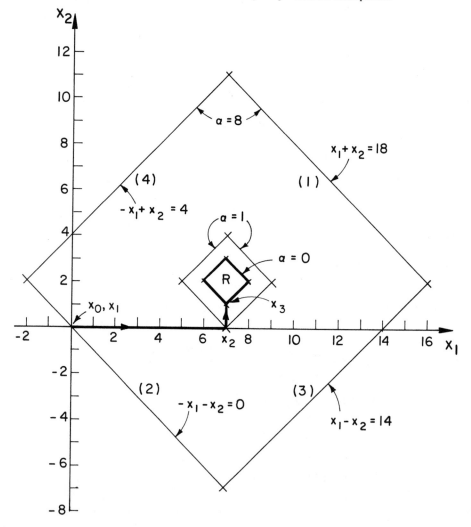

original feasible region is expanded by adding the same amount (α_0) to the right-hand side of each constraint until the origin $(x_0 = 0)$ becomes feasible. Applying Algorithm 2 to the initial point problem shrinks the modified feasible region until it is reduced to the original.

We have now developed a complete capability to solve the LP

$$\min\{c'x \mid Ax \leq b\}. \tag{2.21}$$

It is customary to refer to the two parts of the analysis as follows.[4]

Phase 1:
Formulate an initial point problem for (2.21). Solve it using Algorithm 2. If (2.21) does in fact possess a feasible solution, obtain x_0 as in Case 2. Set $D_0^{-1} = I$ and $J_0 = \{0, \ldots, 0\}$.

Phase 2:
Using the initial data x_0, J_0, and D_0^{-1} obtained in Phase 1, apply Algorithm 2 to solve (2.21).

In the procedure above we have chosen to associate the matrix $D_0^{-1} = I$ and the index set $J_0 = \{0, \ldots, 0\}$ with x_0, thus ignoring the fact that certain constraints are active at x_0. We now show how initial data x_0, J_0, and D_0^{-1}, which reflect these active constraints, can be extracted from the final data obtained by applying Algorithm 2 to the initial point problem. Let $\hat{x}_0$, $\hat{J}_0$, and $\hat{D}_0^{-1}$ denote the final data just described. We use the "hat" notation to emphasize that these are all $(n + 1)$-dimensional quantities. To determine D_0^{-1} from $\hat{D}_0^{-1}$ we require that $m + 1 \in \hat{J}_0$; that is, that the constraint $-\alpha \leq 0$ is associated with one of the columns of $\hat{D}_0^{-1}$. If the gradients of those constraints of (2.21) which are active at x_0 are linearly independent, then it can be shown (Exercise 2.26) that $m + 1 \in \hat{J}_0$. If this is not the case, Procedure Φ may be used to obtain a new matrix $\hat{D}_0^{-1}$ with the desired property (Exercise 2.26). Suppose for simplicity that the constraint $-\alpha \leq 0$ is associated with the last column of $\hat{D}_0^{-1}$. We can partition $\hat{x}_0$, $\hat{J}_0$, and $\hat{D}_0'$ as

$$\hat{x}_0 = \begin{bmatrix} x_0 \\ 0 \end{bmatrix}, \quad \hat{J}_0 = \{J_0, m + 1\},$$

and

$$\hat{D}_0' = \left[\begin{array}{c|c} a_{\alpha_{10}}, \ldots, a_{\alpha_{n0}} & 0 \\ \hline -1, \ldots, -1 & -1 \end{array} \right] = \left[\begin{array}{c|c} D_0' & 0 \\ \hline -e' & -1 \end{array} \right],$$

---

[4] The terms "Phase 1" and "Phase 2" are historically linked with the simplex method (see Chapter 4). They are used here in the same sense.

where $J_0 = \{\alpha_{10}, \ldots, \alpha_{n0}\}$, e is an n-vector of 1s, and

$$D_0' = \left[a_{\alpha_{10}}, \ldots, a_{\alpha_{n0}} \right].$$

Now $\hat{D}_0^{-1}$ is known and we would like to determine D_0^{-1}. From the bordered inverse lemma (see Exercise 2.24), $\hat{D}_0^{-1}$ can be written in partitioned form as

$$\hat{D}_0^{-1} = \left[\begin{array}{c|c} D_0^{-1} & -D_0^{-1}e \\ \hline 0' & -1 \end{array} \right].$$

Therefore, we can obtain D_0^{-1} by striking out the last row and column of $\hat{D}_0^{-1}$.

More generally, suppose that the kth element of $\hat{J}_0$ is $m + 1$, so that the kth column of $\hat{D}_0^{-1}$ is $(0', -1)'$. Arguing as above, we obtain J_0 from $\hat{J}_0$ by deleting the kth element and D_0^{-1} from $\hat{D}_0^{-1}$ by deleting the last row and the kth column.

Applying this process to the final data obtained in Example 2.9 gives (with $k = 2$)

$$x_0 = \begin{bmatrix} 7 \\ 1 \end{bmatrix}, \quad J_0 = \{3, 2\}, \quad \text{and} \quad D_0^{-1} = \begin{bmatrix} 1/2 & -1/2 \\ -1/2 & -1/2 \end{bmatrix}.$$

Computer programs which implement the Phase 1–Phase 2 procedure are given in Appendix B, Sections B.1.4 and B.1.5. The first program (Section B.1.4) begins the Phase 2 problem with $D_0^{-1} = I$ and $J_0 = \{0, \ldots, 0\}$. The second (Section B.1.5) extracts D_0^{-1} and J_0 from $\hat{D}_0^{-1}$ and $\hat{J}_0$, respectively, as described above. The programs are illustrated by applying them to the problem of minimizing $-x_2$ subject to the constraints of Example 2.9. The initial point problem for this is formulated and solved in Example 2.9. The computations for this may be compared with the output for the two Phase 1–Phase 2 programs shown in Figures B.24 and B.27.

2.6 A MODEL PROBLEM WITH EXPLICIT LINEAR EQUALITY CONSTRAINTS

Thus far our model problem has included inequality constraints only. We shall now show that a simple modification of Step 1.2 of Algorithm 2 is all that is needed to handle problems with inequality and equality constraints.

Let the model problem be

$$
\left.
\begin{aligned}
\text{minimize:} \quad & c'x \\
\text{subject to:} \quad & a_i'x \le b_i, \quad i = 1, \ldots, m, \\
& a_i'x = b_i, \quad i = m + 1, \ldots, m + r.
\end{aligned}
\right\} \tag{2.22}
$$

Without loss of generality, we may assume that $b_i \geq 0$ for $i = m + 1$ $, \ldots, m + r$.

First suppose that x_0 is a nondegenerate extreme point of (2.22). Then

$$a_i' x_0 = b_i, \quad i = m + 1, \ldots, m + r. \tag{2.23}$$

For simplicity, assume furthermore that

$$a_i' x_0 = b_i, \quad i = 1, \ldots, n - r; \quad a_i' x_0 < b_i, \quad i = n - r + 1, \ldots, m.$$

Then the appropriate index set and matrix to be associated with x_0 are

$$J_0 = \{1, \ldots, n - r, m + 1, \ldots, m + r\}$$

and

$$D_0' = \left[a_1, \ldots, a_{n-r}, a_{m+1}, \ldots, a_{m+r} \right],$$

respectively. As usual, let

$$D_0^{-1} = \left[c_{10}, \ldots, c_{n0} \right] \tag{2.24}$$

and consider the effect of choosing $s_0 = c_{k0}$ for some $k \in \{n - r + 1$ $, \ldots, n\}$. To be specific, let $k = n$. Then

$$a_{m+r}'(x_0 - \sigma s_0) = a_{m+r}' x_0 - \sigma = b_{m+r}$$

if and only if $\sigma = 0$. Therefore, none of the columns of D_0^{-1} corresponding to gradients of equality constraints is a candidate for the search direction s_0. On the other hand, if $k \in \{1, \ldots, n - r\}$, then by definition of the inverse matrix

$$a_i'(x_0 - \sigma s_0) = a_i' x_0 - \sigma \cdot 0 = b_i$$

for $i = m + 1, \ldots, m + r$ and all σ.

More generally, let x_0 be any feasible point for (2.22). Again we have the equalities (2.23). In addition, let

$$a_i' x_0 = b_i, \quad i = 1, \ldots, \rho < n - r; \quad a_i' x_0 < b_i, \quad i = \rho + 1, \ldots, m.$$

In this case we associate the index set

$$J_0 = \{1, \ldots, \rho, 0, \ldots, 0, m + 1, \ldots, m + r\}$$

and the matrix

$$D_0' = \left[a_1, \ldots, a_\rho, d_{\rho+1}, \ldots, d_{n-r}, a_{m+1}, \ldots, a_{m+r} \right]$$

with x_0. Let D_0^{-1} be as in (2.24) and set $s_0 = c_{k0}$. It follows again that $k \in \{n - r + 1, \ldots, n\}$ and $\sigma \neq 0$ imply that at least one equality constraint is violated, whereas $k \in \{1, \ldots, n - r\}$ implies that $x_0 - \sigma s_0$ satisfies the equality constraints for all σ.

Therefore, all c_{i0} with $m + 1 \leq \alpha_{i0} \leq m + r$ are excluded as potential candidates for s_0. Incorporating this restriction into Step 1.2 of Algorithm 2 we obtain an algorithm which is applicable to a model problem with inequality and equality constraints.

ALGORITHM 3

Model Problem:

$$\min \{ c'x \mid a_i'x \leq b_i, i = 1, \ldots, m; \ a_i'x = b_i, i = m + 1, \ldots, m + r \}$$

Initialization:
Start with any feasible point x_0, $J_0 = \{ \alpha_{10}, \ldots, \alpha_{n0} \}$, and D_0^{-1}, where $D_0' = \left[d_1, \ldots, d_n \right]$ is nonsingular, $d_i = a_{\alpha_{i0}}$ for all i with $1 \leq \alpha_{i0} \leq m + r$, and, each of $m + 1, \ldots, m + r$ is in J_0. Compute $c'x_0$ and set $j = 0$.

Step 1:   Computation of Search Direction s_j.
Same as Step 1 of Algorithm 2.

Step 1.1:
Same as Step 1.1 of Algorithm 2.

Step 1.2:
Compute

$$v_i = c'c_{ij}, \quad \text{for all } i \text{ with } 1 \leq \alpha_{ij} \leq m.$$

Determine the smallest index k such that

$$v_k = \max \{ v_i \mid \text{all } i \text{ with } 1 \leq \alpha_{ij} \leq m \}.$$

If $v_k \leq 0$, stop with optimal solution x_j. Otherwise, set $s_j = c_{kj}$ and go to Step 2.

Step 2:   Computation of Maximum Feasible Step Size σ_j.
Same as Step 2 of Algorithm 2.

Step 3:   Update.
Same as Step 3 of Algorithm 2.

We illustrate Algorithm 3 in

Example 2.10

$$\begin{array}{llll}
\text{minimize:} & - x_1 & & \\
\text{subject to:} & 2x_1 + x_2 \leq 5, & (1) \\
& - x_1 \quad\quad \leq 0, & (2) \\
& x_1 + x_2 = 3. & (3)
\end{array}$$

Note that $m = 2$ and $r = 1$.

Initialization:

$$x_0 = \begin{bmatrix} 1 \\ 2 \end{bmatrix}, \quad D_0^{-1} = \begin{bmatrix} 1 & -1 \\ 0 & 1 \end{bmatrix}, \quad J_0 = \{3,0\}, \quad c'x_0 = -1, \quad j = 0.$$

Iteration 0

Step 1:   Transfer to Step 1.1.

Step 1.1: $|v_2| = \max\{-, |1|\} = 1, \quad k = 2, \quad s_0 = \begin{bmatrix} -1 \\ 1 \end{bmatrix}.$

Step 2: $\sigma_0 = \min\left\{\dfrac{-1}{-1}, -\right\} = 1, \quad l = 1.$

Step 3: $x_1 = \begin{bmatrix} 1 \\ 2 \end{bmatrix} - \begin{bmatrix} -1 \\ 1 \end{bmatrix} = \begin{bmatrix} 2 \\ 1 \end{bmatrix}, \quad c'x_1 = -2,$

$$D_1^{-1} = \begin{bmatrix} -1 & 1 \\ 2 & -1 \end{bmatrix}, \quad J_1 = \{3,1\}, \quad j = 1.$$

Iteration 1

Step 1:   Transfer to Step 1.2.

Step 1.2: $v_2 = \max\{-, -1\} = -1, \quad k = 2.$

$v_2 \leq 0$; stop with optimal solution x_1.

The progress of Algorithm 3 with Example 2.10 is shown in Figure 2.11.

Figure 2.11   Algorithm 3 applied to Example 2.10.

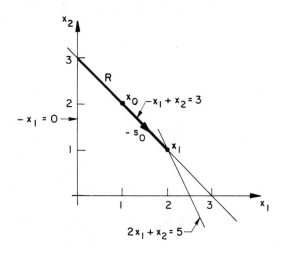

A computer program which implements Algorithm 3 is given in Appendix B, Section B.1.3. The output from applying this program to the problem of Example 2.10 is shown in Figure B.20.

The properties of Algorithm 3 are summarized in the following theorem.

Theorem 2.4.

Let Algorithm 3 be applied to the model problem (2.22) with an arbitrary feasible x_0, and let $x_1, \ldots, x_j, \ldots$ be the sequence of iterates so obtained. If for every j the gradients of those constraints that are active at x_j are linearly independent, Algorithm 3 terminates after a finite number of steps with either an optimal solution or the information that the problem is unbounded from below.

Proof:

The proof is identical with the proof of Theorem 2.3 except in the case if termination occurs with "optimal solution x_j." Let

$$D_j' = \left[d_{1j}, \ldots, d_{nj} \right]$$

with $d_{ij} = a_{\alpha_{ij}}$ for all i with $\alpha_{ij} \geq 1$. Then we have

$$-c = \sum_{i=1}^{n} -(c'c_{ij})d_{ij}, \qquad (2.25)$$

where $c'c_{ij} = 0$ for all i with $\alpha_{ij} = 0$, and $c'c_{ij} \leq 0$ for all i with $1 \leq \alpha_{ij} \leq m$. For all i with $m + 1 \leq \alpha_{ij} \leq m + r$ the multiplier $c'c_{ij}$ can be zero, positive, or negative. Let x be any feasible solution. Then $x = x_j + (x - x_j)$ and

$$a_i'x = a_i'x_j + a_i'(x - x_j) = b_i;$$

i.e.,

$$a_i'(x - x_j) = 0, \quad \text{for } i = m + 1, \ldots, m + r.$$

Furthermore, for $1 \leq \alpha_{ij} \leq m$,

$$a_{\alpha_{ij}}'x = a_{\alpha_{ij}}'x_j + a_{\alpha_{ij}}'(x - x_j) \leq b_i;$$

i.e.,

$$a_{\alpha_{ij}}'(x - x_j) \leq 0.$$

Using (2.25), we have therefore

$$-c'x = -c'x_j - c'(x - x_j) = -c'x_j - \sum_{i=1}^{n} (c'c_{ij})a_{\alpha_{ij}}'(x - x_j) \leq -c'x_j.$$

Thus $c'x_j \leq c'x$ and x_j is indeed an optimal solution. ∎

For an appropriate way to handle linearly dependent gradients of active constraints the reader is again referred to Appendix A.

To complete the specification of Algorithm 3, it remains to give a method for determining the initial data x_0, D_0^{-1}, and J_0. It is convenient to assume that $b_{m+1}, \ldots, b_{m+r}$ are all nonnegative. This assumption may be satisfied by multiplying both sides of an equality constraint by -1, if necessary. Having done this, set

$$d = - \sum_{i=m+1}^{m+r} a_i$$

and consider the following <u>initial point</u> or <u>phase</u> 1 <u>problem</u> for (2.22).

$$\left.\begin{array}{lll} \text{minimize:} & d'x + \alpha & \\ \text{subject to:} & a_i'x - \alpha \leq b_i, & i = 1, \ldots, m, \\ & a_i'x \qquad \leq b_i, & i = m+1, \ldots, m+r, \\ & \qquad -\alpha \leq 0. & \end{array}\right\} \quad (2.26)$$

Observe the following properties:

1. Equation (2.26) is an LP in $n + 1$ variables (α is a scalar) and $m + r + 1$ inequality constraints ($-\alpha \leq 0$ is constraint number $m + r + 1$).
2. A feasible solution for (2.26) is $x_0 = 0$ together with any α_0 satisfying

$$\alpha_0 \geq \max\{0, -b_i \mid i = 1, \ldots, m\}.$$

   (We have assumed that $b_i \geq 0$ for $i = m + 1, \ldots, m + r$.)
3. Equation (2.26) may be solved by applying Algorithm 2 with initial data as in property 2. An optimal solution exists because for any feasible point $(x', \alpha)'$

$$d'x + \alpha = - \sum_{i=m+1}^{m+r} a_i'x + \alpha \geq - \sum_{i=m+1}^{m+r} b_i.$$

Let x^*, α^* be an optimal solution for the initial point problem. There are two cases to be considered.

Case 1: $\quad d'x^* + \alpha^* > - \sum\limits_{i=m+1}^{m+r} b_i$

This implies that (2.22) has no feasible solution. Indeed, if we assume that $\hat{x}$ is any feasible solution for (2.22), then $\hat{x}$ and $\hat{\alpha} = 0$ is feasible for (2.26) and we have

$$d'\hat{x} + \hat{\alpha} = - \sum_{i=m+1}^{m+r} a_i'\hat{x} = - \sum_{i=m+1}^{m+r} b_i.$$

This contradicts the optimality of x^* and α^*.

Case 2: $d'x^* + \alpha^* = -\displaystyle\sum_{i=m+1}^{m+r} b_i$

In this case we claim that x^* is feasible for (2.22). To see this, first observe that because x^* is feasible for (2.26),

$$a_i'x^* \leq b_i, \quad i = m + 1, \ldots, m + r.$$

The definition of d implies that

$$d'x^* \geq -\sum_{i=m+1}^{m+r} b_i.$$

With the condition of Case 2, this implies that $\alpha^* \leq 0$. But the last constraint of (2.26) asserts that $\alpha^* \geq 0$. Thus $\alpha^* = 0$ and x^* satisfies the first m constraints of (2.22). Finally, suppose that $a_k'x^* < b_k$ for some k with $m + 1 \leq k \leq m + r$. Since x^* is feasible for (2.26), this implies that

$$\sum_{i=m+1}^{m+r} a_i'x^* < \sum_{i=m+1}^{m+r} b_i.$$

This contradiction establishes that

$$a_i'x^* = b_i, \quad i = m + 1, \ldots, m + r,$$

and x^* is feasible for (2.22), as required.

In addition to the feasible point x_0, Algorithm 3 requires a matrix D_0^{-1} and an index set J_0 with the property that each of the numbers $m + 1, \ldots, m + r$ is in J_0. We now show how these initial data can be obtained from the final data obtained by applying Algorithm 2 to the initial point problem. Let $\hat{x}_0$, $\hat{J}_0$, and $\hat{D}_0^{-1}$ denote these final data, where we use the "hat" notation again to emphasize that these are all $(n + 1)$-dimensional quantities.

If $\hat{J}_0$ contains each of the numbers $m + 1, \ldots, m + r, m + r + 1$, we have two options. First we can use the initial data $\hat{x}_0$, $\hat{J}_0$, and $\hat{D}_0^{-1}$ and apply Algorithm 3 to the problem

$$\left.\begin{array}{ll}
\text{minimize:} & c'x \\
\text{subject to:} & a_i'x - \alpha \leq b_i, \quad i = 1, \ldots, m, \\
& a_i'x \quad\quad = b_i, \quad i = m + 1, \ldots, m + r, \\
& \alpha = 0.
\end{array}\right\} \quad (2.27)$$

Obviously, solving this problem is equivalent to solving the model problem (2.22). Second, if the kth element of $\hat{J}_0$ is $m + r + 1$ [i.e., if the kth column of $\hat{D}_0'$ corresponds to the gradient of the $(m + r + 1)$th constraint $-\alpha \leq 0$], we can proceed as in Section 2.5 and obtain x_0 from $\hat{x}_0$ by deleting the last element, J_0 from $\hat{J}_0$ by deleting the kth element, and D_0^{-1} from $\hat{D}_0^{-1}$ by deleting the last row and the kth column. Using the initial data x_0, J_0, and D_0^{-1} we can apply Algorithm 3 to (2.22).

The arguments above show that a complete method to solve the LP

$$
\left.
\begin{array}{ll}
\text{minimize:} & c'x \\
\text{subject to:} & a_i'x \leq b_i, \quad i = 1, \ldots, m, \\
& a_i'x = b_i, \quad i = m+1, \ldots, m+r.
\end{array}
\right\}
\qquad (2.28)
$$

can be described as follows.

Phase 1:
Formulate an initial point problem for (2.28). Solve it using Algorithm 2. Let $\hat{x}_0$, $\hat{J}_0$, and $\hat{D}_0^{-1}$ be the final data thus obtained.

Phase 2:
If (2.28) has a feasible solution either apply Algorithm 3 with $\hat{x}_0$, $\hat{J}_0$, and $\hat{D}_0^{-1}$ as initial data to the LP (2.27) or obtain x_0, J_0, and D_0^{-1} from $\hat{x}_0$, $\hat{J}_0$, and $\hat{D}_0^{-1}$, respectively, as described above and apply Algorithm 3 with these initial data to (2.28).

We illustrate the Phase 1–Phase 2 procedure in the following example.

Example 2.11

$$
\begin{array}{lrrrrl}
\text{minimize:} & & -7x_2 & +3x_3 & & \\
\text{subject to:} & -x_1 & -2x_2 & -x_3 & \leq -3, & (1) \\
& -x_1 & & & \leq 1, & (2) \\
& & & -x_3 & \leq 0, & (3) \\
& x_1 & +3x_2 & +2x_3 & = 27, & (4) \\
& 2x_1 & -x_2 & +3x_3 & = 12. & (5)
\end{array}
$$

The Phase 1 problem is

$$
\begin{array}{lrrrrrl}
\text{minimize:} & -3x_1 & -2x_2 & -5x_3 & +\alpha & & \\
\text{subject to:} & -x_1 & -2x_2 & -x_3 & -\alpha & \leq -3, & (1) \\
& -x_1 & & & -\alpha & \leq 1, & (2) \\
& & & -x_3 & -\alpha & \leq 0, & (3) \\
& x_1 & +3x_2 & +2x_3 & & \leq 27, & (4) \\
& 2x_1 & -x_2 & +3x_3 & & \leq 12, & (5) \\
& & & & -\alpha & \leq 0. & (6)
\end{array}
$$

Setting $x_0 = 0$, we solve this problem by using Algorithm 2 as follows.

Initialization:

$$x_0 = \begin{bmatrix} 0 \\ 0 \\ 0 \end{bmatrix}, \quad \alpha_0 = 3 = \max\{0, 3, -1, 0\}, \quad D_0^{-1} = \begin{bmatrix} 1 & 0 & 0 & 0 \\ 0 & 1 & 0 & 0 \\ 0 & 0 & 1 & 0 \\ 0 & 0 & 0 & 1 \end{bmatrix},$$

$$J_0 = \{0, 0, 0, 0\}, \quad d'x_0 + \alpha_0 = 3, \quad j = 0.$$

Iteration 0

Step 1:　　Transfer to Step 1.1.

Step 1.1:　　$|v_3| = \max\{|-3|, |-2|, |-5|, |11|\} = 5, \quad k = 3,$

$$s_0 = \begin{bmatrix} 0 \\ 0 \\ -1 \\ 0 \end{bmatrix}.$$

Step 2:　　$\sigma_0 = \min\left\{ -, -, -, \dfrac{-27}{-2}, \dfrac{-12}{-3}, - \right\} = 4, \quad l = 5.$

Step 3:　　$\begin{bmatrix} x_1 \\ \alpha_1 \end{bmatrix} = \begin{bmatrix} 0 \\ 0 \\ 0 \\ 3 \end{bmatrix} - 4 \begin{bmatrix} 0 \\ 0 \\ -1 \\ 0 \end{bmatrix} = \begin{bmatrix} 0 \\ 0 \\ 4 \\ 3 \end{bmatrix}, \quad d'x_1 + \alpha_1 = -17,$

$$D_1^{-1} = \begin{bmatrix} 1 & 0 & 0 & 0 \\ 0 & 1 & 0 & 0 \\ -2/3 & 1/3 & 1/3 & 0 \\ 0 & 0 & 0 & 1 \end{bmatrix}, \quad J_1 = \{0, 0, 5, 0\}, \quad j = 1.$$

Iteration 1

Step 1:　　Transfer to Step 1.1.

Step 1.1:　　$|v_2| = \max\left\{ |\dfrac{1}{3}|, |\dfrac{-11}{3}|, -, |11| \right\} = \dfrac{11}{3}, \quad k = 2,$

$$s_1 = \begin{bmatrix} 0 \\ -1 \\ -1/3 \\ 0 \end{bmatrix}.$$

Step 2: $\sigma_1 = \min\left\{-,-,-,\dfrac{-19}{-11/3},-,-\right\} = \dfrac{57}{11}, \quad l = 4.$

Step 3: $\begin{bmatrix} x_2 \\ \alpha_2 \end{bmatrix} = \begin{bmatrix} 0 \\ 0 \\ 4 \\ 3 \end{bmatrix} - \dfrac{57}{11}\begin{bmatrix} 0 \\ -1 \\ -1/3 \\ 0 \end{bmatrix} = \begin{bmatrix} 0 \\ 57/11 \\ 63/11 \\ 3 \end{bmatrix},$

$d'x_2 + \alpha_2 = -36,$

$$D_2^{-1} = \begin{bmatrix} 1 & 0 & 0 & 0 \\ 1/11 & 3/11 & -2/11 & 0 \\ -7/11 & 1/11 & 3/11 & 0 \\ 0 & 0 & 0 & 1 \end{bmatrix}, \quad J_2 = \{0,4,5,0\},$$

$j = 2.$

Iteration 2

Step 1:   Transfer to Step 1.1.

Step 1.1: $|v_4| = \max\{\,|0|,-,-,|1|\,\} = 1, \quad k = 4, \quad s_2 = \begin{bmatrix} 0 \\ 0 \\ 0 \\ 1 \end{bmatrix}.$

Step 2: $\sigma_2 = \min\left\{\dfrac{-177/11}{-1},\dfrac{-4}{-1},\dfrac{-96/11}{-1},-,-,\dfrac{-3}{-1}\right\} = 3,$

$l = 6.$

Step 3: $\begin{bmatrix} x_3 \\ \alpha_3 \end{bmatrix} = \begin{bmatrix} 0 \\ 57/11 \\ 63/11 \\ 3 \end{bmatrix} - 3\begin{bmatrix} 0 \\ 0 \\ 0 \\ 1 \end{bmatrix} = \begin{bmatrix} 0 \\ 57/11 \\ 63/11 \\ 0 \end{bmatrix}, \quad d'x_3 + \alpha_3 = -39,$

$$D_3^{-1} = \begin{bmatrix} 1 & 0 & 0 & 0 \\ 1/11 & 3/11 & -2/11 & 0 \\ -7/11 & 1/11 & 3/11 & 0 \\ 0 & 0 & 0 & -1 \end{bmatrix}, \quad J_3 = \{0,4,5,6\},$$

$j = 3.$

Iteration 3

Step 1:   Transfer to Step 1.1.

Step 1.1: $|v_1| = \max\{\,|0|\,, -, -, -\,\} = 0,\;\; k = 1.$ Transfer to Step 1.2.

Step 1.2: $v_2 = \max\{\,-, -1, -1, -1\,\} = -1,\;\; k = 2.$

$v_2 \leq 0$; stop with optimal solution x_3 and α_3.

We obtain the initial data, D_0^{-1}, J_0, and x_0, for the Phase 2 problem by striking out row 4 and column 4 of D_3^{-1}, the last element of J_3, and the last component of $(x_3', \alpha_3)'$ above, respectively. The Phase 2 problem is now solved by applying Algorithm 3.

Initialization:

$$x_0 = \begin{bmatrix} 0 \\ 57/11 \\ 63/11 \end{bmatrix}, \quad D_0^{-1} = \begin{bmatrix} 1 & 0 & 0 \\ 1/11 & 3/11 & -2/11 \\ -7/11 & 1/11 & 3/11 \end{bmatrix}, \quad J_0 = \{0, 4, 5\},$$

$$c'x_0 = \frac{-210}{11}, \quad j = 0.$$

Iteration 0

Step 1:   Transfer to Step 1.1.

Step 1.1: $|v_1| = \max\left\{\,\left|\dfrac{-28}{11}\right|, -, -\,\right\} = \dfrac{28}{11},\;\; k = 1,$

$$s_0 = \begin{bmatrix} -1 \\ -1/11 \\ 7/11 \end{bmatrix}.$$

Step 2: $\sigma_0 = \min\left\{\,-, -, \dfrac{-63/11}{-7/11}\,\right\} = 9,\;\; l = 3.$

Step 3: $x_1 = \begin{bmatrix} 0 \\ 57/11 \\ 63/11 \end{bmatrix} - 9 \begin{bmatrix} -1 \\ -1/11 \\ 7/11 \end{bmatrix} = \begin{bmatrix} 9 \\ 6 \\ 0 \end{bmatrix},\;\; c'x_1 = -42,$

$$D_1^{-1} = \begin{bmatrix} 11/7 & 1/7 & 3/7 \\ 1/7 & 2/7 & -1/7 \\ -1 & 0 & 0 \end{bmatrix}, \quad J_1 = \{3, 4, 5\}, \quad j = 1.$$

Iteration 1

Step 1:   Transfer to Step 1.2.

Step 1.2: $v_1 = \max\{-4, -, -\} = -4$, $k = 1$.

$v_1 \le 0$; stop with optimal solution x_1.

A computer program which implements the above Phase 1–Phase 2 procedure, including the redundancy tests described below, is given in Appendix B, Section B.1.6. The output from applying this program to the problem of Example 2.11 is shown in Figure B.31.

It remains to show that $\hat{J}_0$ contains each of the numbers $m + 1$,..., $m + r$, $m + r + 1$. This is indeed the case if the gradients a_i of those constraints that are active at x_0 are linearly independent. Let Algorithm 2, applied to (2.26), terminate with

$$\hat{x}_0 = \begin{bmatrix} x_0 \\ 0 \end{bmatrix}, \quad \hat{J}_0 = \{\alpha_{10}, \ldots, \alpha_{n0}, \alpha_{n+1,0}\}, \quad \text{and} \quad \hat{D}_0^{-1}.$$

We use the following notation:

$$\hat{c} = \begin{bmatrix} d \\ 1 \end{bmatrix} \quad \text{and} \quad \hat{D}_0' = \begin{bmatrix} d_1, \ldots, d_n, d_{n+1} \\ \delta_1, \ldots, \delta_n, \delta_{n+1} \end{bmatrix},$$

where $d_i = a_{\alpha_{i0}}$ for all i with $1 \le \alpha_{i0} \le m + r$,

$$\delta_i = 0, \quad \text{for all } i \text{ with } m + 1 \le \alpha_{i0} \le m + r, \tag{2.29}$$

and $d_i = 0$, $\delta_i = -1$ if $\alpha_{i0} = m + r + 1$. With

$$\hat{D}_0^{-1} = \begin{bmatrix} \hat{c}_{10}, \ldots, \hat{c}_{n0}, \hat{c}_{n+1,0} \end{bmatrix} \tag{2.30}$$

and

$$v_i = \hat{c}' \hat{c}_{i0}, \quad i = 1, \ldots, n + 1$$

we have

$$\hat{c} = \sum_{i=1}^{n+1} v_i \begin{bmatrix} d_i \\ \delta_i \end{bmatrix}$$

or

$$d = \sum_{i=1}^{n+1} v_i d_i \tag{2.31}$$

and

$$1 = \sum_{i=1}^{n+1} v_i d_i. \tag{2.32}$$

Observing that $d_i = 0$ if $\alpha_{i0} = m + r + 1$,

$$v_i = 0, \quad \text{for all } i \text{ with } \alpha_{i0} = 0, \tag{2.33}$$

and using the definition of d, we obtain from (2.31)

$$-\sum_{k=m+1}^{m+r} a_k = \sum v_i a_{\alpha_{i0}}, \tag{2.34}$$

where the summation on the right-hand side is over all i with $1 \leq \alpha_{i0} \leq m + r$. Because x_0 is a feasible solution for (2.26), every a_i occurring on either side of (2.34) is the gradient of a constraint which is active at x_0. By assumption, these gradients are linearly independent. This implies that each of the vectors $a_{m+1}, \ldots, a_{m+r}$ occurs on the right-hand side of (2.34); that is,

$$m + i \in \hat{J}_0, \quad \text{for } i = 1, \ldots, r.$$

Furthermore, $v_i = -1$ for all i with $m + 1 \leq \alpha_{i0} \leq m + r$ and

$$v_i = 0, \quad \text{for all } i \text{ with } 1 \leq \alpha_{i0} \leq m. \tag{2.35}$$

Using (2.29), (2.33), and (2.35), we note that $v_i \delta_i = 0$ for all i with $0 \leq \alpha_{i0} \leq m + r$. Thus it follows from (2.32) that $\alpha_{i0} = m + r + 1$ for some $1 \leq i \leq n + 1$ (i.e., $m + r + 1$ is in $\hat{J}_0$).

To conclude this section we discuss briefly the case that at least one of the numbers $m + 1, \ldots, m + r, m + r + 1$ is not in $\hat{J}_0$.

First suppose that $m + r + 1 \notin \hat{J}_0$; that is, the gradient of the constraint $-\alpha \leq 0$ is not a column of $\hat{D}'_0$. Using the same notation for $\hat{D}'_0$ as before, we can write $(0', -1)'$ as

$$\begin{bmatrix} 0 \\ -1 \end{bmatrix} = \sum_{i=1}^{n+1} \lambda_i \begin{bmatrix} d_i \\ \delta_i \end{bmatrix}. \tag{2.36}$$

Because of (2.29) there is at least one i with $0 \leq \alpha_{i0} \leq m$ and $\lambda_i \neq 0$. Now determine k such that

$$|(\hat{c}_{k0})_{n+1}| = \max \{ |(\hat{c}_{i0})_{n+1}| \mid \text{all } i \text{ with } 0 \leq \alpha_{i0} \leq m \}.$$

Thus $(\hat{c}_{k0})_{n+1} \neq 0$ because it follows from (2.30) and (2.36) that

$$\hat{c}'_{i0} \begin{bmatrix} 0 \\ 1 \end{bmatrix} = \lambda_i, \quad i = 1, \ldots, n + 1.$$

Thus we can use Procedure Φ to determine the matrix

$$\Phi \left(\hat{D}_0^{-1}, \begin{bmatrix} 0 \\ -1 \end{bmatrix}, k \right).$$

Replacing $\hat{D}_0^{-1}$ with this new matrix if necessary, we may assume that the vector $(0', -1)'$ is one of the columns of $\hat{D}_0'$.

Next suppose that there is a ρ such that $m + 1 \leq \rho \leq m + r$ and $\rho \notin \hat{J}_0$. Let

$$a_\rho = \sum_{i=1}^{n+1} \lambda_i \begin{bmatrix} d_i \\ \delta_i \end{bmatrix}, \quad \text{with } \lambda_i = \hat{c}_{i0}' \begin{bmatrix} a_\rho \\ 0 \end{bmatrix}.$$

If $\lambda_i \neq 0$ for at least one i with $0 \leq \alpha_{i0} \leq m$, determine k such that

$$|\lambda_k| = \max \{ |\lambda_i| \mid \text{all } i \text{ with } 0 \leq \alpha_{i0} \leq m \}.$$

Then $\lambda_k \neq 0$ and

$$\Phi \left(\hat{D}_0^{-1}, \begin{bmatrix} a_\rho \\ 0 \end{bmatrix}, k \right)$$

is well-defined. Now let $\lambda_i = 0$ for all i with $0 \leq \alpha_{i0} \leq m$. Then

$$a_\rho \in \text{span} \{ a_{\alpha_{i0}} \mid \text{all } i \text{ with } m + 1 \leq \alpha_{i0} \leq m + r \}.$$

It follows that the equation $a_\rho' x = b_\rho$ is superfluous and can be deleted.

Continuing this process and deleting equations if necessary, we obtain a matrix $\tilde{D}_0^{-1}$ and a corresponding index set $\tilde{J}_0$ which contains the indices of all remaining equality constraints. Thus we can replace $\hat{J}_0$ and $\hat{D}_0^{-1}$ by $\tilde{J}_0$ and $\tilde{D}_0^{-1}$, respectively.

EXERCISES

2.1. Consider the feasible region defined by

$$\begin{align}
x_1 - 2x_2 + 3x_3 &\leq 2, & (1) \\
-x_1 + x_2 + x_3 &\leq 1, & (2) \\
x_1 + x_2 + x_3 &\leq 4, & (3) \\
- x_2 + 6x_3 &\leq 5. & (4)
\end{align}$$

Let $x_0 = (1, 1, 1)'$. Is x_0 an extreme point? Is x_0 degenerate? State your reasons.

2.2. Determine all extreme points for

$$\begin{align}
- x_1 + 2x_2 &\leq 4, & (1) \\
-2x_1 + x_2 &\leq -1, & (2) \\
x_1 + x_2 &\leq 7, & (3) \\
x_1 &\leq 6, & (4) \\
x_2 &\leq 3, & (5) \\
- x_2 &\leq -1. & (6)
\end{align}$$

Which are degenerate and which are nondegenerate? Why?

2.3. Determine all extreme points for

$$4x_1 + 6x_2 + 6x_3 + 4x_4 = 2, \qquad (1)$$
$$6x_1 + 2x_2 + x_3 + 2x_4 = 3, \qquad (2)$$
$$-x_1 \qquad\qquad\qquad\quad \leq 0, \qquad (3)$$
$$-x_2 \qquad\qquad\quad \leq 0, \qquad (4)$$
$$-x_3 \qquad\quad \leq 0, \qquad (5)$$
$$-x_4 \leq 0. \qquad (6)$$

2.4. Use Algorithm 1 to solve the following problems.

(a) minimize: $-4x_1 - 7x_2$

subject to: $x_1 + 3x_2 \leq 9, \qquad (1)$

$2x_1 + x_2 \leq 8, \qquad (2)$

$-x_1 \qquad\quad \leq 0, \qquad (3)$

$-x_2 \leq 0. \qquad (4)$

Initial point: $x_0 = (0,0)'$.

(b) minimize: $-x_1 - x_2$

subject to: $-x_1 + x_2 \leq 2, \qquad (1)$

$-x_1 + 3x_2 \leq 10, \qquad (2)$

$2x_1 + x_2 \leq 15, \qquad (3)$

$3x_1 - x_2 \leq 15, \qquad (4)$

$-x_1 \qquad\quad \leq 0, \qquad (5)$

$-x_2 \leq 0. \qquad (6)$

Initial point: $x_0 = (0,0)'$.

2.5. Use Algorithm 1 to minimize each of the objective functions **(a)** $x_1 - x_2$ and **(b)** $x_1 - 4x_2$ on the feasible region defined by

$$x_1 - 3x_2 \leq 3, \qquad (1)$$
$$-x_2 \leq 0, \qquad (2)$$
$$-x_1 \qquad\quad \leq 0, \qquad (3)$$
$$-2x_1 + x_2 \leq 2, \qquad (4)$$
$$-x_1 + 2x_2 \leq 7. \qquad (5)$$

Initial point: $x_0 = (0,0)'$.

2.6. Use Algorithm 1 to solve the following problem.

$$
\begin{array}{lrrcll}
\text{minimize:} & -3x_1 & - & x_2 & & \\
\text{subject to:} & & - & x_2 & \le & 0, & (1) \\
& -2x_1 & + & x_2 & \le & 4, & (2) \\
& -x_1 & + & 2x_2 & \le & 5, & (3) \\
& -x_1 & + & 5x_2 & \le & 11, & (4) \\
& 2x_1 & + & x_2 & \le & 11, & (5) \\
& 6x_1 & + & x_2 & \le & 27. & (6)
\end{array}
$$

Initial point: $x_0 = (-2, 0)'$.

2.7. Use Algorithm 1 to solve the following problems.

(a)

$$
\begin{array}{lrrrcll}
\text{minimize:} & -x_1 & - & 2x_2 & - & 3x_3 & \\
\text{subject to:} & x_1 & - & 2x_2 & - & 3x_3 & \le & 12, & (1) \\
& & & x_2 & - & 7x_3 & \le & 6, & (2) \\
& -x_1 & & & & & \le & 0, & (3) \\
& & - & x_2 & & & \le & 0, & (4) \\
& & & & - & x_3 & \le & 0. & (5)
\end{array}
$$

Initial point: $x_0 = (0, 0, 0)'$.

(b)

$$
\begin{array}{lrrrcll}
\text{minimize:} & -2x_1 & - & 3x_2 & - & 4x_3 & \\
\text{subject to:} & 3x_1 & & & + & 4x_3 & \le & 32, & (1) \\
& -5x_1 & + & 16x_2 & + & 4x_3 & \le & 64, & (2) \\
& 25x_1 & + & 7x_2 & + & 10x_3 & \le & 150, & (3) \\
& -x_1 & & & & & \le & 0, & (4) \\
& & - & x_2 & & & \le & 0, & (5) \\
& & & & - & x_3 & \le & 0. & (6)
\end{array}
$$

Initial point: $x_0 = (0, 0, 0)'$.

(c)

$$
\begin{array}{lrrrcll}
\text{minimize:} & -x_1 & - & 2x_2 & - & x_3 & \\
\text{subject to:} & x_1 & + & x_2 & + & 4x_3 & \le & 6, & (1) \\
& -x_1 & + & x_2 & + & 2x_3 & \le & 2, & (2) \\
& 5x_1 & - & x_2 & + & 2x_3 & \le & 6, & (3) \\
& -x_1 & & & & & \le & 0, & (4) \\
& & - & x_2 & & & \le & 0, & (5) \\
& & & & - & x_3 & \le & 0. & (6)
\end{array}
$$

Initial point: $x_0 = (0, 0, 0)'$.

(d)   minimize: $-x_1$

subject to: $-x_1 + 4x_2 - x_3 \leq 2,$ (1)

$6x_1 - 2x_2 - 2x_3 \leq 40,$ (2)

$x_1 + 2x_2 + 3x_3 \leq 6,$ (3)

$-x_1 \qquad\qquad\qquad \leq 0,$ (4)

$-x_2 \qquad\qquad \leq 0,$ (5)

$-x_3 \leq 0.$ (6)

Initial point: $x_0 = (0,0,0)'.$

(e)   minimize: $-14x_1 - 18x_2 - 16x_3 - 80x_4$

subject to: $4.5x_1 + 8.5x_2 + 6x_3 + 20x_4 \leq 6000,$ (1)

$x_1 + x_2 + 4x_3 + 40x_4 \leq 4000,$ (2)

$-x_1 \qquad\qquad\qquad\qquad \leq 0,$ (3)

$-x_2 \qquad\qquad\qquad \leq 0,$ (4)

$-x_3 \qquad\qquad \leq 0,$ (5)

$-x_4 \leq 0.$ (6)

Initial point: $x_0 = (0,0,0,0)'.$

2.8. Use Algorithm 2 to solve the following problems. In each case, use $D_0^{-1} = I$ and $J_0 = \{0, \ldots, 0\}.$

(a)   minimize: $-4x_1 - 2x_2$

subject to: $-x_1 \qquad\qquad \leq -1,$ (1)

$-x_2 \leq -1,$ (2)

$x_1 \qquad\quad \leq 6,$ (3)

$x_1 + x_2 \leq 11,$ (4)

$x_1 + 4x_2 \leq 29.$ (5)

Initial point: $x_0 = (3,3)'.$

(b)   minimize: $-4x_1 - 3x_2$

subject to: $x_1 + 2x_2 \leq 2,$ (1)

$x_1 - 3x_2 \leq 3,$ (2)

$-2x_1 + x_2 \leq 36,$ (3)

$9x_1 + 3x_2 \leq 1,$ (4)

$-x_2 \leq -2.$ (5)

Initial point: $x_0 = (-10,3)'.$

(c)   minimize: $-2x_1 - 3x_2$

subject to:
$$3x_1 + 2x_2 \leq 1, \qquad (1)$$
$$2x_1 + x_2 \leq 3, \qquad (2)$$
$$x_1 + 4x_2 \leq 1, \qquad (3)$$
$$x_1 - x_2 \leq -1, \qquad (4)$$
$$x_1 + x_2 \leq -2. \qquad (5)$$

Initial point: $x_0 = (-6, -5)'$.

(d)   minimize: $-5x_1 - 6x_2 - 6x_3$

subject to:
$$x_1 + 2x_2 + 2x_3 \leq 2, \qquad (1)$$
$$x_1 + 3x_2 + 3x_3 \leq 1, \qquad (2)$$
$$x_1 + 2x_2 - 3x_3 \leq 3, \qquad (3)$$
$$x_1 + 2x_2 - x_3 \leq -1, \qquad (4)$$
$$x_3 \leq -1. \qquad (5)$$

Initial point: $x_0 = (-6, 0, -3)'$.

(e)   minimize: $x_1 - x_2 - 2x_3 - 3x_4$

subject to:
$$-x_1 - x_2 + x_3 - x_4 \leq 5, \qquad (1)$$
$$x_1 \qquad\qquad + x_4 \leq 4, \qquad (2)$$
$$x_1 + x_2 \qquad\qquad \leq 1, \qquad (3)$$
$$2x_2 - 3x_3 - 2x_4 \leq 2, \qquad (4)$$
$$- x_4 \leq 2, \qquad (5)$$
$$2x_1 + 3x_2 - 2x_3 \qquad \leq 3, \qquad (6)$$
$$x_3 + x_4 \leq 4, \qquad (7)$$
$$- x_1 \qquad\qquad \leq 4, \qquad (8)$$
$$-x_1 + 3x_2 + 2x_3 + 2x_4 \leq 17. \qquad (9)$$

Initial point: $x_0 = (-1, 1, 1, -1)'$.

2.9. State the Phase 1 problem for
$$x_1 \qquad - x_3 \leq 6, \qquad (1)$$
$$-x_1 - x_2 - x_3 \leq -2, \qquad (2)$$
$$x_1 - x_2 \qquad \leq 4, \qquad (3)$$
$$x_1 \qquad\qquad \leq 1, \qquad (4)$$
$$x_2 \qquad \leq 1, \qquad (5)$$
$$x_3 \leq 1. \qquad (6)$$

Solve the Phase 1 problem using Algorithm 2 and $x_0 = (0, 0, 0)'$.

2.10. Solve the following problem by using Algorithm 2 and the Phase 1–Phase 2 method of Section 2.5 with $x_0 = (0,0)'$.

$$\text{minimize:} \quad -x_1 - x_2$$

$$
\begin{aligned}
\text{subject to:} \quad -x_1 \quad\quad\quad &\leq 1, \quad\quad (1) \\
-x_2 &\leq 1, \quad\quad (2) \\
-x_1 + 5x_2 &\leq 11, \quad\quad (3) \\
-x_1 - 2x_2 &\leq -1, \quad\quad (4) \\
6x_1 + x_2 &\leq 27. \quad\quad (5)
\end{aligned}
$$

2.11. State the Phase 1 problem for

$$
\begin{aligned}
x_1 + x_2 + x_3 &\leq 10, \quad\quad (1) \\
2x_1 \quad\quad\;\, + 3x_3 &\leq 11, \quad\quad (2) \\
-x_1 \quad\quad\quad &\leq 0, \quad\quad (3) \\
-x_3 &\leq 0, \quad\quad (4) \\
-2x_1 + 2x_2 - 5x_3 &= -13, \quad\quad (5) \\
x_2 \quad\quad\quad &= 0. \quad\quad (6)
\end{aligned}
$$

Solve the Phase 1 problem with $x_0 = (0,0,0)'$. The reader may find it useful to solve this question in two ways and compare the results. First, substitute $x_2 = 0$ into the remaining constraints, giving a two-variable problem. Second, solve the problem as an explicit three-variable problem.

2.12. Solve the following problems by using the Phase 1–Phase 2 method of Section 2.6 with $x_0 = 0$.

(a)   minimize: $8x_2 + 14x_3$

$$
\begin{aligned}
\text{subject to:} \quad x_1 + x_2 + x_3 &\leq 4, \quad\quad (1) \\
4x_1 - 2x_2 - x_3 &\leq 10, \quad\quad (2) \\
x_1 \quad\quad\;\, - x_3 &= 0, \quad\quad (3) \\
x_1 + 2x_2 + 3x_3 &= 6. \quad\quad (4)
\end{aligned}
$$

(b)   minimize: $9x_1 + 8x_2 - x_3 - x_4$

$$
\begin{aligned}
\text{subject to:} \quad x_1 + 2x_2 - x_3 \quad\quad &\leq 1, \quad\quad (1) \\
-x_1 \quad\quad\quad &\leq 0, \quad\quad (2) \\
-x_2 \quad\quad\quad &\leq 0, \quad\quad (3) \\
-x_3 \quad\quad &\leq 0, \quad\quad (4) \\
-x_4 &\leq 0, \quad\quad (5) \\
3x_1 + x_2 \quad\quad - x_4 &= 1. \quad\quad (6)
\end{aligned}
$$

2.13. State the initial point problem for the constraints of
   (a)   Example 1.1,
   (b)   Example 2.3,
   (c)   Example 2.6.
   In each case, use Algorithm 2 with $x_0 = 0$ to solve the initial point problem, and show the progress of the algorithm on a graph of the feasible region.

2.14. Let x be such that

$$a_i'x = b_i, \quad i = 1, \ldots, k.$$

Show that x is uniquely determined if and only if the set $\{a_1, \ldots, a_k\}$ contains n linearly independent vectors.

2.15. For Algorithm 1, suppose that constraints $1, 2, \ldots, n$ are active at x_0, have linearly independent gradients, and the algorithm proceeds by deleting constraint k from the active set. Suppose that l is the index of the new active constraint as determined in Step 2. Show that $a_1, \ldots, a_{k-1}, a_{k+1}, \ldots, a_n$ together with a_l are linearly independent. *Hint:* By definition of σ_0, $a_l'c_{k0} < 0$.

2.16. There is no reason to believe that choosing k as in Step 1 of Algorithm 1 will give the smallest objective function value among all extreme points adjacent to x_0. Give a method for choosing k which does produce the smallest such objective function value. Calculate the number of arithmetic operations in terms of m and n required by your method. Give a graphical example for which this new choice of k drastically reduces the total number of iterations required for solution. Give a second graphical example of a problem for which this new choice of k drastically increases the total number of iterations required for solution.

2.17. Show that $x_0 \in R$ is an extreme point of the set $R = \{x \mid Ax \leq b\}$ if and only if there are no distinct $x_1, x_2 \in R$ such that $x_0 = (1/2)(x_1 + x_2)$.

2.18. Let A be an (m, n) matrix and suppose that the set $R = \{x \mid Ax \leq b\}$ is nonempty. Show that R has extreme points if and only if $\text{rank}(A) = n$.

2.19. Apply Algorithm 1 to the problem

$$\min\{c'x \mid a_i'x \leq b_i, \quad i = 1, \ldots, m\}.$$

Suppose that x_{j-1}, x_j, x_{j+1} are nondegenerate extreme points obtained in three consecutive iterations. Show that no constraint $a_i'x \leq b_i$ can be active at x_j but inactive at both x_{j-1} and x_{j+1}.

2.20. (a)   For the model problem (2.19), observe that α need not be added to those constraints for which $b_i \geq 0$, provided that the initial point $x_0 = 0$ is used. Suppose this is done. Does this affect the computation of x_0, D_0^{-1}, and J_0 from $\hat{x}_0$, $\hat{D}_0^{-1}$, and $\hat{J}_0$, respectively? Apply this approach to the problem of Example 2.9. Draw the appropriately modified Figure 2.10.
   (b)   If $x_0 \neq 0$, α need not be added to any constraint for which $a_i'x_0 \leq b_i$. Apply this approach to the constraints of Example 1.2 using $x_0 = (8, 2)'$ and then $x_0 = (10, -1)'$. In each case, show the progress of Algorithm 2 on a graph of the feasible region.

2.21. Let

$$D' = \left[d_1, d_2, \ldots, d_n \right]$$

be an (n, n) nonsingular matrix and let

$$D^{-1} = \left[c_1, c_2, \ldots, c_n \right].$$

(a) Show that

$$d_i' c_j = \begin{cases} 0, & 1 \le i \le n,\ 1 \le j \le n,\ i \ne j, \\ 1, & i = j,\ 1 \le i \le n. \end{cases}$$

(b) Show that any n-vector d can be written as

$$d = \sum_{i=1}^{n} (d' c_i) d_i.$$

Hint: Consider the linear equations $D'x = d$ and solve for x.

2.22. Justify Procedure Φ; that is, let $H = \Phi(D^{-1}, d, k)$ and show that $H = \hat{D}^{-1}$ by showing that $\hat{D}H = I$ and $\hat{D}$ is nonsingular. *Hint:* Use the fact that $DD^{-1} = D^{-1}D = I$.

2.23. Show that Procedure Φ requires approximately n^2 arithmetic operations. How can Procedure Φ be used to find the inverse of a nonsingular matrix? How many arithmetic operations are required?

2.24. *Bordered Inverse Lemma.* Let A be an (n, n) nonsingular matrix. Let

$$H = \begin{bmatrix} A & u \\ v' & \alpha \end{bmatrix},$$

and let

$$H^{-1} = \begin{bmatrix} B & p \\ q' & \beta \end{bmatrix},$$

provided that H is nonsingular, and where B is an (n, n) matrix, u, v, p, and q are n-vectors, and α and β are scalars. Show that

$$p = -\beta A^{-1} u, \qquad\qquad q' = -\beta v' A^{-1},$$
$$B = A^{-1} + \beta A^{-1} u v' A^{-1}, \qquad\qquad \beta = (\alpha - v' A^{-1} u)^{-1},$$

and

$$A^{-1} = B - \frac{1}{\beta} p q'.$$

There are several points to this exercise. Suppose that A^{-1}, u, v, and α are known. Then H^{-1} can be easily calculated in terms of an (n, n) matrix and the new row and column which define it. Further, suppose that H^{-1} is known and we wish to compute A^{-1}. This can be done using the fifth equation; that is, the inverse of an (n, n) submatrix of an $(n + 1, n + 1)$ matrix can easily be computed in terms of the partitions of the inverse of the larger matrix. Finally, observe that H is nonsingular if and only if $(\alpha - v' A^{-1} u) \ne 0$.

2.25. Let a, c, $a_1, \ldots, a_n$ be n-vectors. Suppose that $c \neq 0$ and $a_1, \ldots, a_n$ are linearly independent.

    (a) Let $a_i'c = 0$, $i = 1, \ldots, n-1$. Show that $a_1, \ldots, a_{n-1}, a$ are linearly independent if and only if $a'c \neq 0$.

    (b) Let $a = \lambda_1 a_1 + \cdots + \lambda_n a_n$. Show that $a_1, \ldots, a_{n-1}, a$ are linearly independent if and only if $\lambda_n \neq 0$.

2.26. Let $\hat{x}_0$, $\hat{J}_0$, and $\hat{D}_0^{-1} = \left[\hat{c}_{10}, \ldots, \hat{c}_{n0}, \hat{c}_{n+1,0} \right]$ be the final data obtained by applying Algorithm 2 to the initial point problem for the LP

$$\min \{ c'x \mid a_i'x \le b_i, \ i = 1, \ldots, m \}.$$

Show that $m + 1 \in \hat{J}_0$ if the vectors a_i, for all $i \in \hat{J}_0$, are linearly independent. If $m + 1 \notin \hat{J}_0$, let $\hat{d} = (0', -1)'$ and determine k such that $|\hat{d}'\hat{c}_{k0}| \ge |\hat{d}'\hat{c}_{i0}|$ for $i = 1, \ldots, n + 1$. Show that the matrix $\Phi(\hat{D}_0^{-1}, \hat{d}, k)$ is defined. Note that the index set associated with the new matrix is obtained from $\hat{J}_0$ by replacing the kth element with $m + 1$.

3

THEORY

In Chapters 1 and 2 we have seen that an optimal solution is characterized by the condition that $-c$ lies in the cone spanned by the gradients of the active constraints. In this chapter we formulate this condition for a general problem and prove that it is both necessary and sufficient for optimality. Since linear programming problems may be formulated in a variety of ways, we derive optimality conditions for several different model problems.

Associated with any linear programming problem (called the primal problem) there is a second linear programming problem called the dual problem. The dual problem is formulated in terms of the data of the primal problem. Various relationships between the primal and dual problems are established. It is shown that once the primal has been solved, the final data may be used directly to find an optimal solution for the dual.

The solution of a linear programming problem in not necessarily unique. We discuss the determination of alternate optimal solutions and give conditions under which the optimal solution is uniquely determined.

The effect of small changes in the problem data for a linear programming problem is examined. In particular, the sensitivity of the optimal solution to small changes in the coefficients of the objective function and the right-hand side of the constraints is discussed.

3.1 OPTIMALITY CONDITIONS

Consider the model problem

$$\min \{ c'x \mid a_i'x \le b_i, \ i = 1, \ldots, m \}. \tag{3.1}$$

Theorem 2.1 shows that if

$$
\left.\begin{aligned}
a_i' x_0 &= b_i, \quad i = 1, \ldots, k, \\
a_i' x_0 &< b_i, \quad i = k + 1, \ldots, m,
\end{aligned}\right\}
\tag{3.2}
$$

$$
-c = \sum_{i=1}^{k} u_i a_i,
\tag{3.3}
$$

$$
u_i \geq 0, \quad i = 1, \ldots, k,
\tag{3.4}
$$

then x_0 is optimal for (3.1). The assumption that it is the first k constraints which are active at x_0 was for simplicity. We next formulate optimality conditions analogous to (3.2) to (3.4) but without the ordering assumption. The conditions (3.2) to (3.4) state that x_0 is a feasible point at which $-c$ lies in the cone spanned by the gradients of the active constraints. Let us associate a multiplier u_i with each constraint i. We can write (3.3) and (3.4) as

$$
-c = \sum_{i=1}^{m} u_i a_i, \quad u_i \geq 0, \quad i = 1, \ldots, m
\tag{3.5}
$$

provided that $u_i = 0$ for each inactive constraint i. A convenient way to write this last condition is

$$
u_i(a_i' x_0 - b_i) = 0, \quad i = 1, \ldots, m.
\tag{3.6}
$$

If $a_i' x_0 < b_i$ (i.e., constraint i is inactive at x_0), then (3.6) forces $u_i = 0$ and the ith term in the sum (3.5) vanishes. If $a_i' x_0 = b_i$ (i.e., constraint i is active at x_0), then (3.6) imposes no restriction on u_i. Summarizing, the optimality conditions for (3.1) are

$$
a_i' x_0 \leq b_i, \quad i = 1, \ldots, m,
\tag{3.7}
$$

$$
-c = \sum_{i=1}^{m} u_i a_i, \quad u_i \geq 0, \quad i = 1, \ldots, m,
\tag{3.8}
$$

$$
u_i(a_i' x_0 - b_i) = 0, \quad i = 1, \ldots, m.
\tag{3.9}
$$

When (3.1) is written more compactly as

$$
\min \{ c' x \mid Ax \leq b \},
\tag{3.10}
$$

the optimality conditions may also be written more compactly as

$$
Ax_0 \leq b,
\tag{3.11}
$$

$$
-c = A' u, \quad u \geq 0,
\tag{3.12}
$$

$$
u'(Ax_0 - b) = 0.
\tag{3.13}
$$

Conditions (3.11) and (3.12) are equivalent matrix counterparts of (3.7) and (3.8), respectively. Expanding (3.13) gives

$$\sum_{i=1}^{m} u_i(a_i' x_0 - b_i) = 0.$$

From (3.11) and (3.12), each term in this sum is nonpositive. Since these terms sum to zero, each term must have value zero; that is,

$$u_i(a_i' x_0 - b_i) = 0, \quad i = 1, \ldots, m.$$

Thus (3.11) to (3.13) are equivalent to (3.7) to (3.9).

Conditions (3.7) to (3.9) are a specialization of the Karush-Kuhn-Tucker conditions [Mangasarian, 1969] for a nonlinear programming problem to the linear case. We will next show that they are both necessary and sufficient for optimality. This is one of the most fundamental results of linear programming.

Theorem 3.1.

The vector x_0 is an optimal solution for the model problem

$$\min \{ c'x \mid Ax \leq b \}$$

if and only if there is an m-vector u which, with x_0, satisfies the optimality conditions

$$Ax_0 \leq b,$$
$$A'u = -c, \quad u \geq 0,$$
$$u'(Ax_0 - b) = 0.$$

Proof:

First suppose that x_0 satisfies the conditions in the statement of the theorem. Let x be any feasible point for the problem. We have

$$
\begin{aligned}
c'(x_0 - x) &= u'Ax - u'Ax_0 \quad \text{(because } c = -A'u) \\
&= u'Ax - u'b \quad \text{[because } u'(Ax_0 - b) = 0] \\
&= u'(Ax - b) \\
&\leq 0,
\end{aligned}
$$

where the last inequality follows from $u \geq 0$ and the feasibility of x. Thus $c'x_0 \leq c'x$ for all feasible x. By definition, this implies that x_0 is indeed optimal.

Next, suppose that x_0 is an optimal solution. Apply Algorithm A (see Appendix A) with the initial point x_0. By Theorem A.1 (see Appendix A), the algorithm will terminate with an optimal solution x_j. Because x_0 is optimal by assumption we have $c'x_j = c'x_0$, which implies that $x_0 = x_1 = \cdots = x_j$. Let

$$D_j^{-1} = \left[c_{1j}, \ldots, c_{nj} \right]$$

and

$$J_j = \{ \alpha_{1j}, \ldots, \alpha_{nj} \}$$

be associated with x_j. Then we have $v_i = c'c_{ij}$ with $v_i = 0$ for all i with $\alpha_{ij} = 0$ and $v_i \leq 0$ for all i with $\alpha_{ij} \geq 1$. Therefore,

$$c = \sum v_i a_{\alpha_{ij}},$$

where the summation is over all i with $\alpha_{ij} \geq 1$. Let $u = (u_1, \ldots, u_m)'$ with

$$u_i = 0, \quad \text{for all } i \notin J_j$$

and

$$u_i = -v_k, \quad \text{if } i \in J_j \text{ and } \alpha_{kj} = i.$$

Then

$$-c = \sum_{i=1}^{m} u_i a_i = A'u \quad \text{and} \quad u \geq 0.$$

Furthermore, for every i either $u_i = 0$ or $i \in J_j$ and thus $a_i'x_0 = b_i$. Therefore, $u_i(a_i'x_0 - b_i) = 0$ for all i [i.e., $u'(Ax_0 - b) = 0$]. Clearly, $Ax_0 \leq b$. Thus the conditions of the theorem are necessary for an optimal solution. ∎

In the proof of Theorem 3.1, the essential idea in proving necessity is to apply Algorithm 2 beginning with optimal solution x_0. However, it may be the case that x_0 is degenerate and the algorithm could cycle. However, Algorithm A is a variant of Algorithm 2 which is guaranteed not to cycle. Upon first reading of the proof, the reader may find it helpful to assume that x_0 is nondegenerate and that Algorithm 2 and Theorem 2.3 are used in place of Algorithm A and Theorem A.1, respectively.

We have derived optimality conditions for the model problem (3.10). It is useful to derive optimality conditions for other model problems. Consider the model problem

$$\min \{ c'x \mid A_1 x \leq b_1, \ A_2 x = b_2 \}, \tag{3.14}$$

which has explicit inequality and equality constraints. The optimality conditions for it may be deduced by first writing it in the same form as (3.10). Because the equality constraints $A_2 x = b_2$ are equivalent to the inequality constraints $A_2 x \leq b_2$ and $-A_2 x \leq -b_2$, (3.14) may be rewritten as

$$\min\{c'x \mid A_1x \leq b_1, \ A_2x \leq b_2, \ -A_2x \leq -b_2\}, \qquad (3.15)$$

which is of the same form as (3.10) with

$$A = \begin{bmatrix} A_1 \\ A_2 \\ -A_2 \end{bmatrix} \quad \text{and} \quad b = \begin{bmatrix} b_1 \\ b_2 \\ -b_2 \end{bmatrix}.$$

Letting

$$u = \begin{bmatrix} u_1 \\ u_2 \\ u_3 \end{bmatrix},$$

the optimality conditions (3.11) to (3.13) for (3.15) are

$$A_1x_0 \leq b_1, \quad A_2x_0 = b_2,$$
$$A_1'u_1 + A_2'(u_2 - u_3) = -c, \quad u_1 \geq 0, \ u_2 \geq 0, \ u_3 \geq 0,$$
$$u_1'(A_1x_0 - b_1) + (u_2 - u_3)'(A_2x_0 - b_2) = 0.$$

Notice that these depend only on $(u_2 - u_3)$ and while both u_2 and u_3 are required to be nonnegative, the components of $(u_2 - u_3)$ can be either positive or negative. Replacing $(u_2 - u_3)$ with u_2 and using $A_2x_0 = b_2$ to simplify the third condition, the optimality conditions for (3.14) are thus

$$A_1x_0 \leq b_1, \quad A_2x_0 = b_2,$$
$$A_1'u_1 + A_2'u_2 = -c, \quad u_1 \geq 0,$$
$$u_1'(A_1x_0 - b_1) = 0.$$

Theorem 3.1 for the model problem (3.14) becomes

Theorem 3.2.

x_0 is optimal for $\min\{c'x \mid A_1x \leq b_1, \ A_2x = b_2\}$ if and only if there are vectors u_1 and u_2 which, with x_0, satisfy the optimality conditions

$$A_1x_0 \leq b_1, \quad A_2x_0 = b_2,$$
$$A_1'u_1 + A_2'u_2 = -c, \quad u_1 \geq 0,$$
$$u_1'(A_1x_0 - b_1) = 0.$$

As a further example, we next derive the optimality conditions for the model problem

$$\min\{c'x \mid Ax = b, \ x \geq 0\}. \qquad (3.16)$$

Rewriting the nonnegativity constraints of (3.16) in the equivalent form $-Ix \leq 0$, (3.16) becomes

$$\min\{c'x \mid Ax = b, \ -Ix \leq 0\},$$

which is of the same form as (3.14). The optimality conditions for (3.16) are thus

$$\left.\begin{array}{c} Ax = b, \ x \geq 0, \\ A'u_1 - u_2 = -c, \ u_2 \geq 0, \\ u_2'x = 0, \end{array}\right\} \tag{3.17}$$

and the appropriate version of Theorem 3.1 is as follows.

Theorem 3.3.
x_0 is optimal for $\min\{c'x \mid Ax = b, \ x \geq 0\}$ if and only if there are vectors u_1 and u_2 which, with x_0, satisfy

$$Ax_0 = b, \ x_0 \geq 0,$$
$$A'u_1 - u_2 = -c, \ u_2 \geq 0,$$
$$u_2'x_0 = 0.$$

We remark that because of the explicit presence of nonnegativity constraints in (3.16), the optimality conditions for (3.16) may be expressed in a more compact (but equivalent) manner by eliminating u_2 and replacing u_1 with u. Doing so in (3.17) gives

$$Ax = b, \ x \geq 0,$$
$$A'u \geq -c,$$
$$(A'u + c)'x = 0.$$

Theorem 3.3 then becomes

Theorem 3.4.
x_0 is optimal for $\min\{c'x \mid Ax = b, \ x \geq 0\}$ if and only if there is a vector u which, with x_0, satisfies

$$Ax_0 = b, \ x_0 \geq 0,$$
$$A'u \geq -c,$$
$$(A'u + c)'x_0 = 0.$$

Table 3.1 summarizes the optimality conditions for a variety of model problems. Theorem 3.1 applies to each model problem with the indicated optimality conditions. The reader is asked to verify the optimality conditions for the remaining model problems of Table 3.1 (Exercise 3.16).

TABLE 3.1 Optimality Conditions for Various Model Problems

model problem: optimality conditions:	$\min\{c'x \mid Ax \leq b\}$ $Ax \leq b$ $A'u = -c,\ u \geq 0$ $u'(Ax - b) = 0$
model problem: optimality conditions:	$\min\{c'x \mid Ax \geq b\}$ $Ax \geq b$ $A'u = c,\ u \geq 0$ $u'(Ax - b) = 0$
model problem: optimality conditions:	$\min\{c'x \mid A_1x \leq b_1,\ A_2x = b_2\}$ $A_1x \leq b_1,\ A_2x = b_2$ $A_1'u_1 + A_2'u_2 = -c,\ u_1 \geq 0$ $u_1'(A_1x - b_1) = 0$
model problem: optimality conditions:	$\min\{c'x \mid Ax \geq b,\ x \geq 0\}$ $Ax \geq b,\ x \geq 0$ $A'u_1 + u_2 = c,\ u_1 \geq 0,\ u_2 \geq 0$ $u_1'(Ax - b) = 0,\ u_2'x = 0$ OR $Ax \geq b,\ x \geq 0$ $A'u \leq c,\ u \geq 0$ $u'(Ax - b) = 0,\ x'(A'u - c) = 0$
model problem: optimality conditions:	$\min\{c'x \mid Ax = b,\ x \geq 0\}$ $Ax = b,\ x \geq 0$ $A'u_1 - u_2 = -c,\ u_2 \geq 0$ $u_2'x = 0$ OR $Ax = b,\ x \geq 0$ $A'u \geq -c$ $(A'u + c)'x = 0$
model problem: optimality conditions:	$\min\{c'x \mid A_1x \leq b_1,\ A_2x = b_2,\ x \leq b_3,\ x \geq b_4\}$ $A_1x \leq b_1,\ A_2x = b_2,\ x \leq b_3,\ x \geq b_4$ $A_1'u_1 + A_2'u_2 + u_3 - u_4 = -c,\ u_1 \geq 0,\ u_3 \geq 0,\ u_4 \geq 0$ $u_1'(A_1x - b_1) + u_3'(x - b_3) + u_4'(-x + b_4) = 0$
model problem: optimality conditions:	$\min\{c_1'x_1 + c_2'x_2 \mid A_1x_1 + A_2x_2 \leq b,\ x_2 \geq 0\}$ $A_1x_1 + A_2x_2 \leq b,\ x_2 \geq 0$ $A_1'u = -c_1,\ A_2'u \geq -c_2,\ u \geq 0$ $u'(A_1x_1 + A_2x_2 - b) - (A_2'u + c_2)'x_2 = 0$

Necessary and sufficient conditions for the existence of an optimal solution for an LP are given in the following theorem.

Theorem 3.5.

There exists an optimal solution for $\min\{c'x \mid Ax \le b\}$ if and only if $R \equiv \{x \mid Ax \le b\}$ is nonnull and $c'x$ is bounded from below on R.

Proof:

Since R is nonnull, there is an $x_0 \in R$. Applying Algorithm A to the given problem beginning with x_0, termination with the message that the problem is unbounded from below cannot occur. Consequently, from Theorem A.1, termination with an optimal solution must occur in a finite number of steps. ■

3.2 DUALITY

For any given LP there is a second LP, formulated in terms of the data of the given one, called the dual problem. It turns out that solving the given LP automatically determines a solution to the dual LP. This, and other properties of the dual, are developed in this section.

We motivate the definition of the dual as follows. According to Theorem 3.1, solving

$$\min\{c'x \mid Ax \le b\}, \tag{3.18}$$

is equivalent to finding a pair x, u such that

$$Ax \le b, \tag{3.19}$$

$$A'u = -c, \quad u \ge 0, \tag{3.20}$$

$$u'(Ax - b) = 0. \tag{3.21}$$

Now (3.20) implies that $u'Ax = -c'x$, so that (3.19) to (3.21) are equivalent to

$$Ax \le b, \tag{3.22}$$

$$A'u = -c, \quad u \ge 0, \tag{3.23}$$

$$c'x = -b'u. \tag{3.24}$$

Observe that (3.22) expresses linear constraints on x, (3.23) expresses linear constraints on u, and (3.24) is a single linear constraint relating both x and u. A reasonable question to ask is the following: Is there a second LP, whose variables are the components of u, for which (3.22) to (3.24) are the optimality conditions with multiplier vector x? The answer is yes, and the second LP is

$$\max\{-b'u \mid A'u = -c, \quad u \ge 0\}. \tag{3.25}$$

In order to deduce the optimality conditions for (3.25), we first write it as the equivalent minimization problem

$$\min \{ b'u \mid A'u = -c, \; u \geq 0 \}. \tag{3.26}$$

According to Theorem 3.4, the optimality conditions for (3.26) are

$$\left. \begin{array}{r} A'u = -c, \; u \geq 0, \\ (A')'x \geq -b, \\ x'(A')u = -b'u. \end{array} \right\} \tag{3.27}$$

Replacing x with $-x$, and utilizing $(A')' = A$, (3.27) reduces to the conditions

$$Ax \leq b,$$

$$A'u = -c, \; u \geq 0,$$

$$u'(Ax - b) = 0,$$

which are identical to (3.19) to (3.21).

Because of the special relationship between (3.18) and (3.26), we make the following definition. For the <u>primal problem</u>

$$\min \{ c'x \mid Ax \leq b \},$$

the <u>dual problem</u> is

$$\max \{ -b'u \mid A'u = -c, \; u \geq 0 \}.$$

Example 3.1

Formulate the dual of

$$\begin{array}{rll} \text{minimize:} & -5x_1 + 2x_2 & \\ \text{subject to:} & -2x_1 + x_2 \leq 2, & (1) \\ & x_1 + 2x_2 \leq 14, & (2) \\ & 4x_1 + 3x_2 \leq 36, & (3) \\ & -x_1 \leq 0, & (4) \\ & -x_2 \leq 0. & (5) \end{array}$$

(This is the problem of Example 2.2.) By definition, the dual is

$$\begin{array}{rll} \text{maximize:} & -2u_1 - 14u_2 - 36u_3 & \\ \text{subject to:} & -2u_1 + u_2 + 4u_3 - u_4 = 5, & (1) \\ & u_1 + 2u_2 + 3u_3 - u_5 = -2, & (2) \\ & u_i \geq 0, \; i = 1, \ldots, 5. \end{array}$$

We note that the dual problem is defined in terms of the specified primal problem. Any other model LP may be defined as the primal problem, and its dual may be deduced by first transforming the new primal into the form of the primal above and then applying the definition of the dual. This procedure is illustrated in the proof of the following theorem and again later in this section.

A simple property of a primal-dual pair is

Theorem 3.6.

The dual of the dual is the primal.

Proof:

For the primal problem, the dual is

$$\max \{ -b'u \mid A'u = -c, \; u \geq 0 \}. \tag{3.28}$$

To formulate the dual of (3.28), we must first rewrite (3.28) in the form of the primal problem; that is, as a minimization problem with less-than-or-equal-to constraints. The procedure is quite similar to deriving the optimality conditions for different model problems (see discussion prior to Theorem 3.2 and following). Proceeding as indicated, the dual problem (3.28) is equivalent to

$$\min \{ b'u \mid A'u \leq -c, \; -A'u \leq c, \; -Iu \leq 0 \}, \tag{3.29}$$

which is of the same form as the primal problem with c, A, and b replaced with

$$b, \quad \begin{bmatrix} A' \\ -A' \\ -I \end{bmatrix}, \quad \text{and} \quad \begin{bmatrix} -c \\ c \\ 0 \end{bmatrix}, \quad \text{respectively.}$$

By definition, the dual of (3.29) is

$$\max \{ c'(x_1 - x_2) \mid A(x_1 - x_2) - x_3 = -b, \; x_1 \geq 0, \; x_2 \geq 0, \; x_3 \geq 0 \}.$$

Although both x_1 and x_2 are required to be nonnegative, their difference can take on any sign. Replacing $-(x_1 - x_2)$ with x, eliminating x_3, and writing the resulting problem in the equivalent minimization form gives

$$\min \{ c'x \mid Ax \leq b \};$$

that is, the dual of (3.28) is the primal problem. ∎

The two most important results of duality theory are the Weak and Strong Duality Theorems, as follows.

Theorem 3.7 (Weak Duality Theorem).

Let x be any feasible solution for the primal problem

$$\min \{ c'x \mid Ax \leq b \},$$

and let u be any feasible solution for the dual problem

$$\max \{ -b'u \mid A'u = -c, \ u \geq 0 \}.$$

Then $c'x \geq -b'u$.

Proof:

$$
\begin{aligned}
c'x + b'u &= -u'Ax + b'u \quad &\text{(because } u \text{ is dual feasible)} \\
&= u'(b - Ax) \\
&\geq 0. \quad &\text{(because } u \geq 0 \text{ and } Ax \leq b)
\end{aligned}
$$

Therefore, $c'x \geq -b'u$. ∎

Theorem 3.7 asserts that the primal objective function value is bounded from below by the dual objective function value, and conversely, the dual objective function value is bounded from above by the primal objective function value. Consequently, if there are primal and dual feasible points such that the two objective function values are equal, then the primal feasible point must be optimal for the primal and the dual feasible point must be optimal for the dual. More formally, we have

Corollary 3.1.

Let x_0 and u_0 be feasible for the primal and dual problems, respectively. If $c'x_0 = -b'u_0$, then x_0 and u_0 are optimal solutions for the primal and dual, respectively.

Proof:

Let x and u be any feasible solutions for the primal and dual problems, respectively. From the Weak Duality Theorem (Theorem 3.7),

$$c'x \geq -b'u_0 = c'x_0.$$

It follows from definition that x_0 is optimal for the primal.

Similarly, from the Weak Duality Theorem (Theorem 3.7),

$$-b'u \leq c'x_0 = -b'u_0,$$

and again by definition, u_0 is optimal for the dual. ∎

Example 3.2

For Example 3.1, $x = (6,4)'$ and $u = (0,5,0,0,12)'$ are feasible for the primal and dual, respectively. Furthermore,

$$c'x = -22 \quad \text{and} \quad -b'u = -70.$$

Thus $c'x \geq -b'u$, in agreement with the Weak Duality Theorem (Theorem 3.7).

For the same example, $x_0 = (9,0)'$ and $u_0 = \left(0,0,\frac{5}{4},0,\frac{23}{4} \right)'$ are also feasible for the primal and dual, respectively. Furthermore,

$$c'x_0 = -45 \quad \text{and} \quad -b'u_0 = -45.$$

Corollary 3.1 asserts that x_0 is optimal for the primal and u_0 is optimal for the dual.

Theorem 3.8 (Strong Duality Theorem).

If x_0 is an optimal solution for the primal problem, then there exists an optimal solution u_0 for the dual problem and $c'x_0 = -b'u_0$.

Proof:

From Theorem 3.1, x_0 is optimal for the primal if and only if there exists an m-vector u_0 which, with x_0, satisfies

$$Ax_0 \le b,$$

$$A'u_0 = -c, \quad u_0 \ge 0,$$

$$u_0'(Ax_0 - b) = 0.$$

From the second optimality condition, u_0 is feasible for the dual. Furthermore, the third condition implies that $u_0'Ax_0 = u_0'b$. The second implies that $c'x_0 = -u_0'Ax_0$, so that $c'x_0 = -b'u_0$. It now follows from Corollary 3.1 that u_0 is optimal for the dual problem. ■

Let x_0 be optimal for the primal. The proof of Theorem 3.8 uses the result from Theorem 3.1 that x_0 satisfies the optimality conditions and proceeds by demonstrating that the vector u_0 of the optimality conditions is, in fact, optimal for the dual. For these reasons, each of the three optimality conditions is frequently referred to by the parenthetical names following:

$$Ax \le b, \qquad \text{(primal feasibility)}$$

$$A'u = -c, \quad u \ge 0, \qquad \text{(dual feasibility)}$$

$$u'(Ax - b) = 0. \qquad \text{(complementary slackness)}$$

We have referred to u_i as the multiplier associated with constraint i. It is also referred to as the dual variable associated with constraint i.

Theorem 3.9 (Converse Duality Theorem).

If u_0 is an optimal solution for the dual problem, there exists an optimal solution x_0 for the primal problem and $c'x_0 = -b'u_0$.

Proof:

The theorem follows directly from the Strong Duality Theorem (Theorem 3.8) and Theorem 3.6. ■

Theorem 3.10.

If both the primal and dual problems have feasible solutions, they both have optimal solutions.

Proof:

By the Weak Duality Theorem (Theorem 3.7), $c'x$ is bounded from below on the nonnull feasible region for the primal. It follows from Theorem 3.5 that the primal possess an optimal solution.

The remainder of the theorem follows from Theorem 3.8. ∎

Suppose Algorithm 3 has been applied to

$$\left.\begin{array}{ll} \text{minimize:} & c'x \\[4pt] \text{subject to:} & a_i'x \ \leq\ b_i, \ i = 1,\ldots,m, \\[4pt] & a_i'x \ =\ b_i, \ i = m+1,\ldots,m+r. \end{array}\right\} \qquad (3.30)$$

We next show that if termination occurs with "optimal solution x_j," an optimal solution for the dual of (3.30) is immediately available at no additional computational cost. First, write (3.30) in the equivalent form

$$\min\{c'x \mid A_1x \leq b_1, \ A_2x = b_2\}.$$

The dual problem is

$$\max\{-b_1'u_1 - b_2'u_2 \mid A_1'u_1 + A_2'u_2 = -c, \ u_1 \geq 0\}. \qquad (3.31)$$

Let

$$D_j' = \left[d_{1j},\ldots,d_{nj}\right], \quad D_j^{-1} = \left[c_{1j},\ldots,c_{nj}\right],$$

and

$$J_j = \{\alpha_{1j},\ldots,\alpha_{nj}\}.$$

Then

$$d_{ij} = a_{\alpha_{ij}}, \quad \text{for all } i \text{ with } 1 \leq \alpha_{ij} \leq m+r.$$

Let v be an n-vector and consider the linear equations

$$D_j'v = c.$$

Then

$$v'D_j = c'.$$

Multiplying on the right by D_j^{-1} gives

$$v' = c'D_j^{-1},$$

or, in component form,

$$v_i = c'c_{ij}, \ i = 1,\ldots,n.$$

Termination can occur only at Step 1.2, so that from Step 1.1 either $\alpha_{ij} \geq 1$ for $i = 1,\ldots,n$, or

$$v_i = 0, \quad \text{for all } i \text{ with } \alpha_{ij} = 0.$$

In either case, the definition of v implies that

$$c = \sum_{i=1}^{n} v_i d_{ij}$$

$$= \sum v_i a_{\alpha_{ij}},$$

where the last sum is over all i with $1 \leq \alpha_{ij} \leq m + r$. Now define the $(m + r)$-vector u according to

$$\left.\begin{array}{rl} u_{\alpha_{ij}} &= -c' c_{ij}, \quad \text{all } i \text{ with } 1 \leq \alpha_{ij} \leq m + r, \\ u_i &= 0, \qquad \text{all } i \notin J_j. \end{array}\right\} \qquad (3.32)$$

Then

$$-c = \sum_{i=1}^{m+r} u_i a_i, \quad u_i \geq 0, \ i = 1, \ldots, m,$$

and

$$u_i(a_i' x_j - b_i) = 0, \quad i = 1, \ldots, m.$$

Thus x_j together with u satisfy the optimality conditions for (3.30). As in the proof of the Strong Duality Theorem (Theorem 3.8), u is an optimal solution for the dual problem (3.31).

We have shown that u defined by (3.32) is an optimal solution for the dual problem. Its computation involves little additional computational effort. Indeed, just prior to termination, Step 1.2 of Algorithm 3 has already computed the quantities $v_i = c' c_{ij}$ for all i with $1 \leq \alpha_{ij} \leq m$ so only those with $m + 1 \leq \alpha_{ij} \leq m + r$ need be computed. Since in many applications the optimal solution for the dual may be as important as that for the primal, most computer programs for the solution of an LP normally include the optimal dual solution together with the primal. The optimal dual solution is also important in performing sensitivity analysis, as we shall see in Section 3.4.

We remark that Algorithms 1 and 2 are special cases of Algorithm 3, so that for any of these three algorithms, an optimal solution for the dual may be obtained from (3.32).

Example 3.3

In Example 3.1, we formulated the dual of the problem of Example 2.2. The primal problem was solved by Algorithm 1 in Example 2.2. The final data was

$$x_2 = \begin{bmatrix} 9 \\ 0 \end{bmatrix}, \quad D_2^{-1} = \begin{bmatrix} 1/4 & 3/4 \\ 0 & -1 \end{bmatrix}, \quad \text{and} \quad J_2 = \{3, 5\}.$$

In Step 1 of the last iteration, we had

$$v_1 = \frac{-5}{4}, \quad v_2 = \frac{-23}{4},$$

so that, from (3.32),

$$u = \left(0, 0, \frac{5}{4}, 0, \frac{23}{4}\right)'$$

is an optimal solution for the dual problem.

Computer programs are given in Appendix B, Section B.1 which implement Algorithms 1, 2, and 3. A subroutine is given in Appendix B, Section B.1.7 which computes an optimal solution for the dual problem using (3.32). Figure B.34 shows the output from this subroutine after the problem of Example 2.2 was solved using Algorithm 1. This may be compared with Example 3.3.

Consider the case when Algorithm 3 starts with an extreme point. A dual "point" $u = u_j$ may be computed as in (3.32). Then, by construction,

$$\sum_{i=1}^{m+r} u_i a_i = -c;$$

that is, u satisfies the first part of the dual feasibility conditions. Furthermore, the pair (x_j, u) satisfy complementary slackness. However, prior to termination, $c'c_{kj} > 0$ in Step 1.2 so that the remaining part of the dual feasibility condition

$$u_i \geq 0, \quad i = 1, \ldots, m$$

is satisfied if and only if x_j is optimal. Thus Algorithm 3 can be described as a method which at every iteration satisfies primal feasibility and complementary slackness and iteratively works toward satisfying the nonnegativity portion of dual feasibility. At each iteration, Step 1.2 implicitly chooses ρ such that

$$u_\rho = \min\{u_i \mid i = 1, \ldots, m\}$$

and (explicitly using k such that $\alpha_{kj} = \rho$) deletes constraint ρ from the active set. Thus the constraint with the "most negative" dual variable is dropped from the active set.

We have defined the dual for the model primal problem $\min\{c'x \mid Ax \leq b\}$. The dual of any other primal problem may be deduced by first writing it in the same form as our model primal problem, using the definition to form the dual, and then simplifying. The process is quite similar to deriving optimality conditions for various model problems as was done in Section 3.1. We illustrate the procedure as follows.

Consider the model problem

$$\min\{c'x \mid Ax = b, \ x \geq 0\}. \tag{3.33}$$

This can be written as the equivalent maximization problem

$$\max\{-c'x \mid Ax = b, \ x \geq 0\}.$$

This is of the same form as the dual problem in our definition of a primal-dual pair with c, A, b, and x replaced with b, A', $-c$, and u, respectively. According to Theorem 3.6 the dual of (3.33) is

$$\min\{-b'u \mid A'u \leq c\},$$

or, equivalently,

$$\max\{b'u \mid A'u \leq c\}.$$

By replacing u with $-u$ in the above, the dual of (3.33) may also be written as

$$\max\{-b'u \mid A'u \geq -c\}.$$

In this formulation, the feasibility condition is identical to the dual feasibility part of the optimality conditions for (3.33) as given in Table 3.1.

As a second example, consider the model problem

$$\min\{c'x \mid Ax \geq b\}. \tag{3.34}$$

To formulate its dual, we first write the inequality constraints in the form of our model primal LP. This gives

$$\min\{c'x \mid (-A)x \leq (-b)\}.$$

With A and b replaced with $(-A)$ and $(-b)$, respectively, the definition of the dual now asserts that the dual of (3.34) is

$$\max\{b'u \mid -A'u = -c, \ u \geq 0\},$$

or, equivalently,

$$\max\{b'u \mid A'u = c, \ u \geq 0\}.$$

Using the procedure above, the dual of any model primal problem can be deduced. Table 3.2 summarizes a number of primal-dual pairs. The reader is asked to verify the remaining entries (Exercise 3.21). Appropriate versions of Theorems 3.6 through 3.10 and Corollary 3.1 hold for any correctly formulated primal-dual pair.

TABLE 3.2   Dual Problems for Various Primal Problem Formulations

primal problem:	$\min\{c'x \mid Ax \le b\}$
dual problem:	$\max\{-b'u \mid A'u = -c, \ u \ge 0\}$
primal problem:	$\min\{c'x \mid Ax \ge b\}$
dual problem:	$\max\{b'u \mid A'u = c, \ u \ge 0\}$
primal problem:	$\min\{c'x \mid A_1x \le b_1, \ A_2x = b_2\}$
dual problem:	$\max\{-b_1'u_1 - b_2'u_2 \mid A_1'u_1 + A_2'u_2 = -c, \ u_1 \ge 0\}$
primal problem:	$\min\{c'x \mid Ax \ge b, \ x \ge 0\}$
dual problem:	$\max\{b'u \mid A'u \le c, \ u \ge 0\}$
primal problem:	$\min\{c'x \mid Ax = b, \ x \ge 0\}$
dual problem:	$\max\{b'u \mid A'u \le c\}$
primal problem:	$\min\{c'x \mid A_1x \le b_1, \ A_2x = b_2, \ x \le b_3, \ x \ge b_4\}$
dual problem:	$\max\{-b_1'u_1 - b_2'u_2 - b_3'u_3 + b_4'u_4 \mid$
	$\quad A_1'u_1 + A_2'u_2 + u_3 - u_4 = -c,$
	$\quad u_1 \ge 0, \ u_3 \ge 0, \ u_4 \ge 0\}$
primal problem:	$\min\{c_1'x_1 + c_2'x_2 \mid A_1x_1 + A_2x_2 \le b, \ x_2 \ge 0\}$
dual problem:	$\max\{-b'u \mid A_1'u = -c_1, \ A_2'u \ge -c_2, \ u \ge 0\}$

3.3   UNIQUENESS; ALTERNATE OPTIMAL SOLUTIONS

Let x_0 be an optimal solution for our model LP

$$\min\{c'x \mid Ax \le b\}.$$

If x_0 is uniquely determined, we refer to x_0 as *the* optimal solution. Lacking this knowledge, we must necessarily refer to x_0 as *an* optimal solution. We have already observed an example of alternate optimal solutions in Example 2.6. Another is the following.

Example 3.4

$$
\begin{aligned}
\text{minimize:} \quad & -2x_1 - x_2 \\
\text{subject to:} \quad & x_1 + 3x_2 \le 9, \quad (1) \\
& 2x_1 + x_2 \le 8, \quad (2) \\
& -x_1 \quad\quad\ \le -1, \quad (3) \\
& \quad\ -x_2 \le -1. \quad (4)
\end{aligned}
$$

Suppose that we apply Algorithm 3 to this problem with the initial data

$$x_0 = \begin{bmatrix} 3 \\ 2 \end{bmatrix}, \quad D_0^{-1} = \begin{bmatrix} -1/5 & 3/5 \\ 2/5 & -1/5 \end{bmatrix}, \quad \text{and} \quad J_0 = \{1, 2\}.$$

Algorithm 3 immediately terminates in Step 1.2 with the information that x_0 is optimal. From (3.32), an optimal solution for the dual is $u = (0, 1, 0, 0)'$. The geometry for this problem is shown in Figure 3.1. It is clear from the figure that all points on the line segment joining x_0 and $(7/2, 1)'$ are also optimal solutions. In this case, $-c = (2, 1)'$ is parallel (in fact, equal) to the gradient of the active constraint $2x_1 + x_2 = 8$, so that the dual feasibility condition is satisfied at all points along the line segment.

More generally, suppose that Algorithm 3 has been applied to

$$\min\{c'x \mid a_i'x \le b_i, \ i = 1, \ldots, m\}. \tag{3.35}$$

Let x_0 be the optimal solution for the primal, let u be the optimal solution for the dual obtained from (3.32), and let D_0^{-1} and J_0 be the associated data. Suppose that for some $i \in J_0$, $u_i = 0$; that is, constraint i is in the active set and its multiplier has value zero. Let k be such that $\alpha_{k0} = i$. Then a_i is column k of D_0'. By construction of u_i in (3.32),

$$c'c_{k0} = 0. \tag{3.36}$$

Now define $s_0 = c_{k0}$ and observe that from (3.36),

$$c'(x_0 - \sigma s_0) = c'x_0, \quad \text{for all } \sigma \ge 0.$$

Thus $x_0 - \sigma s_0$ is also an optimal solution for all σ for which $x_0 - \sigma s_0$ remains feasible. The largest such value of σ is the maximum feasible step

Figure 3.1 Geometry of Example 3.4.

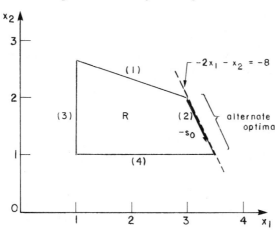

size σ_0, and may be computed as in Step 2 of Algorithm 3. Let l be the index of the restricting constraint for σ_0 and assume that $\sigma_0 > 0$. Then $x_0 - \sigma s_0$ is optimal for all σ with $0 \leq \sigma \leq \sigma_0$. Constraint i, which is active at x_0, becomes inactive at $x_0 - \sigma s_0$ for $\sigma > 0$ and constraint l, which is inactive at $x_0 - \sigma s_0$ for $\sigma < \sigma_0$, becomes active at $x_0 - \sigma_0 s_0$.

Returning to Example 3.4, we have

$$i = 1, \quad k = 1, \quad s_0 = \begin{bmatrix} -1/5 \\ 2/5 \end{bmatrix}, \quad \sigma_0 = \frac{5}{2}, \quad \text{and} \quad l = 4.$$

As in the analysis above, we have that

$$\begin{bmatrix} 3 \\ 2 \end{bmatrix} - \sigma \begin{bmatrix} -1/5 \\ 2/5 \end{bmatrix}$$

is optimal for all σ with $0 \leq \sigma \leq 5/2$.

The complementary slackness part of the optimality conditions for

$$\min \{ c'x \mid a_i'x \leq b_i, \ i = 1, \ldots, m \}$$

states that

$$u_i (a_i'x_0 - b_i) = 0, \quad i = 1, \ldots, m.$$

It is natural to think of this condition as meaning that the ith dual variable must be zero if constraint i is inactive at x_0. However, it is also possible for u_i to have value zero when constraint i is active (i.e., both terms in the above product may be zero). This was the case with $i = 1$ in Example 3.4. We saw that alternate optima were a consequence of this condition. To conclude uniqueness of an optimal solution, we must preclude this condition. Consequently, we say that x_0 satisfies the <u>strict</u> <u>complementary</u> <u>slackness</u> <u>condition</u> for (3.35) if

$$a_i'x_0 = b_i \quad \text{implies} \quad u_i > 0, \quad i = 1, \ldots, m;$$

that is, if all multipliers associated with active constraints are strictly positive.

A useful criterion for the uniqueness can be formulated in terms of the strict complementary slackness condition as follows.

Theorem 3.11.

Let x_0 be an optimal solution for (3.35). If x_0 is a nondegenerate extreme point and x_0 satisfies the strict complementary slackness condition, then x_0 is the unique optimal solution for (3.35).

Proof:

By hypothesis, there are exactly n constraints active at x_0 and their gradients are linearly independent. Assume, without loss of generality, that the active constraints are the first n. Let $D_0' = \begin{bmatrix} a_1, \ldots, a_n \end{bmatrix}$. Then D_0 is nonsingular. Let $D_0^{-1} = \begin{bmatrix} c_1, \ldots, c_n \end{bmatrix}$. The first n multipliers associated with x_0 are the solution of the linear equations

$$D_0' \begin{bmatrix} u_1 \\ u_2 \\ \vdots \\ u_n \end{bmatrix} = -c,$$

so that[1]

$$u_i = -c'c_i, \quad i = 1, \ldots, n. \tag{3.37}$$

Let x be any feasible point for (3.35) with $x \neq x_0$. Since $c_1, \ldots, c_n$ are linearly independent, there are numbers $\rho_1, \ldots, \rho_n$ such that

$$x - x_0 = \rho_1 c_1 + \rho_2 c_2 + \cdots + \rho_n c_n. \tag{3.38}$$

Let k be any integer such that $1 \leq k \leq n$. Then

$$
\begin{aligned}
a_k'(x - x_0) &= a_k'x - a_k'x_0 \\
&= a_k'x - b_k \quad \text{(because } a_k'x_0 = b_k) \\
&\leq 0. \quad \text{(because } a_k'x \leq b_k)
\end{aligned}
$$

Furthermore, taking the inner product of both sides of (3.38) with a_k, it follows by definition of the inverse matrix that

$$\rho_k = a_k'(x - x_0). \tag{3.39}$$

From the above, we now have that

$$\rho_i \leq 0, \quad \text{for } i = 1, \ldots, n.$$

Now, from (3.37) and (3.38),

$$
\begin{aligned}
c'(x - x_0) &= \sum_{i=1}^n \rho_i c'c_i \\
&= -\sum_{i=1}^n \rho_i u_i.
\end{aligned}
\tag{3.40}
$$

The strict complementary slackness condition asserts that

$$u_i > 0, \quad \text{for } i = 1, \ldots, n.$$

Furthermore, because x_0 is a nondegenerate extreme point and $x \neq x_0$, there is at least one i, $1 \leq i \leq n$, such that $\rho_i < 0$. Consequently, from (3.40),

$$c'x_0 < c'x$$

for all feasible x with $x \neq x_0$. This implies that x_0 is the unique optimal solution. ∎

---

[1] The analysis here is identical to that preceding (3.32).

Example 3.5

From (3.32), the multipliers for the optimal solution obtained in Example 2.2 are

$$u = \left(0, 0, \frac{5}{4}, 0, \frac{23}{4}\right)'.$$

Only constraints (3) and (5) are active at the optimal solution so that it is a nondegenerate extreme point. Furthermore, by inspection of the multiplier vector above, the strict complementary slackness condition is satisfied. Theorem 3.11 asserts that the optimal solution is unique.

We complete this section by extending Theorem 3.11 to a model problem having both inequality and equality constraints. Consider the model problem

$$\min\{c'x \mid A_1x \leq b_1, \ A_2x = b_2\}. \tag{3.41}$$

Let x_0 be an optimal solution. We say that x_0 satisfies the <u>strict complementary slackness condition</u> for (3.41) if each multiplier associated with an active inequality constraint is strictly positive. No requirement is imposed on the multipliers for the equality constraints. The proof of the following is similar to that of Theorem 3.11 and is left as an exercise (Exercise 3.26).

Theorem 3.12.

Let x_0 be an optimal solution for (3.41). If x_0 is a nondegenerate extreme point and x_0 satisfies the strict complementary slackness condition, then x_0 is the unique optimal solution for (3.41).

Note that Theorem 3.11 may be used to ascertain uniqueness for the primal problem

$$\min\{c'x \mid Ax \leq b\}$$

and Theorem 3.12 may be used to determine uniqueness for the dual problem

$$\max\{-b'u \mid A'u = -c, \ u \geq 0\}.$$

3.4 SENSITIVITY ANALYSIS

The data for a linear programming problem are usually not known with certainty. For example, the data may be measurements, made to a certain tolerance, of physical data. In this section, we investigate the effect on the optimal solution of small changes in the coefficients of the objective function and the right-hand side of the constraint functions. We introduce this type of analysis with the following example.

Example 3.6

minimize: $-\ x_1\ -\ x_2$

subject to: $x_1\ +\ 3x_2\ \le\ \ 9,$ (1)

$\qquad\qquad 2x_1\ +\ x_2\ \le\ \ 8,$ (2)

$\qquad -\ x_1\qquad\quad\ \le\ -1,$ (3)

$\qquad\qquad\ -\ x_2\ \le\ -1.$ (4)

We have seen in Example 1.1 that the optimal solution is $x_0 = (3,2)'$. It is easy to verify that

$$u\ =\ \left(\frac{1}{5},\frac{2}{5},0,0\right)'$$

is optimal for the dual problem. Furthermore, x_0 is a nondegenerate extreme point and the strict complementary slackness condition is satisfied. Theorem 3.11 asserts that x_0 is the unique optimal solution. Let

$$c_0\ =\ \begin{bmatrix} -1 \\ -1 \end{bmatrix}.$$

The solution above is optimal for $c = c_0$.

Suppose that c is allowed to vary in some neighborhood of c_0. From Figure 3.2, it is clear that x_0 remains optimal for all c in the cone spanned by a_1 and a_2. The multipliers do, however, depend on c. Let

$$c\ =\ \begin{bmatrix} c_1 \\ c_2 \end{bmatrix}.$$

Figure 3.2   Dependence of the optimal solution on c.

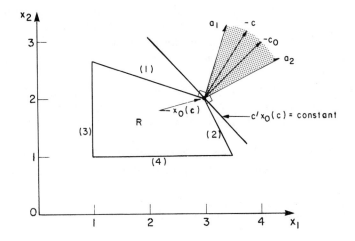

Set $D_0' = [a_1, a_2]$ and $J_0 = \{1, 2\}$. As in Example 3.4,

$$D_0^{-1} = \begin{bmatrix} -1/5 & 3/5 \\ 2/5 & -1/5 \end{bmatrix}.$$

As in (3.32), the multipliers may be obtained as explicit functions of c:

$$u = \begin{bmatrix} \frac{1}{5}c_1 - \frac{2}{5}c_2 \\ -\frac{3}{5}c_1 + \frac{1}{5}c_2 \\ 0 \\ 0 \end{bmatrix}.$$

This may be expressed equivalently in matrix notation as

$$u \equiv u(c) = \begin{bmatrix} -(D_0^{-1})' \\ \vdots \\ 0 \end{bmatrix} c. \tag{3.42}$$

Now $x_0 \equiv x_0(c)$ remains optimal for all c such that

$$u(c) \geq 0. \tag{3.43}$$

Equation (3.43) can be expressed equivalently as

$$(D_0^{-1})'c \leq 0.$$

In the case at hand, x_0 is optimal for all $c = (c_1, c_2)'$ such that

$$\left. \begin{array}{rcrcl} c_1 & - & 2c_2 & \geq & 0, \\ -3c_1 & + & c_2 & \geq & 0. \end{array} \right\} \tag{3.44}$$

Now x_0 is independent of c, and this is a special case of a linear function of c. Furthermore, the optimal solution for the dual is the linear function of c given by (3.42). Therefore, we have shown that the optimal solutions for both the primal and dual are linear functions of c for all c in the neighborhood of c_0 defined by (3.44).

Define $b_0 = (9, 8, -1, -1)'$ and $b = (b_1, b_2, b_3, b_4)'$. We continue the example by keeping $c = c_0$ fixed and examining the behavior of the optimal solution as it depends on b, while b varies in some neighborhood of b_0.[2]

Let $x_0(b)$ and $u(b)$ denote the optimal primal and dual solutions as they depend on b, respectively. It seems reasonable that for b sufficiently close to b_0, the active constraints at the optimal solution will remain unchanged.

---

[2] The alert reader will appreciate that by Theorem 3.6, this part of the example is the dual of the first part.

Then

$$D_0 x_0(b) = \begin{bmatrix} b_1 \\ b_2 \end{bmatrix},$$

$$x_0(b) = D_0^{-1} \begin{bmatrix} b_1 \\ b_2 \end{bmatrix}$$

$$= \begin{bmatrix} -\frac{1}{5}b_1 + \frac{3}{5}b_2 \\ \frac{2}{5}b_1 - \frac{1}{5}b_2 \end{bmatrix} \tag{3.45}$$

$$= \begin{bmatrix} D_0^{-1}, 0 \end{bmatrix} \begin{bmatrix} b_1 \\ b_2 \\ b_3 \\ b_4 \end{bmatrix}. \tag{3.46}$$

Since $c = c_0$ remains fixed, $u(b) = u$ remains unchanged. Thus $x_0(b)$ and $u(b)$ are optimal for the primal and dual, respectively, for all b for which $x_0(b)$ remains feasible. Now by construction, the active constraints (1) and (2) remain active for all b. However, b must be restricted so that constraints (3) and (4) remain satisfied. Substituting $x_0(b)$ from (3.45) into constraints (3) and (4) gives

$$\left. \begin{aligned} \frac{1}{5}b_1 - \frac{3}{5}b_2 - b_3 \quad\quad &\leq 0, \\ -\frac{2}{5}b_1 + \frac{1}{5}b_2 \quad\quad - b_4 &\leq 0. \end{aligned} \right\} \tag{3.47}$$

$x_0(b)$ and $u(b)$ are therefore primal and dual optimal, respectively, for all b satisfying (3.47). Therefore, $x_0(b)$ and $u(b)$ are linear functions of b for all b in this neighborhood of b_0. This part of the example is illustrated in Figure 3.3.

As the final part of this example, we investigate the behavior of the optimal objective function value as it depends first on c and then on b.

Because $c'x_0(c) = c'x_0$, for all c satisfying (3.44), we have[3]

$$\frac{\partial}{\partial c_i}(c'x_0(c)) \bigg|_{c=c_0} = (x_0)_i, \quad i = 1, 2.$$

---

[3] We have used the notation $(x_0)_i$ to denote the ith component of the vector x_0.

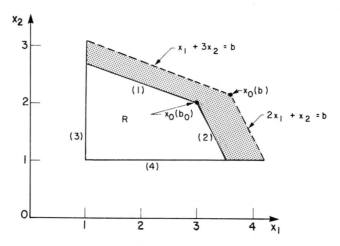

Figure 3.3 Dependence of the optimal solution on b.

From (3.42) and (3.46),

$$c'x_0(b) = c'D_0^{-1} \begin{bmatrix} b_1 \\ b_2 \end{bmatrix}$$

$$= -b'u,$$

where we have used the fact that $u_3 = u_4 = 0$. Thus

$$\frac{\partial}{\partial b_i}(c'x_0(b)) \bigg|_{b = b_0} = -u_i(b_0), \quad i = 1, \ldots, 4.$$

The generalization of these properties is contained in the following theorem.

Theorem 3.13.

Let $x_0(b_0, c_0)$ and $u(b_0, c_0)$ be optimal solutions for

$$\min\{c'x \mid a_i'x \leq b_i, \ i = 1, \ldots, m\}$$

and its dual, respectively, for $b = b_0$ and $c = c_0$. Assume that $x_0(b_0, c_0)$ is a nondegenerate extreme point and that the strict complementary slackness condition is satisfied. Let $x_0(b, c)$ and $u(b, c)$ denote the optimal solutions for the primal and dual, respectively, as they depend on b and c. Then

(a) $x_0(b,c)$ and $u(b,c)$ are linear functions of b and c for all b and c in some nonnull neighborhood of b_0 and c_0, which includes b_0 and c_0 in its interior,

(b) $\left. \dfrac{\partial}{\partial c_i}(c'x_0(b,c)) \right|_{\substack{b\,=\,b_0 \\ c\,=\,c_0}} = (x_0(b_0,c_0))_i, \quad i = 1, \ldots, n,$

(c) $\left. \dfrac{\partial}{\partial b_i}(c'x_0(b,c)) \right|_{\substack{b\,=\,b_0 \\ c\,=\,c_0}} = -(u_0(b_0,c_0))_i, \quad i = 1, \ldots, m.$

Proof:

Assume that the constraints active at $x_0(b_0, c_0)$ have been renumbered, if necessary, so that the first n are active. Let D_0' and D_0^{-1} be as in Theorem 3.11. Then, as in (3.46) and (3.42), respectively,

$$x_0(b,c) = \left[D_0^{-1}, 0 \right] b, \tag{3.48}$$

$$u(b,c) = \begin{bmatrix} -(D_0^{-1})' \\ \vdots \\ 0 \end{bmatrix} c. \tag{3.49}$$

$x_0(b,c)$ and $u(b,c)$ are optimal for all b, c such that primal and dual feasibility are satisfied; that is, all b and c satisfying

$$(a_i' D_0^{-1}) \begin{bmatrix} b_1 \\ \vdots \\ b_n \end{bmatrix} \le b_i, \quad i = n+1, \ldots, m, \tag{3.50}$$

and

$$(D_0^{-1})'c \le 0. \tag{3.51}$$

Equations (3.50) and (3.51) impose linear inequality constraints on b and c, respectively. Because $x_0(b_0, c_0)$ is a nondegenerate extreme point, this region is nonnull and (b_0, c_0) lies in its interior. Furthermore, the optimal solutions for the primal and dual, $x_0(b,c)$ and $u(b,c)$, respectively, are the linear functions of b and c defined by (3.48) and (3.49), respectively, for all b and c in this region. This proves part (a).

To establish parts (b) and (c), observe that from (3.48), the optimal objective function value is the linear function of c:

$$c'(x_0(b,c)) = c'\left[D_0^{-1}, 0 \right] b. \tag{3.52}$$

Taking the partial derivative of both sides of (3.52) with respect to c_i and evaluating the result at $b = b_0$, $c = c_0$, verifies (b). Taking the partial derivative of both sides of (3.52) with respect to b_i, and using (3.49), proves part (c). ■

Example 3.7

Consider Example 3.6 further. Suppose that the problem reflects an economic model where the objective function represents cost and the right-hand side represents upper bounds on scarce resources. Consider the case when additional units of b_1 may be purchased at a unit cost of $\$p$. If p is very large, it will be unattractive to purchase any additional units. However, if p is relatively small, it may be profitable to purchase some units. We address the question of determining values of p for which more units should be purchased and the additional question of how many.

Let b_1 be replaced with $b_1 + \delta b_1$. Purchasing δb_1 additional units costs $p\delta b_1$ dollars. Balancing this is the fact that the optimal objective function is now

$$-5 - u_1\delta b_1.$$

The overall cost is the above plus $p\delta b_1$:

$$-5 - u_1\delta b_1 + p\delta b_1 = -5 + (p - u_1)\delta b_1.$$

Thus the overall cost is strictly reduced, unchanged, or increased according to whether $p < u_1$, $p = u_1$, or $p > u_1$. The break-even price is precisely the dual variable associated with constraint (1); $u_1 = 1/5$.

Because of this price interpretation, the optimal dual variables are frequently called shadow prices in the context of an economic model.

We continue by assuming that $p < 1/5$, so that it is advantageous to purchase additional units of b_1. The question is now: How many? We may use (3.45) to obtain the optimal solution as an explicit function of δb_1. Substituting $9 + \delta b_1$ for b_1 and $b_2 = 8$ gives

$$x_0(\delta b_1) = \begin{bmatrix} 3 - \frac{1}{5}\delta b_1 \\ 2 + \frac{2}{5}\delta b_1 \end{bmatrix}.$$

According to (3.47), $x_0(\delta b_1)$ remains feasible for all δb_1 satisfying[4]

$$\frac{-5}{2} \le \delta b_1 \le 10.$$

---

[4] $\delta b_1 < 0$ would mean that it is more profitable to *sell* δb_1 units at $\$p$ per unit.

EXERCISES

3.1. Consider the following problem.

$$
\begin{aligned}
\text{minimize:} \quad & -x_1 - x_2 \\
\text{subject to:} \quad & -x_1 \qquad\qquad \leq -2, && (1) \\
& -4x_1 + x_2 \leq -6, && (2) \\
& -2x_1 + x_2 \leq -2, && (3) \\
& -4x_1 + 3x_2 \leq -2, && (4) \\
& -x_1 + 2x_2 \leq \;\;\; 2, && (5) \\
& x_1 + 2x_2 \leq \;\;\; 6, && (6) \\
& 4x_1 + 3x_2 \leq 14. && (7)
\end{aligned}
$$

(a) Let $x_0 = (2, 2)'$. Show graphically that x_0 is optimal.

(b) Consider the statement "$-c$ is in the cone spanned by the gradients of constraints i and k." Find all i and k such that this statement is true. Numerically obtain the corresponding multipliers.

(c) Discuss the unusual aspects of this problem with respect to the proof of Theorem 3.1.

(d) Write down the dual problem and determine an optimal solution for it. Is this optimal solution unique?

3.2. Show that $(2, 2)'$ is an optimal solution for the problem

$$
\begin{aligned}
\text{minimize:} \quad & -x_1 - x_2 \\
\text{subject to:} \quad & -2x_1 + x_2 \leq -2, && (1) \\
& 2x_1 + x_2 \leq \;\;\; 6, && (2) \\
& x_1 + x_2 \geq \;\;\; 1, && (3) \\
& x_1 \qquad\qquad \geq \;\;\; 1. && (4)
\end{aligned}
$$

Write down the dual problem and determine an optimal solution for it.

3.3. State the dual of

$$
\begin{aligned}
\text{minimize:} \quad & -6x_1 - 4x_2 - 8x_3 \\
\text{subject to:} \quad & x_1 + x_2 + x_3 \leq \;\;\; 4, && (1) \\
& x_1 + 2x_2 - x_3 \leq -1, && (2) \\
& 2x_1 \qquad\quad + 3x_3 \leq \;\;\; 8, && (3) \\
& 2x_2 + 4x_3 \leq 10, && (4) \\
& x_1 + 2x_2 + 3x_3 \leq \;\;\; 7. && (5)
\end{aligned}
$$

Let $x_0 = (1, 0, 2)'$.

(a) Use complementary slackness to find an associated solution u_0 for the dual.

(b) Use Corollary 3.1 to show that x_0 and u_0 are optimal for the primal and dual, respectively.

(c) Is x_0 the unique optimal solution for the primal? Why?

(d) Is u_0 the unique optimal solution for the dual? Why?

3.4. State the dual of

$$\text{minimize:} \quad -2u_1 - 14u_2 - 36u_3$$

$$\text{subject to:} \quad -2u_1 + u_2 + 4u_3 - u_4 \qquad = \quad 5, \qquad (1)$$

$$u_1 + 2u_2 + 3u_3 \qquad - u_5 = -2, \qquad (2)$$

$$-1 \leq u_i \leq 3, \quad i = 1, \ldots, 5.$$

3.5. Consider the problem

$$\min \{ c'x \mid Ax \leq b \}$$

where

$$A = \begin{bmatrix} 3 & 1 \\ 4 & 3 \\ 1 & 2 \end{bmatrix}, \quad b = \begin{bmatrix} -3 \\ -6 \\ -2 \end{bmatrix}, \quad \text{and} \quad x_0 = \frac{-1}{5} \begin{bmatrix} 6 \\ 2 \end{bmatrix}.$$

For each of the following vectors c, determine if x_0 is an optimal solution, and if so, if it is unique.

$$c = \begin{bmatrix} -5 \\ -5 \end{bmatrix}, \quad c = \begin{bmatrix} -3 \\ -1 \end{bmatrix}, \quad c = \begin{bmatrix} -12 \\ -9 \end{bmatrix}.$$

3.6. Consider the problem

$$\text{minimize:} \quad -4x_1 - 5x_2 - 4x_3$$

$$\text{subject to:} \quad x_1 \qquad\quad + x_3 \leq 4, \qquad (1)$$

$$2x_1 + x_2 - x_3 \leq 3, \qquad (2)$$

$$x_2 + x_3 \leq 5, \qquad (3)$$

$$- x_2 + x_3 \leq 2, \qquad (4)$$

$$- x_1 + x_2 - x_3 \leq -2, \qquad (5)$$

$$x_1 + x_2 + x_3 \leq 7. \qquad (6)$$

Let $x_0 = (1, 2, 3)'$.

(a) Is x_0 an extreme point?

(b) Show that x_0 is optimal.

(c) State the dual problem. What is an optimal solution for it?

(d) Determine all alternate optimal solutions for the primal problem.

3.7. Consider the problem

$$\text{minimize:} \quad -5x_1 + 2x_2$$

$$\text{subject to:} \quad -2x_1 + x_2 \leq 2, \qquad (1)$$

$$x_1 + 2x_2 \leq 14, \qquad (2)$$

$$4x_1 + 3x_2 \leq 36, \qquad (3)$$

$$- x_1 \qquad\quad \leq 0, \qquad (4)$$

$$- x_2 \leq 0. \qquad (5)$$

Let $c_0 = (-5, 2)'$ and $b_0 = (2, 14, 36, 0, 0)'$.

(a) What are the optimal solutions for the primal and dual when $c = c_0$ and $b = b_0$?

(b) For $b = b_0$, determine the optimal primal and dual solutions as explicit functions of c. For what values of c do these solutions remain optimal?

(c) For $c = c_0$, determine the optimal primal and dual solutions as explicit functions of b. For what values of b do these solutions remain optimal?

3.8. Consider the problem

$$\text{minimize:} \quad -2x_1 - x_2$$

$$\text{subject to:} \quad x_1 + x_2 \leq 4, \quad (1)$$

$$-x_1 + x_2 \leq 0, \quad (2)$$

$$-x_2 \leq -2. \quad (3)$$

(a) Obtain optimal primal and dual solutions for this problem.

(b) Does Theorem 3.13 apply to this problem? What happens when $b_1 = 4$ is decreased?

3.9. What happens in Example 3.7 when δb_1 is increased beyond 10? Obtain the optimal solution for all $\delta b_1 \geq 10$. Illustrate with a graph.

3.10. Give a similar analysis to that in Example 3.7 for the case when $b_1 = 9$ and b_2 is replaced with $8 + \delta b_2$.

3.11. Why does Theorem 3.13 not apply to the problem of Exercise 3.1? Nevertheless, obtain $x_0(b, c)$ and $u(b, c)$ from (3.48) and (3.49), respectively. Also determine the constraints (3.50) and (3.51) on b and c. Do b_0 and c_0 lie in the interior of this region? What happens as b_2 is increased from -6?

3.12. State the dual of the problem of Example 2.4. Use the final data from Algorithm 3 and (3.32) to find an optimal solution for the dual problem.

3.13. Repeat Exercise 3.12 for the problem of Example 2.5.

3.14. Repeat Exercise 3.12 for the problem of Example 2.6.

3.15. Repeat Exercise 3.12 for the problem of Example 2.10.

3.16. Verify the remaining entries of Table 3.1.

3.17. Assume that $\text{rank}(A) = n$ and that $\min\{c'x \mid Ax \leq b\}$ is bounded from below. Prove that for any feasible point x_0 there exists an extreme point x_1 such that $c'x_1 \leq c'x_0$.

3.18. Prove

Farkas' Lemma.

Let A be an (m, n) matrix, and c an n-vector. Then either
there exists an n-vector x with $Ax \leq 0$ and $c'x < 0$,

or,

there exists an m-vector u with $A'u = -c$ and $u \geq 0$,
but never both.

Hint: Use the primal problem

$$\min\{c'x \mid Ax \leq 0\},$$

its dual, and the duality theorems of Section 3.2.

3.19. Let x_0 be optimal for $\min\{c'x \mid Ax \le b\}$.

    (a) Show directly that

$$As \le 0 \quad \text{implies} \quad c's \ge 0 \quad \text{for all } n\text{-vectors } s.$$

    (b) Assume that x_0 is a nondegenerate extreme point. Show directly that x_0 satisfies the optimality conditions. *Hint:* Assume, without loss of generality, that the active constraints are the first n. Define D_0' and D_0^{-1} as usual. Consider the effect of using any column of D_0^{-1} as a search direction s and use part (a).

    (c) Why does the argument of part (b) break down if x_0 is degenerate? Explain in terms of the problem of Exercise 3.1.

    (d) Use Farkas' Lemma to provide an alternate proof of the necessity part of Theorem 3.1. *Hint:* Use part (a).

Remark: A traditional development of the necessary conditions for optimality for a linear programming problem has been to first prove Farkas' Lemma directly (a somewhat lengthy process) and then to show necessity using Farkas' Lemma as above. An alternate approach is to develop a parallel version of Algorithm A, resolving degeneracy by means of a lexicographic ordering [Dantzig, Orden, and Wolfe, 1955], and then prove sufficiency by the constructive method of the proof of Theorem 3.1 as we have done. We have chosen to use Bland's rules to resolve degeneracy because of the simplicity of their statement.

3.20. Show that if the primal problem is unbounded from below, the dual problem has no feasible solution. *Hint:* Use the Weak Duality Theorem (Theorem 3.7).

3.21. Verify the remaining entries of Table 3.2.

3.22. The purpose of this exercise is to provide an alternate derivation of the notions of primal and dual linear programming problems.

    (a) Show that $\min\{c'x\}$ is unbounded from below if $c \ne 0$, and has an optimal solution if and only if $c = 0$.

    (b) Let c and x be n-vectors, b and u be m-vectors, and let A be an (m, n) matrix. Let $A = [a_1, \ldots, a_m]$. Assume that A, b, and c are given. Define

$$L(x, u) = c'x + u'(Ax - b),$$

$$L_*(x) = \max\{L(x, u) \mid u \ge 0\},$$

$$L^*(u) = \min\{L(x, u)\}.$$

Define the "primal" problem to be $\min\{L_*(x)\}$, and the "dual" problem to be $\max\{L^*(u) \mid u \ge 0\}$. Show that

$$L_*(x) = \begin{cases} c'x, & \text{if } Ax \le b, \\ +\infty, & \text{if } a_i'x > b_i \text{ for at least one } i, \end{cases}$$

so that the "primal" problem is just

$$\min\{c'x \mid Ax \le b\};$$

that is, the "primal" is identical to our usual primal problem. Also show that the "dual" problem is equivalent to our usual dual problem

$$\max\{-b'u \mid A'u = -c, \ u \ge 0\}.$$

(c) Give an alternate proof of the Weak Duality Theorem (Theorem 3.7) by showing (in just one line!) that

$$L_*(x) \geq L^*(u).$$

3.23. Formulate the optimality conditions and the dual problem for
   (a) $\min \{ c'x \mid A_1 x = b_1, \quad A_2 x \leq b_2, \quad A_3 x \geq b_3, \quad x \geq 0 \}$,
   (b) $\min \{ c_1' x_1 + c_2' x_2 \mid A_{11} x_1 + A_{12} x_2 \leq b_1, \quad A_{21} x_1 + A_{22} x_2 = b_2,$
   $x_1 \geq 0, \quad x_2 \leq b_3 \}$.

3.24. Suppose that the feasible region for the problem

$$\min \{ c'x \mid Ax \leq b \}$$

is bounded. Show that every optimal solution is a convex combination of the optimal extreme points.

3.25. What is the optimal solution for

$$\min \{ c'x \mid a'x = 1, \quad x \geq 0 \},$$

where $c_i < 0$, $a_i > 0$, $i = 1, \ldots, n$? State the dual problem and find an optimal solution for it.

3.26. Complete the proof of Theorem 3.12.

3.27. Show that the two problems

$$\min \{ c'x \mid Ax \geq b, \quad x \geq 0 \} \quad \text{and} \quad \max \{ b'u \mid A'u \leq c, \quad u \geq 0 \}$$

with $b \leq 0$ and $c \geq 0$ have optimal solutions.

3.28. Is it possible for neither the primal problem $\min \{ c'x \mid Ax \leq b \}$ nor its dual problem $\max \{ -b'u \mid A'u = -c, \quad u \geq 0 \}$ to possess a feasible solution? *Hint:* Consider the following primal problem [Dantzig, 1963]:

minimize: $-x_1 - x_2$

subject to: $-x_1 \qquad\qquad \leq \quad 0,$ (1)

$\qquad\qquad\qquad - x_2 \leq \quad 0,$ (2)

$\qquad\quad x_1 - x_2 = \quad -5,$ (3)

$\qquad\quad x_1 - x_2 = \quad\;\; 5.$ (4)

3.29. Suppose that the problem

$$\min \{ c'x \mid Ax \leq b \}$$

has a unique optimal solution $\hat{x}$. Show that the dual problem has a unique optimal solution if and only if the gradients of the constraints active at $\hat{x}$ are linearly independent.

3.30. Consider the problem (3.30) and its equivalent formulation

$$\min \{ c'x \mid A_1 x \leq b_1, \quad A_2 x = b_2 \}.$$

Suppose that Algorithm 3 has been applied to this problem to obtain an optimal solution x_0. Let $J_0 = \{ \alpha_1, \ldots, \alpha_n \}$ and let $D_0^{-1} = \begin{bmatrix} c_1, \ldots, c_n \end{bmatrix}$ be the associated data. Assume that x_0 is an extreme point. Let t denote the number of columns c_i of D_0^{-1} for which

$$1 \leq \alpha_i \leq m \quad \text{and} \quad c'c_i = 0.$$

Let E_0 be the (n,t) matrix whose columns consist of precisely those c_i for which

$$1 \leq \alpha_i \leq m \quad \text{and} \quad c'c_i = 0.$$

Let S be the entire family of optimal solutions for (3.30).

(a) Prove

$$S = \{x \mid x = x_0 + E_0w, \ A_1E_0w \leq b_1 - A_1x_0\}.$$

(b) If $t > 1$, then (3.30) has alternate optimal solutions and it may be desirable to find one that minimizes some other linear function (e.g., $d'x$). Show that this problem can be formulated in t variables.

(c) Using $x_0 = (-3, 1, 1)'$ and $J_0 = \{4, 2, 3\}$, find S for the problem

$$\text{minimize:} \qquad x_1 + x_2 + x_3$$

$$
\begin{array}{lllll}
\text{subject to:} & x_1 & & & \leq 1, & (1) \\
& & x_2 & & \leq 1, & (2) \\
& & & x_3 \leq 1, & (3) \\
& -x_1 & - x_2 & - x_3 \leq 1. & & (4)
\end{array}
$$

3.31. Research Problem. Analyze the behavior of the optimal solution for the problem

$$\min\{c'x \mid Ax \leq b\}$$

under small changes in one of the columns of A.

4

THE SIMPLEX METHOD

This chapter is devoted to the simplex method of linear programming. This is the original solution procedure for a linear programming problem developed by George B. Dantzig in the late 1940s and the early 1950s. The appearance of the simplex method is different from that of the algorithms developed in Chapter 2. This is due to the use of a different model problem, and is a surface difference only. Under the appropriate assumptions, all of these algorithms generate the same sequence of points.

4.1 THE REVISED SIMPLEX METHOD

We derive the revised simplex method from Algorithm 3. Beginning with an extreme point, the steps of Algorithm 3 can be expressed as follows.

Step 1:
Compute the multipliers for the active inequality constraints. If they are all nonnegative, stop: the current solution is optimal. Otherwise, choose the active inequality constraint with the smallest multiplier[1] and compute the search direction that deletes it from the active set.

Step 2:
If the objective function can be decreased indefinitely by moving along the search direction, the problem is unbounded from below: stop. Oth-

---

[1] We think of this as being the "most negative" multiplier.

erwise, move as far as possible along the search direction until some previously inactive constraint restricts further decrease.

Step 3:

Modify the active set by deleting the constraint found in Step 1 from
the active set and adding the constraint determined in Step 2. Update
any quantities required for further computation and return to Step 1.

The simplex method performs precisely these steps and is thus
equivalent to Algorithm 3. However, the algebraic formulation of each step
differs from Algorithm 3 for two reasons. First, the model problem assumed
by the simplex method is different from that for Algorithm 3. Second, the
steps of the simplex method are designed to take advantage of the algebraic
structure of the model problem.

The simplex method uses the model problem

$$\min\{c'x \mid Ax = b, \ x \geq 0\}, \tag{4.1}$$

where A is an (m, n) matrix, c and x are n-vectors, and b is an m-vector.[2]
Note that the only inequality constraints of the model problem are the nonnegativity constraints $x \geq 0$. Assume that rank$(A) = m$. From Theorem 3.3,
the optimality conditions for (4.1) are

$$Ax = b, \ x \geq 0, \tag{4.2}$$

$$A'u - v = -c, \ v \geq 0, \tag{4.3}$$

$$v'x = 0. \tag{4.4}$$

We have used u to denote the m-vector of multipliers associated with the
equality constraints and v to denote the n-vector of multipliers associated
with the nonnegativity constraints. Suppose that a nondegenerate extreme
point for (4.1) is known. We next formulate one iteration of Algorithm 3 for
(4.1) in such a way as to take advantage of the algebraic structure of (4.1).
First observe that n of the constraints of (4.1) must be active. Since m of
these must be the m equality constraints, the remaining $n - m$ must be active inequality constraints. Thus any extreme point of (4.1) must necessarily
have $n - m$ components with value zero. The m positive components are
called basic variables.[3] The $n - m$ components having value zero are called
nonbasic variables. The extreme point is called a basic feasible solution.

Let $A = \left[A_1, A_2, \ldots, A_n \right]$ (i.e., A_i denotes the ith *column* of A).
Suppose for simplicity that the first m components of our extreme point are

---

[2] Note this change of notation. Previously, we have used m to denote the number of inequality constraints. In this chapter, m denotes the number of *equality* constraints.

[3] More generally, the basic variables need only be nonnegative. If one such variable has
value zero, then at least $n + 1$ constraints are active and the corresponding point is a degenerate
extreme point.

the basic variables and consequently, the last $n - m$ are nonbasic. The current extreme point may then be partitioned as

$$\begin{bmatrix} x_B \\ x_{NB} \end{bmatrix},$$

where the "B" and "NB" subscripts are to be read as "basic" and "nonbasic," respectively. Let

$$B = \begin{bmatrix} A_1, A_2, \ldots, A_m \end{bmatrix};$$

that is, B is the (m, m) submatrix of A consisting of the first m columns of A, in their natural order. B is called the basis matrix. Let

$$N = \begin{bmatrix} A_{m+1}, A_{m+2}, \ldots, A_n \end{bmatrix},$$

so that N is the $(m, (n - m))$ submatrix of A consisting of the last $n - m$ columns of A in their natural order. Thus the equality constraints of (4.1) may be written as

$$B x_B + N x_{NB} = b.$$

The specific value of the current extreme point is found by setting $x_{NB} = 0$. This gives

$$B x_B = b.$$

Because

$$\begin{bmatrix} x_B \\ x_{NB} \end{bmatrix}$$

is a nondegenerate extreme point, it follows that B is nonsingular (Exercise 4.22). Therefore,

$$x_B = B^{-1} b.$$

Step 1 of Algorithm 3 requires the computation of the multipliers for the active inequality constraints. We do this as follows. Let c_B denote the first m components of c and let c_{NB} denote the last $n - m$ components. Partition the multiplier vector for the inequality constraints, v, in a similar manner. Thus

$$c = \begin{bmatrix} c_B \\ c_{NB} \end{bmatrix} \quad \text{and} \quad v = \begin{bmatrix} v_B \\ v_{NB} \end{bmatrix}.$$

Because $x_{NB} = 0$, the objective function value at the current extreme point is

$$(c_B', c_{NB}') \begin{bmatrix} x_B \\ x_{NB} \end{bmatrix} = c_B' x_B + c_{NB}' x_{NB} = c_B' x_B.$$

With the "B" and "NB" notation, the first part of the dual feasibility conditions (4.3) and the complementary slackness condition (4.4) become

$$\begin{bmatrix} B' \\ N' \end{bmatrix} u - \begin{bmatrix} v_B \\ v_{NB} \end{bmatrix} = \begin{bmatrix} -c_B \\ -c_{NB} \end{bmatrix},$$

and

$$v_B' x_B + v_{NB}' x_{NB} = 0,$$

respectively. Equating corresponding partitions gives

$$B'u - v_B = -c_B, \tag{4.5}$$

$$N'u - v_{NB} = -c_{NB}, \tag{4.6}$$

$$v_B' x_B = 0, \tag{4.7}$$

$$v_{NB}' x_{NB} = 0. \tag{4.8}$$

Because $x_B > 0$, (4.7) implies that $v_B = 0$. Substituting this in (4.5) gives

$$B'u = -c_B,$$

or

$$u = -(B^{-1})' c_B. \tag{4.9}$$

Having obtained u from (4.9), the multipliers for the active nonnegativity constraints are obtained directly from (4.6):

$$v_{NB} = c_{NB} + N'u. \tag{4.10}$$

Because we have assumed that the basic variables are the first m, (4.10) also may be written in more conventional notation as

$$v_i = c_i + A_i'u, \quad i = m + 1, \ldots, n. \tag{4.11}$$

Having computed v_{NB}, we proceed by computing k such that

$$v_k = \min\{v_i \mid i = m + 1, \ldots, n\}. \tag{4.12}$$

If $v_k \geq 0$, the current point is optimal. Otherwise, we wish to delete the kth nonnegativity constraint from the active set. This is done by increasing the kth variable, x_k. Then the kth variable, which is presently nonbasic, is to become basic (i.e., x_k is increased from its lower bound of zero). Suppose for simplicity that $k = m + 1$. Then the kth component of the search direction is -1.[4] Let s_B denote the part of the search direction associated with the current basic variables. Also, let s_{NB} denote the part of the search direction associated with the nonbasic variables other than the kth. The complete

---

[4] The "-1" component is used because we wish to *increase* x_k.

search direction is thus

$$s_j = \begin{bmatrix} s_B \\ -1 \\ s_{NB} \end{bmatrix}.$$

Other than the nonbasic variable which is to become basic, the nonbasic variables remain at their lower bounds (i.e., their nonnegativity constraints remain active). Therefore, $s_{NB} = 0$. The changing components are

$$\begin{bmatrix} x_B \\ 0 \end{bmatrix} - \sigma \begin{bmatrix} s_B \\ -1 \end{bmatrix},$$

where σ denotes the step size. Since this must satisfy the equality constraints,

$$Bx_B + 0 \cdot A_k - \sigma Bs_B + \sigma A_k = b.$$

But $Bx_B = b$, so that this reduces to

$$s_B = B^{-1}A_k. \tag{4.13}$$

As σ is increased from zero, the basic components of the new point, $x_B - \sigma s_B$, must remain nonnegative. Therefore,

$$\sigma s_B \leq x_B. \tag{4.14}$$

Using (4.13) and (4.9), the objective function value at the new point

$$\begin{bmatrix} x_B \\ 0 \\ 0 \end{bmatrix} - \sigma \begin{bmatrix} s_B \\ -1 \\ 0 \end{bmatrix} \tag{4.15}$$

is

$$c' \left(\begin{bmatrix} x_B \\ 0 \\ 0 \end{bmatrix} - \sigma \begin{bmatrix} s_B \\ -1 \\ 0 \end{bmatrix} \right) = c_B' x_B - \sigma c_B' s_B + \sigma c_k$$

$$= c_B' x_B + \sigma(c_k - c_B' B^{-1} A_k)$$

$$= c_B' x_B + \sigma(c_k + u' A_k)$$

$$= c_B' x_B + \sigma v_k. \tag{4.16}$$

Suppose that $s_B \leq 0$. Then σ may be increased indefinitely and (4.15) is feasible for all $\sigma \geq 0$. Since $v_k < 0$, it follows from (4.16) that the objective function value tends to $-\infty$ as $\sigma \rightarrow +\infty$ (i.e., the problem is unbounded from below).

Suppose next that $(s_B)_i > 0$ for at least one i. Then from (4.14), the largest value of σ for which

$$\begin{bmatrix} x_B \\ 0 \\ 0 \end{bmatrix} - \sigma \begin{bmatrix} s_B \\ -1 \\ 0 \end{bmatrix}$$

remains feasible is

$$\sigma_B = \min\left\{ \frac{(x_B)_i}{(s_B)_i} \;\middle|\; \text{all } i \text{ with } (s_B)_i > 0 \right\}. \tag{4.17}$$

Let l be such that

$$\sigma_B = \frac{(x_B)_l}{(s_B)_l}. \tag{4.18}$$

Then the basic variable x_l has been reduced to zero at the new extreme point,

$$\begin{bmatrix} x_B \\ 0 \\ 0 \end{bmatrix} - \sigma_B \begin{bmatrix} s_B \\ -1 \\ 0 \end{bmatrix}.$$

From (4.16), its objective function value is

$$c_B' x_B + \sigma_B v_k,$$

which differs from the previous objective function value by $\sigma_B v_k$. Thus the objective function value is strictly reduced provided that $\sigma_B > 0$.

Variable k has changed from nonbasic to basic and variable l has changed from basic to nonbasic. The simplex method continues by modifying the basis matrix to reflect this change in basic/nonbasic variables. B is changed by replacing column l with A_k. Procedure Φ can be used to compute the new B^{-1}. However, it is important to observe that B has been modified by a *column* replacement, whereas Procedure Φ is formulated for *row* replacement. Since for any nonsingular matrix H, $(H')^{-1} = (H^{-1})'$, the new basis inverse is

$$\left[\Phi((B^{-1})', A_k, l) \right]'.$$

We have derived one iteration of the simplex method under the assumption that the first m variables are basic and that the incoming basic variable is the $(m + 1)$st. We next remove this simplifying ordering assumption and allow for the possibility that the basic variables are any of the n variables.

The situation is similar to that for Algorithm 1. In that case, we introduced the ordered index set J_j which gave the column numbers of the active constraints in D'_j. We require the analogous information for the simplex method. Because the inequality constraints of the model problem

$$\min \{ c'x \mid Ax = b, \ x \geq 0 \}$$

are precisely n in number and each is a constraint on exactly one of the variables, an ordered index set associated with the basic variables is appropriate. As previously, we use the "B" and "NB" notation for the basic and nonbasic variables, respectively. Define the ordered index set

$$I_B = \{ \beta_1, \beta_2, \ldots, \beta_m \},$$

where

$$B = \left[A_{\beta_1}, A_{\beta_2}, \ldots, A_{\beta_m} \right].$$

For example, suppose that $m = 4$, $n = 16$, variables 2, 3, 7, and 14 are basic, and

$$B = \left[A_7, A_3, A_{14}, A_2 \right].$$

Then

$$I_B = \{ 7, 3, 14, 2 \}. \tag{4.19}$$

It is helpful to remember that the conceptual notion of this index set is the *complement* of the one used in Chapter 2. Here the index set reflects the inactive constraints. Basic variables refer to inactive constraints and nonbasic variables refer to active constraints. Because the inequality constraints of the model problem (4.1) may be associated with the problem variables in an obvious way, the basic/nonbasic notation included in the index set I_B captures all the relevant information of the active set.

With the index set, we now modify our previous computations as follows. Assume that a basic feasible solution is known. Let B denote the basis matrix and let I_B be as defined above. Let c_B denote the m-vector of components of c, ordered according to I_B; for example, for I_B as in (4.19),

$$c_B = (c_7, c_3, c_{14}, c_2)'.$$

Let u denote the multipliers for the equality constraints and let v_B be the m-vector of components of v, ordered according to I_B. Using the complementary slackness condition (4.4), we have $v_B = 0$. Consequently, u must satisfy

$$B'u = -c_B.$$

Thus $u = -(B^{-1})'c_B$, as before. Having obtained u, the multipliers for the nonbasic variables (more correctly, the multipliers for the active constraints) are

$$v_i = c_i + A_i'u, \quad \text{for all } i \notin I_B.$$

In Step 1, we compute k such that

$$v_k = \min\{c_i + A_i'u \mid \text{all } i \notin I_B\}.$$

If $v_k \geq 0$, the current solution is optimal and we terminate. Otherwise, we compute

$$s_B = B^{-1}A_k,$$

and proceed with the computation of the maximum feasible step size.

The critical change there is that the index l associated with the computation of σ_B no longer refers to variable l. Let x_B and s_B denote the components of the basic variables and the basic portion of the search direction, respectively, *ordered according to* I_B. Then the feasibility requirement defining σ_B is

$$\begin{bmatrix} x_{\beta_1} \\ \vdots \\ x_{\beta_m} \end{bmatrix} - \sigma \begin{bmatrix} s_{\beta_1} \\ \vdots \\ s_{\beta_m} \end{bmatrix} \geq 0. \tag{4.20}$$

The component form of (4.20) is

$$(x_B)_i - \sigma(s_B)_i \geq 0,$$

so that σ_B is still correctly determined by (4.17). However, the "l" determined by (4.18) is not, in general, the index of the basic variable, which is reduced to zero. Rather, it is the basic variable x_{β_l}, which is reduced to zero. Thus, variable k becomes basic and variable β_l becomes nonbasic. Since the new basis matrix is now

$$\left[A_{\beta_1}, \ldots, A_{\beta_{l-1}}, A_k, A_{\beta_{l+1}}, \ldots, A_{\beta_m}\right],$$

the update of B^{-1} is as described previously and the index set $I_B = \{\beta_1, \beta_2, \ldots, \beta_m\}$ must be modified by replacing β_l with k.

These computations give us the revised[5] simplex method as follows.

REVISED SIMPLEX METHOD

Model Problem: $\min\{c'x \mid Ax = b, \ x \geq 0\}$

Initialization:
Start with basic index set $I_B = \{\beta_1, \ldots, \beta_m\}$, basis matrix

---

[5] The qualifier "revised" will be explained subsequently.

$$B = \left[A_{\beta_1}, \ldots, A_{\beta_m} \right],$$

B^{-1}, and basic feasible solution $x_B = B^{-1}b$ and $x_{NB} = 0$. Compute $c_B'x_B$.

Step 1:   Computation of Search Direction s_B.
Compute $u = -(B^{-1})'c_B$, the smallest index k, and v_k such that

$$v_k = \min\{c_i + A_i'u \mid i \notin I_B\}.$$

If $v_k \geq 0$, stop with optimal solution $x_B = B^{-1}b$ and $x_{NB} = 0$.
If $v_k < 0$, set $s_B = B^{-1}A_k$ and go to Step 2.

Step 2:   Computation of Maximum Feasible Step Size σ_B.
If $s_B \leq 0$, print "the problem is unbounded from below" and stop.
If $(s_B)_i > 0$ for at least one i, compute σ_B and the smallest index l such that

$$\sigma_B = \frac{(x_B)_l}{(s_B)_l} = \min\left\{ \frac{(x_B)_i}{(s_B)_i} \mid \text{all } i \text{ with } (s_B)_i > 0 \right\},$$

and go to Step 3.

Step 3:   Update.

Replace B^{-1} with $\left[\Phi((B^{-1})', A_k, l) \right]'$, x_B with $B^{-1}b$, and β_l with k. Compute $c_B'x_B$ and go to Step 1.

Primal feasibility and complementary slackness are satisfied at each step of the revised simplex method and termination occurs in Step 1 only if dual feasibility is also satisfied. In this case

$$u = -(B^{-1})'c_B,$$
$$v_i = 0, \qquad \text{all } i \in I_B,$$
$$v_i = c_i + A_i'u, \qquad \text{all } i \notin I_B,$$

is an optimal solution for the dual problem.

We remark that the components of the basic feasible solution x_B, x_{NB} may be determined also from the formulas

$$x_{\beta_i} = (B^{-1}b)_i, \qquad i = 1, \ldots, m,$$
$$x_i = 0, \qquad \text{all } i \notin I_B.$$

We illustrate the revised simplex method in the following examples.

Example 4.1

minimize: $2x_1 + 14x_2 + 36x_3$

subject to: $-2x_1 + x_2 + 4x_3 - x_4 \qquad\qquad = 5,$ (1)

$\qquad\qquad -x_1 - 2x_2 - 3x_3 \qquad\quad + x_5 = 2,$ (2)

$\qquad\qquad x_i \geq 0, \ i = 1, \ldots, 5.$

Initialization:

$$I_B = \{2, 5\}, \quad B = \begin{bmatrix} 1 & 0 \\ -2 & 1 \end{bmatrix}, \quad B^{-1} = \begin{bmatrix} 1 & 0 \\ 2 & 1 \end{bmatrix},$$

$$x_B = \begin{bmatrix} 1 & 0 \\ 2 & 1 \end{bmatrix} \begin{bmatrix} 5 \\ 2 \end{bmatrix} = \begin{bmatrix} 5 \\ 12 \end{bmatrix}, \quad c_B' x_B = 70.$$

Iteration 0

Step 1:[6] $u = -\begin{bmatrix} 1 & 2 \\ 0 & 1 \end{bmatrix} \begin{bmatrix} 14 \\ 0 \end{bmatrix} = \begin{bmatrix} -14 \\ 0 \end{bmatrix},$

$$v_3 = \min\{30, -, -20, 14, -\} = -20, \quad k = 3, \quad s_B = \begin{bmatrix} 4 \\ 5 \end{bmatrix}.$$

Step 2: $\sigma_B = \min\left\{\dfrac{5}{4}, \dfrac{12}{5}\right\} = \dfrac{5}{4}, \quad l = 1.$

Step 3: $B \leftarrow \begin{bmatrix} 4 & 0 \\ -3 & 1 \end{bmatrix}, \quad B^{-1} \leftarrow \left[(\Phi(B^{-1})', \begin{bmatrix} 4 \\ -3 \end{bmatrix}, 1) \right]' = \begin{bmatrix} 1/4 & 0 \\ 3/4 & 1 \end{bmatrix},$

$$x_B \leftarrow \begin{bmatrix} 5/4 \\ 23/4 \end{bmatrix}, \quad I_B \leftarrow \{3, 5\}, \quad c_B' x_B = 45.$$

Iteration 1

Step 1: $u = \begin{bmatrix} -9 \\ 0 \end{bmatrix},$

$$v_2 = \min\{20, 5, -, 9, -\} = 5, \quad k = 2.$$

$v_2 \geq 0$; stop with optimal solution $x = \left(0, 0, \dfrac{5}{4}, 0, \dfrac{23}{4}\right)'.$

Note that $u = (-9, 0)'$ together with $v = (20, 5, 0, 9, 0)'$ is an optimal solution for the dual.

---

[6] We use a dash to indicate the entries associated with basic variables.

A computer program which implements the revised simplex method is given in Appendix B, Section B.2.1. The output from applying this program to the problem of Example 4.1 is shown in Figure B.43.

Example 4.2

$$\begin{aligned}
\text{minimize:} \quad & -5x_1 + 2x_2 \\
\text{subject to:} \quad & -2x_1 + x_2 + x_3 &&&&= 2, && (1) \\
& x_1 + 2x_2 && + x_4 &&= 14, && (2) \\
& 4x_1 + 3x_2 &&&& + x_5 = 36, && (3) \\
& x_i \geq 0, \quad i = 1, \ldots, 5.
\end{aligned}$$

Initialization:

$$I_B = \{1, 2, 5\}, \quad B = \begin{bmatrix} -2 & 1 & 0 \\ 1 & 2 & 0 \\ 4 & 3 & 1 \end{bmatrix}, \quad B^{-1} = \begin{bmatrix} -2/5 & 1/5 & 0 \\ 1/5 & 2/5 & 0 \\ 1 & -2 & 1 \end{bmatrix},$$

$$x_B = B^{-1} \begin{bmatrix} 2 \\ 14 \\ 36 \end{bmatrix} = \begin{bmatrix} 2 \\ 6 \\ 10 \end{bmatrix}, \quad c'_B x_B = 2.$$

Iteration 0

Step 1: $u = \begin{bmatrix} -12/5 \\ 1/5 \\ 0 \end{bmatrix},$

$$v_3 = \min\left\{ -, -, \frac{-12}{5}, \frac{1}{5}, - \right\} = \frac{-12}{5}, \quad k = 3,$$

$$s_B = \begin{bmatrix} -2/5 \\ 1/5 \\ 1 \end{bmatrix}.$$

Step 2: $\sigma_B = \min\left\{ -, \dfrac{6}{1/5}, \dfrac{10}{1} \right\} = 10, \quad l = 3.$

Step 3: $B \leftarrow \begin{bmatrix} -2 & 1 & 1 \\ 1 & 2 & 0 \\ 4 & 3 & 0 \end{bmatrix}, \quad B^{-1} \leftarrow \begin{bmatrix} 0 & -3/5 & 2/5 \\ 0 & 4/5 & -1/5 \\ 1 & -2 & 1 \end{bmatrix}, \quad x_B \leftarrow \begin{bmatrix} 6 \\ 4 \\ 10 \end{bmatrix},$

$$I_B \leftarrow \{1, 2, 3\}, \quad c'_B x_B = -22.$$

Iteration 1

Step 1: $u = \begin{bmatrix} 0 \\ -23/5 \\ 12/5 \end{bmatrix}$,

$$v_4 = \min\left\{-, -, -, \frac{-23}{5}, \frac{12}{5}\right\} = \frac{-23}{5}, \quad k = 4,$$

$$s_B = \begin{bmatrix} -3/5 \\ 4/5 \\ -2 \end{bmatrix}.$$

Step 2: $\sigma_B = \left\{-, \dfrac{4}{4/5}, -\right\} = 5, \quad l = 2.$

Step 3: $B \leftarrow \begin{bmatrix} -2 & 0 & 1 \\ 1 & 1 & 0 \\ 4 & 0 & 0 \end{bmatrix}$, $B^{-1} \leftarrow \begin{bmatrix} 0 & 0 & 1/4 \\ 0 & 1 & -1/4 \\ 1 & 0 & 1/2 \end{bmatrix}$, $x_B \leftarrow \begin{bmatrix} 9 \\ 5 \\ 20 \end{bmatrix}$,

$$I_B \leftarrow \{1, 4, 3\}, \quad c_B' x_B = -45.$$

Iteration 2

Step 1: $u = \begin{bmatrix} 0 \\ 0 \\ 5/4 \end{bmatrix}$,

$$v_5 = \min\left\{-, \frac{23}{4}, -, -, \frac{5}{4}\right\} = \frac{5}{4}, \quad k = 5.$$

$v_5 \geq 0$; stop with optimal solution $x = (9, 0, 20, 5, 0)'$.

Note that $u = \left(0, 0, \dfrac{5}{4}\right)'$ together with $v = \left(0, \dfrac{23}{4}, 0, 0, \dfrac{5}{4}\right)'$ is an optimal solution for the dual problem.

The simplex method requires that a given linear program be written in the form of (4.1); that is, all variables must be nonnegative and any other constraints must be written as equality constraints. Any model problem can be expressed in this form by introducing additional variables as follows.

$$a_i' x \leq b_i \quad \Longleftrightarrow \quad a_i' x + x_{n+i} = b_i, \quad x_{n+i} \geq 0, \tag{4.21}$$

$$a_i' x \geq b_i \quad \Longleftrightarrow \quad a_i' x - x_{n+i} = b_i, \quad x_{n+i} \geq 0. \tag{4.22}$$

In (4.21), x_{n+i} is called a <u>slack variable</u> and in (4.22), it is called a <u>surplus variable</u>.

In addition, any unrestricted variable x_i may be replaced with $x_i^+ - x_i^-$, where $x_i^+ \geq 0$ and $x_i^- \geq 0$. [In fact, this last transformation need only be done conceptually. A simple modification of the revised simplex method allows unrestricted variables to be handled directly. See Exercise 4.16, part (b).] With additional variables, any linear programming problem may be written in the model form required by the simplex method.

We conclude this section by incorporating Bland's anticycling rules into the revised simplex method. Consider a typical iteration. If $\sigma_B > 0$ in Step 2, then because $v_k < 0$ in Step 1, it follows from (4.16) that the new objective function value is strictly less than the present value. As the revised simplex method continues, the objective function value is never increased. It follows that the present basic feasible solution will never be repeated. Each basic feasible solution is associated with a basis matrix. Since there are at most

$$\binom{n}{m}$$

basis matrices, termination in a finite number of steps is assured.

Suppose that there are ties for l in (4.18). Then every associated basic variable in the new extreme point is reduced to zero. At least $n - m + 1$ inequality constraints have become active. Together with the m equality constraints, at least $n + 1$ constraints are now active and the new extreme point is thus degenerate. For the next iteration, (4.18) shows that the basic variables having value zero may result in $\sigma_B = 0$, in which case, the previous finite termination argument no longer applies. Indeed, a zero step size may be obtained for several iterations, and although the basis matrices differ, the basic feasible solutions remain unchanged. If a positive step size is eventually obtained, the objective function will be strictly reduced and the degenerate extreme point will not be encountered further. However, there is no guarantee that this will be the case, and it is theoretically possible for a basis matrix to be repeated. The sequence of iterations between the first and the second occurrences of the same basis matrix would then be repeated. This cycle of iterations would continue indefinitely.

We argue that cycling is unlikely to occur in numerical computations. Cycling requires that ties for l in (4.18) be resolved in precisely the same way for corresponding iterations of all cycles. However, round-off errors tend to produce numerically different quantities in (4.18). This has the effect of varying the choice of l from one potential cycle to the next, thus making cycling unlikely.

Cycling is still a theoretical possibility. The potential problem can be overcome, as in Chapter 2, by using Bland's anticycling rules whenever a degenerate extreme point is encountered and by continuing to use them until either a strictly positive step size or an optimal solution is obtained. Bland's rules are special rules for choosing k and l in Steps 1 and 2, respectively, and

have been formulated in terms of active set concepts in Chapter 2. By identi-
fying nonbasic variables with active constraints, Bland's rules can be formulated for the revised simplex method as follows. In Step 1, v_i is the multiplier
for the ith active constraint for all $i \notin I_B$. In Step 2, constraint β_i becomes
active for all i with

$$\sigma_B = \frac{(x_B)_i}{(s_B)_i}.$$

Bland's rules for the revised simplex method are thus:

Step 1: Let k be the smallest index such that $k \notin I_B$ and $v_k < 0$.

Step 2: Let l be such that β_l is the smallest index with

$$\sigma_B = \frac{(x_B)_l}{(s_B)_l}.$$

It is shown in Appendix A that the use of Bland's rules implies that no
active set will ever be repeated. For the revised simplex method, this means
that no basis will ever be repeated. Thus we have

Theorem 4.1.

Let the revised simplex method be applied to

$$\min\{c'x \mid Ax = b, \ x \geq 0\}$$

with Bland's rules being used whenever a degenerate basic feasible solution is encountered. Then in a finite number of steps, either an optimal
solution is obtained or it is determined that the problem is unbounded
from below.

4.2 THE PHASE 1–PHASE 2 REVISED SIMPLEX METHOD

The revised simplex method requires an initial basic feasible solution for

$$\min\{c'x \mid Ax = b, \ x \geq 0\}. \tag{4.23}$$

In this section we show either how such a solution may be obtained, or that
none exists. Suppose that the equality constraints of (4.23) have been multiplied by -1, if necessary, so that the resulting b is nonnegative. Let y be an
m-vector and consider the problem

$$\left. \begin{array}{rlrl} \text{minimize:} & & e'y & \\ \text{subject to:} & Ax + & y & = b, \\ & x \geq 0, \ y \geq 0, & & \end{array} \right\} \tag{4.24}$$

where e is an m-vector of 1s. (4.24) is called the <u>phase 1 problem</u> for (4.23). The components of y are called <u>artificial variables</u>. The original problem (4.23) is called the <u>phase 2 problem</u>. Observe the following.

1. Equation (4.24) is a problem in $n + m$ variables, and all variables are restricted to be nonnegative.
2. A basic feasible solution for (4.24) is $y_B = b$, $x = 0$.
3. Equation (4.24) may be solved by the revised simplex method with initial basic feasible solution as in point 2, $B = B^{-1} = I$ and $I_B = \{n + 1, \ldots, n + m\}$.

Let x^*, y^* be an optimal solution for (4.24). There are two cases to be considered.

Case 1: $y^* \neq 0$
    This implies that (4.23) has no feasible solution, for suppose to the contrary that x is feasible for (4.23). Then x together with $y = 0$ is feasible for (4.24) and $e'0 = 0 < e'y^*$, in contradiction to the optimality of x^*, y^*.

Case 2: $y^* = 0$
    In this case x^* is a feasible solution for (4.23). Provided that none of the artificial variables are basic, the final basis inverse and associated index set may then be used as initial data in applying the revised simplex method to (4.23). The case of artificial variables remaining in the basis is considered further following Example 4.3.

Example 4.3

$$\text{minimize:} \qquad x_1 + x_2$$
$$\text{subject to:} \qquad x_1 + x_2 \leq 5, \qquad (1)$$
$$x_1 - x_2 \geq 1, \qquad (2)$$
$$x_1 \geq 0, \; x_2 \geq 0.$$

We first convert this problem to the model form required by the simplex method by adding a slack variable x_3 to the first constraint and a surplus variable x_4 to the second constraint. The problem is now:

$$\text{minimize:} \qquad x_1 + x_2$$
$$\text{subject to:} \qquad x_1 + x_2 + x_3 \qquad\quad = 5, \qquad (1)$$
$$x_1 - x_2 \qquad - x_4 = 1, \qquad (2)$$
$$x_i \geq 0, \; i = 1, \ldots, 4.$$

Appending artificial variables to the two equality constraints, the Phase 1 problem is

minimize: $\qquad\qquad\qquad\qquad\qquad\qquad\qquad y_1 + y_2$

subject to: $\qquad x_1 + x_2 + x_3 \qquad\quad + y_1 \qquad\quad = 5, \qquad$ (1)

$\qquad\qquad x_1 - x_2 \qquad\quad - x_4 \qquad\quad + y_2 = 1, \qquad$ (2)

$\qquad\qquad x_i \geq 0, \ i = 1, \ldots, 4, \ y_1 \geq 0, \ y_2 \geq 0.$

Applying the revised simplex method to this Phase 1 problem gives the following iterations.

Initialization:

$$I_B = \{5,6\}, \quad B^{-1} = \begin{bmatrix} 1 & 0 \\ 0 & 1 \end{bmatrix}, \quad x_B = \begin{bmatrix} 5 \\ 1 \end{bmatrix}, \quad c_B' x_B = 6.$$

Iteration 0

Step 1: $u = (-1, -1)',$

$$v_1 = \min\{-2, 0, -1, 1, -, -\} = -2, \quad k = 1, \quad s_B = \begin{bmatrix} 1 \\ 1 \end{bmatrix}.$$

Step 2: $\sigma_B = \min\left\{\dfrac{5}{1}, \dfrac{1}{1}\right\} = 1, \quad l = 2.$

Step 3: $B^{-1} \leftarrow \begin{bmatrix} 1 & -1 \\ 0 & 1 \end{bmatrix}, \quad x_B \leftarrow \begin{bmatrix} 4 \\ 1 \end{bmatrix}, \quad I_B \leftarrow \{5, 1\}, \quad c_B' x_B = 4.$

Iteration 1

Step 1: $u = (-1, 1)',$

$$v_2 = \min\{-, -2, -1, -1, -, 2\} = -2, \quad k = 2,$$

$$s_B = \begin{bmatrix} 2 \\ -1 \end{bmatrix}.$$

Step 2: $\sigma_B = \min\left\{\dfrac{4}{2}, -\right\} = 2, \quad l = 1.$

Step 3: $B^{-1} \leftarrow \begin{bmatrix} 1/2 & -1/2 \\ 1/2 & 1/2 \end{bmatrix}, \quad x_B \leftarrow \begin{bmatrix} 2 \\ 3 \end{bmatrix}, \quad I_B \leftarrow \{2, 1\}, \quad c_B' x_B = 0.$

Iteration 2

Step 1: $u = (0,0)'$,

$$v_3 = \min\{-,-,0,0,1,1\} = 0, \quad k = 3.$$

$v_3 \geq 0$; stop with optimal solution $x = (3,2,0,0,0,0)'$.

We now solve the Phase 2 problem by applying the revised simplex method to the original problem beginning with the basic feasible solution just obtained.

Initialization:

$$I_B = \{2,1\}, \quad B^{-1} = \begin{bmatrix} 1/2 & -1/2 \\ 1/2 & 1/2 \end{bmatrix}, \quad x_B = \begin{bmatrix} 2 \\ 3 \end{bmatrix}, \quad c_B' x_B = 5.$$

Iteration 0

Step 1: $u = (-1,0)'$,

$$v_3 = \min\{-,-,-1,0\} = -1, \quad k = 3, \quad s_B = \begin{bmatrix} 1/2 \\ 1/2 \end{bmatrix}.$$

Step 2: $\sigma_B = \min\left\{ \dfrac{2}{1/2}, \dfrac{3}{1/2} \right\} = 4, \quad l = 1.$

Step 3: $B^{-1} \leftarrow \begin{bmatrix} 1 & -1 \\ 0 & 1 \end{bmatrix}, \quad x_B \leftarrow \begin{bmatrix} 4 \\ 1 \end{bmatrix}, \quad I_B \leftarrow \{3,1\}, \quad c_B' x_B = 1.$

Iteration 1

Step 1: $u = (0,-1)'$,

$$v_4 = \min\{-,2,-,1\} = 1, \quad k = 4.$$

$v_4 \geq 0$; stop with optimal solution $x = (1,0,4,0)'$.

Note that $u = (0,-1)'$ together with $v = (0,2,0,1)'$ is an optimal solution for the dual problem.

A computer program which implements the Phase 1-Phase 2 revised simplex method is given in Appendix B, Section B.2.2. The output from applying this program to the problem of Example 4.3 is shown in Figure B.47.

Consider the Phase 1 problem (4.24) further. Suppose that $b \geq 0$ and the kth column of A is the ith unit vector. Then there is no need to introduce an artificial variable y_i in row i. The initial basic feasible solution for the Phase 1 problem can be formed by making x_k basic with value b_i and a column of the initial basis matrix is then the ith unit vector.

Following this approach, the Phase 1 problem for the problem of Example 4.3 can be written using just one artificial variable as follows.

minimize: y_1

subject to: $x_1 + x_2 + x_3 \qquad\qquad = 5,$ (1)

$x_1 - x_2 \qquad - x_4 + y_1 = 1,$ (2)

$x_i \geq 0, \quad i = 1, \ldots, 4, \quad y_1 \geq 0.$

Consider Case 2 of the Phase 1 analysis further. Since $y^* = 0$, x^* is indeed feasible for (4.23). However, x^* may or may not be an extreme point. If all artificial variables are nonbasic, x^* is indeed an extreme point and the final basis inverse and associated index set may be used as initial data for the Phase 2 problem. However, the possibility remains that some artificial variables remain in the basis. These artificial variables necessarily have value zero, thus implying that (x^*, y^*) is a degenerate optimal solution. In particular, suppose that y_l is basic in row l and let B^{-1} denote the final basis inverse from Phase 1.

Let k be such that

$$|(B^{-1}A_k)_l| = \max\{ |(B^{-1}A_i)_l| \mid i = 1, \ldots, n \}. \qquad (4.25)$$

If $(B^{-1}A_k)_l = 0$, then

$$(B^{-1}A_i)_l = 0, \quad \text{for all } i = 1, \ldots, n,$$

and the lth equality constraint of (4.23) is redundant (Exercise 4.17). If $(B^{-1}A_k)_l \neq 0$, x_k may be made a basic variable and y_l a nonbasic variable. B^{-1} is replaced with

$$\left[\Phi((B^{-1})', A_k, l) \right]'$$

and I_B is updated by replacing β_l with k. In this case, one artificial has left the basis. In at most m steps, an extreme point for (4.23) will be located.

This process may be repeated. At each iteration, either one artificial variable leaves the basis or one redundant equality constraint of (4.23) is deleted.

4.3   THE TABLEAU FORM OF THE SIMPLEX METHOD

The simplex method was developed in the late 1940s and early 1950s. At that time, digital computers were in a very primitive state of development and were not widely available to the scientific and industrial communities. Consequently, the simplex method was originally presented in a form suitable for hand calculation. Now, however, digital computers are widely available to the general public. Indeed, at the time of this writing, microcomputers are commonly available to children in elementary schools and are not uncommon in many private homes. As a result, linear programming problems are normally solved using digital computers. Typically, they are solved by hand only by students writing examinations in university linear programming courses.

Nonetheless, much of the theoretical development of linear programming was formulated in terms of the procedure used for early hand calculation, and this *tableau* format is presented in this and the following sections.

We introduce the procedure by means of the following example.

Example 4.4

$$\text{minimize:} \quad -2x_1 - 3x_2 \qquad\qquad\qquad\qquad = z \qquad (0)^0$$
$$\text{subject to:} \quad -2x_1 + x_2 + x_3 \qquad\qquad\quad = 2, \qquad (1)^0$$
$$x_1 + 2x_2 \qquad + x_4 \qquad\quad = 14, \qquad (2)^0$$
$$4x_1 + 3x_2 \qquad\qquad\quad + x_5 = 36, \qquad (3)^0$$
$$x_i \geq 0, \quad i = 1, \dots, 5.$$

We have introduced an additional variable, z, which is equal to the objective function value. In this example, an initial basic feasible solution is obvious: the basic variables and their values are

$$x_3 = 2, \quad x_4 = 14, \quad \text{and} \quad x_5 = 36.$$

The nonbasic variables, x_1 and x_2, have value zero. Because the basic variables do not appear in the objective function (i.e., the coefficients of x_3, x_4, and x_5 in the objective function all have value zero), and because the coefficient of x_2 is -3, the objective function can be decreased by increasing x_2 from its present value of zero. Note that the objective function could also be decreased by increasing x_1 since its objective function coefficient is -2, but the rate of decrease is greater by increasing x_2.

Feasibility with respect to the equality constraints resulting from an increase in x_2 is accounted for by appropriate changes in the basic variables x_3, x_4, and x_5. Keeping x_1 at zero, the largest increase in x_2 is

$$\min\left\{\frac{2}{1}, \frac{14}{2}, \frac{36}{3}\right\} = 2,$$

at which point x_3 is reduced to zero. Now x_2 replaces x_3 as a basic variable.

To continue, it is necessary to transform the given problem into an equivalent one for which the same analysis can be performed. This requires that each basic variable appear with coefficient unity, in just one equality constraint, and that the basic variables be eliminated from the objective function. Since x_3 was basic in row 1, this transformation can be performed by solving equation $(1)^0$ for x_2 and eliminating x_2 from the objective function and both constraints $(2)^0$ and $(3)^0$. Doing this gives an equivalent representation of the problem as

$$
\begin{array}{llllll}
\text{minimize:} & -8x_1 & +3x_3 & & = z + 6 & (0)^1 \\
\text{subject to:} & -2x_1 + x_2 + x_3 & & & = 2, & (1)^1 \\
& 5x_1 & -2x_3 + x_4 & & = 10, & (2)^1 \\
& 10x_1 & -3x_3 & +x_5 & = 30, & (3)^1
\end{array}
$$

$$x_i \geq 0, \quad i = 1, \ldots, 5.$$

The current basic feasible solution can be read easily from the above. The nonbasic variables x_1 and x_3 are set to zero, and the basic variable values are

$$x_2 = 2, \quad x_4 = 10, \quad \text{and} \quad x_5 = 30.$$

Furthermore, the objective function value is $z = -6$, which is indeed reduced from its previous value of 0.

The process continues by observing that the coefficient of x_1 in the above reformulated problem is -8. Since this is negative, further progress can be made by increasing x_1 from its present value of 0. The largest feasible increase in x_1 is

$$\min\left\{ -, \frac{10}{5}, \frac{30}{10} \right\} = 2,$$

at which point x_4 is reduced to zero. Now x_1 replaces x_4 as a basic variable. The process continues by solving for x_1 in equation $(2)^1$ and using this to eliminate x_1 from the objective function equation, and equations $(1)^1$ and $(3)^1$. Doing so gives

$$
\begin{array}{llllll}
\text{minimize:} & -0.2x_3 & +1.6x_4 & & = z + 22 & (0)^2 \\
\text{subject to:} & x_2 + 0.2x_3 & +0.4x_4 & & = 6, & (1)^2 \\
& x_1 & -0.4x_3 + 0.2x_4 & & = 2, & (2)^2 \\
& & x_3 - 2x_4 & +x_5 & = 10, & (3)^2
\end{array}
$$

$$x_i \geq 0, \quad i = 1, \ldots, 5.$$

The current basic feasible solution is again read directly from this reformulated problem. The nonbasic variables are $x_3 = x_4 = 0$, and the basic variables are

$$x_2 = 6, \quad x_1 = 2, \quad \text{and} \quad x_5 = 10.$$

The objective function value is $z = -22$.

The coefficient of x_3 is negative and the objective function can be further decreased by increasing x_3. The largest feasible increase in x_3 is

$$\min\left\{ \frac{6}{1/5}, -, \frac{10}{1} \right\} = 10,$$

and x_5 is reduced to zero. Now x_3 replaces x_5 as a basic variable. Solving for x_3 in $(3)^2$ and eliminating x_3 from $(0)^2$, $(1)^2$, and $(2)^2$ gives

minimize: $\qquad\qquad\qquad\qquad 1.2x_4 + 0.2x_5 = z + 24 \qquad (0)^3$

subject to: $\qquad\quad x_2 \qquad + 0.8x_4 - 0.2x_5 = 4, \qquad\quad (1)^3$

$\qquad\qquad x_1 \qquad\qquad - 0.6x_4 + 0.4x_5 = 6, \qquad\quad (2)^3$

$\qquad\qquad\qquad\quad x_3 - 2x_4 + x_5 = 10, \qquad\quad (3)^3$

$$x_i \geq 0, \ i = 1, \ldots, 5.$$

The coefficients of the objective function are now nonnegative, from which we can conclude that the optimal solution is the current basic feasible solution $x_4 = x_5 = 0$ and

$$x_2 = 4, \quad x_1 = 6, \quad \text{and} \quad x_3 = 10.$$

The optimal objective function value is -24.

At each iteration of the example above, we solved for the new basic variable in the row associated with the new nonbasic variable and eliminated the new basic variable from the remaining equations, including the objective function equation. The procedure used is a single step of the Gauss-Jordan elimination procedure for solving linear equations. We formalize this as follows. Suppose that the problem is

minimize: $\quad c_1 x_1 + \cdots + c_k x_k + \cdots + c_n x_n = z \qquad (0)$

subject to: $\quad a_{11} x_1 + \cdots + a_{1k} x_k + \cdots + a_{1n} x_n = b_1 \qquad (1)$

$$\vdots \qquad\qquad \vdots \qquad\qquad \vdots \qquad \vdots$$

$\qquad\qquad a_{l1} x_1 + \cdots + a_{lk} x_k + \cdots + a_{ln} x_n = b_l \qquad (l)$

$$\vdots \qquad\qquad \vdots \qquad\qquad \vdots \qquad \vdots$$

$\qquad\quad a_{m1} x_1 + \cdots + a_{mk} x_k + \cdots + a_{mn} x_n = b_m \qquad (m)$

$$x_i \geq 0, \ i = 1, \ldots, n.$$

Suppose that we wish to solve equation (l) for x_k and eliminate x_k from the remaining equations, including the objective function equation. Assume that $a_{lk} \neq 0$. We wish to transform

$$\begin{bmatrix} c_k \\ a_{1k} \\ a_{2k} \\ \vdots \\ a_{lk} \\ \vdots \\ a_{mk} \end{bmatrix} = \begin{bmatrix} c_k \\ \\ \\ A_k \\ \\ \\ \end{bmatrix}$$

into the $(l + 1)$st unit vector. This can be accomplished by premultiplying the $m + 1$ linear equations in $n + 1$ unknowns by the matrix

$$
P = \begin{bmatrix}
1 & 0 & 0 & \cdots & -c_k/a_{lk} & \cdots & 0 \\
0 & 1 & 0 & \cdots & -a_{1k}/a_{lk} & \cdots & 0 \\
0 & 0 & 1 & \cdots & -a_{2k}/a_{lk} & \cdots & 0 \\
\vdots & \vdots & \vdots & & \vdots & & \vdots \\
0 & 0 & 0 & \cdots & 1/a_{lk} & \cdots & 0 \\
\vdots & \vdots & \vdots & & \vdots & & \vdots \\
0 & 0 & 0 & \cdots & -a_{mk}/a_{lk} & \cdots & 1
\end{bmatrix}. \tag{4.26}
$$

P is called a <u>pivot matrix</u> and a_{lk} is called the <u>pivot element</u>. Note that P differs from the $(m + 1, m + 1)$ identity matrix only in that column $l + 1$ is replaced with

$$
\begin{bmatrix}
-c_k/a_{lk} \\
-a_{1k}/a_{lk} \\
-a_{2k}/a_{lk} \\
\vdots \\
1/a_{lk} \\
\vdots \\
-a_{mk}/a_{lk}
\end{bmatrix}.
$$

It is straightforward to verify that

$$
P \begin{bmatrix} c_k \\ A_k \end{bmatrix} = e_{l+1}, \tag{4.27}
$$

where e_{l+1} denotes the $(l + 1)$st $(m + 1)$-dimensional unit vector (Exercise 4.18).

Let the given LP be formulated as above. The tableau form of the simplex method proceeds as follows.

Step 1:
Compute the smallest index k such that

$$
c_k = \min\{c_i \mid i = 1, \ldots, m\}.
$$

If $c_k \geq 0$, the current basic feasible solution is optimal. Stop.
If $c_k < 0$, go to Step 2.

Step 2:
If $a_{ik} \leq 0$ for $i = 1, \ldots, m$, the problem is unbounded from below. Stop.

Otherwise, compute the smallest index l such that

$$\frac{b_l}{a_{lk}} = \min\left\{\frac{b_i}{a_{ik}} \mid \text{all } i \text{ with } a_{ik} > 0\right\},$$

and go to Step 3.

Step 3:
Pivot on row l, column k; that is, premultiply the $m + 1$ linear equations by P, where P is defined by (4.26). Go to Step 1.

Let P_j denote the pivot matrix used at iteration j. Then the jth reformulated problem is obtained from the original by premultiplying by

$$H \equiv P_j P_{j-1} \cdots P_2 P_1.$$

Since each pivot matrix is nonsingular, so is H and the original problem may be recaptured by premultiplying the jth reformulation by H^{-1}. H is related to the basis inverse matrix, B^{-1}, and the multipliers for the equality constraints, u, according to

$$H = \begin{bmatrix} 1 & u' \\ 0 & B^{-1} \end{bmatrix}, \quad H^{-1} = \begin{bmatrix} 1 & c_B' \\ 0 & B \end{bmatrix}. \tag{4.28}$$

Furthermore, the cost coefficient for each nonbasic variable i in each reformulated problem is precisely (Exercise 4.19)

$$v_i = c_i + A_i' u. \tag{4.29}$$

A typical transformed problem can be written compactly as shown in Table 4.1.

TABLE 4.1   Simplex Tableau

Index	Cost	$\hat{c}_1$	$\hat{c}_2$	$\cdots$	$\hat{c}_k$	$\cdots$	$\hat{c}_n$	$-c_B' x_B$
	c_B	$B^{-1}A_1$	$B^{-1}A_2$	$\cdots$	$B^{-1}A_k$	$\cdots$	$B^{-1}A_n$	$B^{-1}b$
β_1	c_{β_1}	$(B^{-1}A_1)_1$	$(B^{-1}A_2)_1$	$\cdots$	$(B^{-1}A_k)_1$	$\cdots$	$(B^{-1}A_n)_1$	$(B^{-1}b)_1$
β_2	c_{β_2}	$(B^{-1}A_1)_2$	$(B^{-1}A_2)_2$	$\cdots$	$(B^{-1}A_k)_2$	$\cdots$	$(B^{-1}A_n)_2$	$(B^{-1}b)_2$
$\vdots$	$\vdots$	$\vdots$	$\vdots$		$\vdots$		$\vdots$	$\vdots$
β_l	c_{β_l}	$(B^{-1}A_1)_l$	$(B^{-1}A_2)_l$	$\cdots$	$(B^{-1}A_k)_l$	$\cdots$	$(B^{-1}A_n)_l$	$(B^{-1}b)_l$
$\vdots$	$\vdots$	$\vdots$	$\vdots$		$\vdots$		$\vdots$	$\vdots$
β_m	c_{β_m}	$(B^{-1}A_1)_m$	$(B^{-1}A_2)_m$	$\cdots$	$(B^{-1}A_k)_m$	$\cdots$	$(B^{-1}A_n)_m$	$(B^{-1}b)_m$

Using this tableau format, the computations for Example 4.4 are as follows (blanks are to be taken as zeros).

TABLEAU 0

		1	2	3	4	5	z
Index	Cost	−2	−3				0
3	0	−2	1	1			2
4	0	1	2		1		14
5	0	4	3			1	36

TABLEAU 1

		1	2	3	4	5	z
Index	Cost	−8		3			6
2	−3	−2	1	1			2
4	0	5		−2	1		10
5	0	10		−3		1	30

TABLEAU 2

		1	2	3	4	5	z
Index	Cost			−0.2	1.6		22
2	−3		1	0.2	0.4		6
1	−2	1		−0.4	0.2		2
5	0			1	−2	1	10

TABLEAU 3

		1	2	3	4	5	z
Index	Cost				1.2	0.2	24
2	−3		1		0.8	−0.2	4
1	−2	1			−0.6	0.4	6
3	0			1	−2	1	10

The entire tableau is updated at each step of this original form of the simplex method. Premultiplication of any tableau by the pivot matrix requires approximately $m + 1$ arithmetic operations for each column, so that $n(m + 1)$ operations are required in all. Actually, only $(n - m)(m + 1)$ are required since it is known *a priori* that m of the transformed columns will be unit vectors. However, not all of the transformed tableau is required. Indeed, a single iteration requires only that the following be computed: update of B^{-1}, $u' = -c_B'B^{-1}$, v_i, all $i \notin I_B$, and $s_B = B^{-1}A_k$. These require a total of $mn + 2m^2$ arithmetic operations, which is significantly smaller than $(n - m)(m + 1)$ when n is large relative to m. Restricting the computations to only those that are actually required gives the revised simplex method. Hence the qualifier "revised."

Comparison of the simplex method and the revised simplex method with respect to the number of arithmetic operations gives only part of the story. The simplex method requires $(m + 1)(n + 1)$ memory locations to store the entire updated tableau at each iteration. In contrast, the revised simplex method requires only m^2 for B^{-1} plus multiples of m and n for other quantities. In this respect the revised simplex method is superior to the simplex method when n is large relative to m.

Although we have formulated the revised simplex method explicitly in terms of B^{-1}, the computations need not be performed in this manner. Indeed, numerical analysts argue that explicit computation of B^{-1} may introduce unnecessary round-off errors and requires more computation than necessary. The quantities of interest satisfy

$$Bx_B = b, \quad Bs_B = A_k, \quad \text{and} \quad B'u = -c_B.$$

These are all sets of m simultaneous linear equations with coefficient matrix B or B'. Some factorization of B is then computed (e.g., LU or QR). This factorization is designed to make the solution of the equations above numerically stable and computationally efficient [Bartels and Golub, 1969]. Suitable updating schemes are used to modify the factorization in response to a change of a column in B.

4.4 ALGEBRAIC EQUIVALENCE OF ALGORITHM 3 AND THE REVISED SIMPLEX METHOD[7]

The reader may have wondered about the algebraic relationship between the revised simplex method and Algorithm 3. The two methods look different. However, since we derived the revised simplex method from Algorithm 3, it follows that the two methods are equivalent. In this section we examine in detail the algebraic relationship between the two methods and discuss their applicability to various model problems.

---

[7] This section may be omitted on first reading.

The model problem for Algorithm 3 is

$$\min\{c'x \mid A_1x \le b_1, \ A_2x = b_2\}, \tag{4.30}$$

while that for the revised simplex method is

$$\min\{c'x \mid Ax = b, \ x \ge 0\}. \tag{4.31}$$

Suppose first that we apply Algorithm 3 to (4.31). The nonnegativity constraints are multiplied by -1 to obtain

$$\min\{c'x \mid Ax = b, \ -Ix \le 0\}. \tag{4.32}$$

Suppose that an initial extreme point is available. Suppose for simplicity that the last $n - m$ nonnegativity constraints are active. Partitioning A as in the revised simplex method, we take

$$D_0' = \begin{bmatrix} B' & 0 \\ N' & -I \end{bmatrix}$$

and

$$J_0 = \{n + 1, \ldots, n + m, m + 1, m + 2, \ldots, n\},$$

where the nonnegativity constraints are numbered $1, 2, \ldots, n$ and the equality constraints are numbered $n + 1, \ldots, n + m$. From Exercise 4.21, part (b), D_0^{-1} may be written in partitioned form as

$$D_0^{-1} = \begin{bmatrix} B^{-1} & B^{-1}N \\ 0 & -I \end{bmatrix}.$$

Because the first m columns of D_0' are gradients of equality constraints, Algorithm 3 does not consider any of these columns as candidates for s_0. Indeed, Step 1 of Algorithm 3 transfers to Step 1.2 and computes k such that

$$c' \begin{bmatrix} B^{-1}N \\ -I \end{bmatrix}_k = \max\left\{ c' \begin{bmatrix} B^{-1}N \\ -I \end{bmatrix}_i \mid i = m + 1, \ldots, n \right\}. \tag{4.33}$$

Letting c_B denote the first m components of c, and, c_i and c_k denote the ith and kth components of c, respectively, (4.33) becomes

$$c_B'B^{-1}A_k - c_k = \max\{c_B'B^{-1}A_i - c_i \mid i = m + 1, \ldots, n\}.$$

With $u' = -c_B'B^{-1}$, (4.33) reduces to

$$c_k + u'A_k = \min\{c_i + u'A_i \mid i = m + 1, \ldots, n\},$$

which is identical to Step 1 of the revised simplex method.

If $c_k + u'A_k < 0$, Algorithm 3 continues by setting s_0 equal to the kth column of D_0^{-1}. Thus

$$s_0 = \begin{bmatrix} B^{-1}A_k \\ 0 \end{bmatrix} - e_k,$$

where e_k is the kth unit vector of dimension n.

The computations for Step 2 of Algorithm 3 proceed as follows. Because only the first m nonnegativity constraints are inactive,

$$(x_0)_i = (B^{-1}b)_i, \quad i = 1, \ldots, m.$$

Because the gradients of the first m nonnegativity constraints are $-e_1$, $-e_2, \ldots, -e_m$, the maximum feasible step size is

$$\sigma_0 = \frac{-e_l' \begin{bmatrix} B^{-1}b \\ 0 \end{bmatrix} - 0}{-e_l' \left[\begin{bmatrix} B^{-1}A_k \\ 0 \end{bmatrix} - e_k \right]}$$

$$= \min \left\{ \frac{-e_i' \begin{bmatrix} B^{-1}b \\ 0 \end{bmatrix} - 0}{-e_i' \left[\begin{bmatrix} B^{-1}A_k \\ 0 \end{bmatrix} - e_k \right]} \;\middle|\; \text{all } i \text{ with } -e_i' \left[\begin{bmatrix} B^{-1}A_k \\ 0 \end{bmatrix} - e_k \right] < 0 \right\}.$$

Simplifying gives

$$\sigma_0 = \frac{(B^{-1}b)_l}{(B^{-1}A_k)_l} = \min \left\{ \frac{(B^{-1}b)_i}{(B^{-1}A_k)_i} \;\middle|\; \text{all } i \text{ with } (B^{-1}A_k)_i > 0 \right\},$$

which is identical to the computation of σ_B in Step 2 of the revised simplex method.

Step 3 of Algorithm 3 obtains D_1' from D_0' by replacing column $m + k$ with $-e_k$ and computes D_1^{-1} accordingly. From Exercise 4.21, part (b), the computations with D_1^{-1} may be performed with the new B^{-1} given by $\left[\Phi((B^{-1})', A_k, l) \right]'$.

We have thus shown algebraically that the computations of Algorithm 3 coincide with those of the revised simplex method when both algorithms are applied to the common problem (4.31) beginning with a common extreme point.

We next perform the converse analysis. By applying the revised simplex method to (4.32), we obtain one iteration of Algorithm 3. In order to focus on the most important points, we omit explicit consideration of equality constraints and apply the revised simplex method to

$$\min \{ c'x \mid Ax \le b \}. \tag{4.34}$$

We first augment the problem with slack variables giving

$$\min \{ c'x \mid Ax + y = b, y \geq 0 \},$$

or, equivalently,

$$\min \left\{ (c',0') \begin{bmatrix} x \\ y \end{bmatrix} \mid [A, I] \begin{bmatrix} x \\ y \end{bmatrix} = b, \ y \geq 0 \right\}, \qquad (4.35)$$

where y_i is the slack variable for constraint i. Let m denote the number of inequality constraints in (4.34). Then y is an m-vector, (4.35) has $n + m$ variables, m equality constraints, and m nonnegativity constraints. The "x" of (4.31) is now the composite vector

$$\begin{bmatrix} x \\ y \end{bmatrix},$$

so that the variables are numbered 1 to n for the components of x and $n + 1, \ldots, n + m$ for the components of y. The components of x are unrestricted in sign. It follows from Exercise 4.16, part (b), that the components of x are always basic. Consequently, at any extreme point of (4.35), n of the m components of y must be nonbasic. Let

$$\begin{bmatrix} x_0 \\ y_0 \end{bmatrix}$$

denote the initial extreme point for (4.35) and suppose for simplicity that the first n components of y_0 are nonbasic. Then we may take

$$I_B = \{ 1, 2, \ldots, n, 2n + 1, \ldots, m \}.$$

The initial basis matrix for (4.35) is

$$B = \begin{bmatrix} & I \\ A & 0 \end{bmatrix},$$

where I denotes the (n, n) identity matrix. Partitioning A as

$$A = \begin{bmatrix} A_0 \\ \tilde{A}_0 \end{bmatrix},$$

where A_0 denotes the first n rows of A and $\tilde{A}_0$ denotes the last $m - n$ rows, we have

$$B = \begin{bmatrix} A_0 & 0 \\ \tilde{A}_0 & I \end{bmatrix}.$$

Since

$$\begin{bmatrix} x_0 \\ y_0 \end{bmatrix}$$

is an extreme point for the $(n + m)$-dimensional problem (4.35), it follows that x_0 is an extreme point for the n-variable problem (4.34). Also, the first n constraints of (4.34) are active at x_0. Assuming that

$$\begin{bmatrix} x_0 \\ y_0 \end{bmatrix}$$

is nondegenerate, A_0 must be nonsingular. It then follows from Exercise 4.21, part (b), that

$$B^{-1} = \begin{bmatrix} A_0^{-1} & 0 \\ -\tilde{A}_0 A_0^{-1} & I \end{bmatrix}.$$

With $c_B = (c', 0')'$, Step 1 of the revised simplex method computes

$$u = - \begin{bmatrix} (A_0^{-1})' & -(A_0^{-1})'\tilde{A}_0' \\ 0' & I \end{bmatrix} \begin{bmatrix} c \\ 0 \end{bmatrix} = \begin{bmatrix} -(A_0^{-1})'c \\ 0 \end{bmatrix},$$

and because the first n variables are always basic, k is chosen such that

$$v_k = \min \{ 0 + e_i'u \mid i = 1, \ldots, n \},$$

$$= \min \left\{ - \left[(A_0^{-1})'c \right]_i \mid i = 1, \ldots, n \right\}.$$

But in the notation of Algorithm 3, $A_0 = D_0$, so k is chosen such that

$$-v_k = -\max \left\{ \left[(D_0^{-1})'c \right]_i \mid i = 1, \ldots, n \right\}.$$

Letting $D_0^{-1} = \left[c_{10}, \ldots, c_{n0} \right]$, this becomes

$$c'c_{k0} = \max \{ c'c_{i0} \mid i = 1, \ldots, n \},$$

which is identical to Step 1.2 of Algorithm 3. As in Step 1.2, let $s_0 = c_{k0}$. In Step 1 of the revised simplex method, the "A_k" for (4.35) is

$$\begin{bmatrix} e_k \\ 0 \end{bmatrix},$$

where the "0" is an $(m - n)$-vector. Consequently, $B^{-1}A_k$ is the kth column of B^{-1}. Letting $\tilde{A}_0' = \begin{bmatrix} a_{n+1}, \ldots, a_m \end{bmatrix}$, we have

$$s_B = \begin{bmatrix} s_0 \\ -a'_{n+1}s_0 \\ -a'_{n+2}s_0 \\ \vdots \\ -a'_m s_0 \end{bmatrix}.$$

Step 2 of the revised simplex method tests the sign of the components of s_B to determine if the problem is unbounded from below. Since the first n variables of (4.35) are unrestricted, the sign test is applied only to the last $m - n$ components of s_B. Thus if

$$a_i' s_0 \geq 0, \quad \text{for } i = n + 1, \ldots, m,$$

both the revised simplex method and Algorithm 3 conclude that the problem is unbounded from below. Otherwise, Step 2 of the revised simplex method continues by computing σ_B. Partitioning b in a manner compatible with A gives

$$b = \begin{bmatrix} b_0 \\ \tilde{b}_0 \end{bmatrix}.$$

Thus

$$x_B = \begin{bmatrix} A_0^{-1} & 0 \\ -\tilde{A}_0 A_0^{-1} & I \end{bmatrix} \begin{bmatrix} b_0 \\ \tilde{b}_0 \end{bmatrix} = \begin{bmatrix} A_0^{-1} b_0 \\ -\tilde{A}_0 A_0^{-1} b_0 + \tilde{b}_0 \end{bmatrix}.$$

Since the components of x are unrestricted in sign, the first n components of x_B do not affect the computation of σ_B. Furthermore, $x_0 = x_B = A_0^{-1} b_0 = D_0^{-1} b_0$, so

$$\sigma_B = \frac{(x_B)_l}{(s_B)_l}$$

$$= \min \left\{ \frac{(-\tilde{A}_0 A_0^{-1} b_0 + \tilde{b}_0)_i}{-a'_{n+i} s_0} \,\middle|\, \text{all } i = n + 1, \ldots, m \text{ with } -a'_{n+i} s_0 > 0 \right\}$$

$$= \min \left\{ \frac{a'_{n+i} x_0 - b_i}{a'_{n+i} s_0} \,\middle|\, \text{all } i = n + 1, \ldots, m \text{ with } a'_{n+i} s_0 < 0 \right\}.$$

Thus $\sigma_B = \sigma_0$ as calculated in Step 2 of Algorithm 3.

      Review of the analysis above shows that in applying the revised simplex method explicitly to (4.35) and implicitly to (4.34), all computations may be performed using $A_0^{-1} = D_0^{-1}$ rather than B^{-1}. Step 3 of the revised simplex method continues by replacing the $(n + k)$th column of B, the $(n + k)$th m-dimensional unit vector, with the $(n + l)$th m-dimensional unit vector. Proceeding as above, all computations of the next iteration of the revised simplex method may be performed with

$$A_1^{-1} = D_1^{-1} = \Phi(D_0^{-1}, a_k, l).$$

In the previous analysis, we have shown that in applying the revised simplex method to (4.35), y_i being nonbasic is equivalent to constraint i of (4.34) being active, and y_i being basic is equivalent to constraint i of (4.34) being inactive.

For a numerical example of the equivalence of Algorithm 3 and the revised simplex method, the reader should observe that the problem of Example 4.2 is precisely that of Example 1.2 reformulated in the canonical form of the simplex method with slack variables x_3, x_4, and x_5 (which play the rôle of y_1, y_2, and y_3) in the previous analysis. The problem of Example 1.2 was solved by Algorithm 1 (which is equivalent to Algorithm 3 for this example) in Example 2.2. The reader will find it instructive to compare the computations for Example 2.2 with those for Example 4.2.

4.5 WHICH ALGORITHM SHOULD BE USED?

We have shown that Algorithm 3 and the revised simplex method are equivalent. Application of Bland's rules precludes the possibility of cycling in either algorithm, so that both algorithms are equivalent in this respect. There remains only the question of which algorithm to apply to a particular problem.

Suppose that we want to solve

$$\min\{c'x \mid Ax \le b\}. \tag{4.36}$$

Let n and m denote the number of variables and constraints, respectively, in (4.36). For Algorithm 3, the dominant portion of the computational work is in updating D_j^{-1}. From Exercise 2.23, each update requires approximately n^2 arithmetic operations. The dominant portion of the computational work per iteration for the revised simplex method is in updating B^{-1}. To apply the revised simplex method to (4.36), (4.36) must first be transformed into the canonical form required by that method. Doing so gives (4.35), and the number of resulting equality constraints is m. Updating B^{-1} for the revised simplex method requires approximately m^2 arithmetic operations per iteration. Thus Algorithm 3 is superior to the revised simplex method if $m > n$, equal if $m = n$, and inferior if $m < n$.

However, an additional possibility must be accounted for. Both Algorithm 3 and the revised simplex method compute optimal solutions for both the primal and dual problems. Since the dual of the dual problem is the primal problem, the primal may be solved indirectly by solving the dual. The dual of (4.36) is

$$\max\{-b'u \mid A'u = -c,\ u \geq 0\}. \tag{4.37}$$

Updating the basis inverse in applying the revised simplex method directly to (4.37) requires approximately n^2 arithmetic operations. Since (4.37) has m variables, each update of Algorithm 3 requires approximately m^2 arithmetic operations. Thus application of the revised simplex method directly to (4.37) is superior to Algorithm 3 if $n < m$, equal if $n = m$, and inferior if $n > m$. Note that this is the opposite result obtained for applying the two algorithms to (4.36).

Next, suppose that we want to solve

$$\min\{c'x \mid Ax = b,\ x \geq 0\}, \tag{4.38}$$

where A is (m, n). Continuing the previous analysis, the basis inverse in applying the revised simplex method to (4.38) is (m, m), while the D_j^{-1} for Algorithm 3 is (n, n). Thus, in terms of the computational work per iteration, the revised simplex method is superior to Algorithm 3 if $m < n$, equal if $m = n$, and inferior if $m > n$.

The dual of (4.38) is

$$\max\{b'u \mid A'u \leq c\}. \tag{4.39}$$

In applying Algorithm 3 to (4.39), D_j^{-1} is (m, m), while the basis inverse used in applying the revised simplex method to (4.39) is (n, n). Thus Algorithm 3 is superior to the revised simplex method if $m < n$, equal if $m = n$, and inferior if $m > n$.

Let us call D_j^{-1} the <u>working</u> <u>matrix</u> for Algorithm 3 and B^{-1} the <u>working</u> <u>matrix</u> for the revised simplex method. The analysis above is summarized in Tables 4.2 and 4.3.

Our analysis can be summarized by saying that it is only equally appropriate to apply both Algorithm 3 and the revised simplex method to a problem if $m = n$. If it is more appropriate to solve the primal problem directly using one of the two algorithms, it is equally appropriate to solve the problem indirectly by applying the other algorithm to the dual problem.

Implicit in this analysis is the assumption that approximately the same number of iterations are required to solve both the primal and dual problems. There is, in general, no way of knowing whether or not this assumption will be satisfied.

TABLE 4.2

Problem to Be Solved Dimension of A	$\min\{c'x \mid Ax \leq b\}$ (m,n)		
	Working Matrix Size		
	Primal	Dual	
Algorithm 3 Revised simplex method	n^2 m^2	m^2 n^2	

TABLE 4.3

Problem to Be Solved Dimension of A	$\min\{c'x \mid Ax = b,\ x \geq 0\}$ (m,n)		
	Working Matrix Size		
	Primal	Dual	
Algorithm 3 Revised simplex method	n^2 m^2	m^2 n^2	

Our analysis is correct as far as it goes. However, it may be appropriate to account for other factors. Suppose that the problem to be solved is

$$\min\{c'x \mid Ax \leq b\}. \tag{4.40}$$

Suppose that A is (m,n) and m is very large compared to n. Then we have argued that Algorithm 3 should be applied directly to (4.40), or the revised simplex method should be applied to the dual. However, an exact optimal solution may not be required. A suboptimal, but suitably good solution may be satisfactory and can be obtained by terminating Algorithm 3 when in Step 1.2, $c'c_{kj}$ is in fact positive but numerically small. If we choose to solve (4.40) indirectly by applying the revised simplex method to the dual of (4.40), a *feasible* solution for (4.40) is obtained only after an *exact* optimal solution has been obtained for the dual; that is, termination with a suboptimal, but suitably good, solution is not possible by solving the dual. It is precisely this situation that arises in solving integer linear programs by solving a sequence of (continuous) linear programs.

A numerical example of the application of both Algorithm 3 and the revised simplex method is given in Exercises 4.9 and 4.10.

EXERCISES

4.1. Consider the problem

$$\min \{ c'x \mid Ax = b, \; x \geq 0 \},$$

where

$$A = \begin{bmatrix} 6 & 1 & 0 & 0 & 1 & 5 \\ 2 & 3 & 1 & 4 & 0 & 2 \\ -1 & 1 & 0 & 2 & 2 & 1 \end{bmatrix} \quad \text{and} \quad b = \begin{bmatrix} 9 \\ 17 \\ 13 \end{bmatrix}.$$

Determine the basic feasible solution associated with the basis matrix

$$B = \begin{bmatrix} 1 & 0 & 1 \\ 3 & 1 & 0 \\ 1 & 0 & 2 \end{bmatrix}.$$

For which of the following vectors c is this basic feasible solution optimal?

$$c = (-5, -2, -1, 2, -3, -10)',$$
$$c = (30, 1, 0, -5, -2, 20)',$$
$$c = (-10, -1, 1, 6, -3, -15)'.$$

Are there alternate optima? *Hint:* Use Theorem 3.12.

4.2. Solve the following problems using the revised simplex method and the initial data indicated.

(a)   minimize: $- x_1 - x_2$

    subject to: $x_1 + 3x_2 + x_3 \qquad = 9,$ (1)

                 $2x_1 + x_2 \qquad + x_4 = 8,$ (2)

                 $x_i \geq 0, \; i = 1, \ldots, 4.$

$$I_B = \{2, 4\}, \quad B = \begin{bmatrix} 3 & 0 \\ 1 & 1 \end{bmatrix}, \quad B^{-1} = \begin{bmatrix} 1/3 & 0 \\ -1/3 & 1 \end{bmatrix}.$$

(b)   minimize: $-x_1 - x_2$

    subject to: $x_1 + x_2 - x_3 = 1,$ (1)

                 $x_i \geq 0, \; i = 1, 2, 3.$

$$I_B = \{1\}, \quad B = [1], \quad B^{-1} = [1].$$

(c)   minimize: $- x_2$

    subject to: $2x_1 + 3x_2 + x_3 \qquad = 21,$ (1)

                $- x_1 + x_2 \qquad + x_4 \qquad = 2,$ (2)

                $x_1 - x_2 \qquad + x_5 = 3,$ (3)

                $x_i \geq 0, \; i = 1, \ldots, 5.$

$$I_B = \{3,4,5\}, \quad B = B^{-1} = \begin{bmatrix} 1 & 0 & 0 \\ 0 & 1 & 0 \\ 0 & 0 & 1 \end{bmatrix}.$$

(d)   minimize: $x_1 + 15x_2 - 3x_3 + 20x_4 - x_5 + 3x_6 + 25x_7$

subject to: $2x_1 + x_2 + x_3 \qquad + x_5 - x_6 \qquad\qquad = 0,$ (1)

$\qquad\qquad - x_1 \qquad + x_3 \qquad - x_5 + 2x_6 + \quad x_7 = 6,$ (2)

$\qquad -2x_1 \qquad + 2x_3 + \quad x_4 + x_5 + 3x_6 \qquad\qquad = 9,$ (3)

$$x_i \geq 0, \ i = 1, \ldots, 7.$$

$$I_B = \{2,7,4\}, \quad B = B^{-1} = \begin{bmatrix} 1 & 0 & 0 \\ 0 & 1 & 0 \\ 0 & 0 & 1 \end{bmatrix}.$$

(e)   minimize: $-14x_1 - 18x_2 - 16x_3 - 80x_4$

subject to: $4.5x_1 + 8.5x_2 + 6x_3 + 20x_4 + x_5 \qquad\qquad = 6000,$ (1)

$\qquad\qquad x_1 + \quad x_2 + 4x_3 + 40x_4 \qquad + x_6 = 4000,$ (2)

$$x_i \geq 0, \ i = 1, \ldots, 6.$$

$$I_B = \{5,6\}, \quad B = B^{-1} = \begin{bmatrix} 1 & 0 \\ 0 & 1 \end{bmatrix}.$$

4.3. Use the Phase 1–Phase 2 revised simplex method to solve the following problems.

(a)   minimize: $10x_1 + x_2 - 7x_3 - 5x_4 - \quad x_5$

subject to: $x_1 - x_2 \qquad + \quad x_4 + \quad x_5 = 1,$ (1)

$\qquad\qquad x_2 + \quad x_3 + 2x_4 + 2x_5 = 7,$ (2)

$\qquad\qquad - x_2 \qquad + 3x_4 - \quad x_5 = 4,$ (3)

$$x_i \geq 0, \ i = 1, \ldots, 5.$$

(b)   minimize: $x_1 - 4x_2 + x_3 + 13x_4 + 23x_5$

subject to: $2x_1 + \quad x_2 + x_3 + \quad 4x_4 - \quad x_5 = 6,$ (1)

$\qquad\qquad x_1 + 2x_2 + x_3 + \qquad x_4 - 4x_5 = 3,$ (2)

$$x_i \geq 0, \ i = 1, \ldots, 5.$$

(c)   minimize: $9x_1 + \quad x_2 - 10x_3 + 10x_4$

subject to: $5x_1 + 2x_2 + \quad 3x_3 + 6x_4 + x_5 \qquad\qquad = 10,$ (1)

$\qquad\qquad 3x_1 + \quad x_2 - \quad 2x_3 + \quad x_4 \qquad\qquad\qquad = 2,$ (2)

$\qquad\qquad x_1 - 4x_2 + \quad 7x_3 + 4x_4 \qquad + x_6 = 1,$ (3)

$\qquad\qquad\qquad 2x_2 - \quad 5x_3 - 3x_4 \qquad\qquad\qquad = 5,$ (4)

$$x_i \geq 0, \ i = 1, \ldots, 6.$$

4.4. Show that the problem

$$\min\{c'x \mid Ax = b, \ x \geq 0\}$$

with

$$A = \begin{bmatrix} 6 & 1 & 0 & 0 & 1 & -5 \\ 2 & 3 & 1 & 4 & 0 & -10 \\ -1 & 1 & 0 & 2 & 2 & -7 \end{bmatrix}, \quad b = \begin{bmatrix} 9 \\ 17 \\ 13 \end{bmatrix},$$

and

$$c = (-5, -1, 1, 10, -3, -25)'$$

has no optimal solution. Does an optimal solution exist if the last component of c is replaced with 20?

4.5. Consider the following problem.

$$
\begin{array}{llll}
\text{minimize:} & -x_1 - x_2 & & \\
\text{subject to:} & -x_1 + x_2 \leq & 3, & (1) \\
 & -x_1 + 3x_2 \leq & 13, & (2) \\
 & 2x_1 + x_2 \leq & 16, & (3) \\
 & 2x_1 - 3x_2 \leq & 8, & (4) \\
 & -2x_1 - 3x_2 \leq & -6. & (5)
\end{array}
$$

(a) Write the problem in the standard form of the simplex method.
(b) Use the Phase 1–Phase 2 revised simplex method to solve the problem. Note that only one artificial variable is required for the Phase 1 problem.

4.6. (a) Write the initial point problem for the problem of Exercise 4.5. Note that α need only appear in the last constraint (see Exercise 2.20).
(b) Use Algorithm 3 to solve the initial point problem and then the stated problem.
(c) Compare these results with those for Exercise 4.5.

4.7. Use the tableau form of the simplex method to solve each of the problems of Exercise 4.2.

4.8. Consider the problem

$$
\begin{array}{ll}
\text{minimize:} & -x_1 - x_2 - x_3 \\
\text{subject to:} & x_1 + x_2 + x_3 \leq -1. \qquad (1)
\end{array}
$$

(a) What is an optimal solution for the problem? Is it unique?
(b) Transform the problem into the canonical form required by the simplex method.
(c) State the Phase 1 problem. Solve it using the revised simplex method. Is the resulting point an extreme point for the given problem? Explain.
(d) State the initial point problem for the above. Solve it using Algorithm 3. Compare the results with those obtained in part (c).

4.9. Consider the problem

$$\text{minimize:} \quad -x_1$$

$$
\begin{array}{llll}
\text{subject to:} & 2x_1 - 5x_2 \le & 1, & (1) \\
& x_1 - 2x_2 \le & 1, & (2) \\
& x_1 - x_2 \le & 3, & (3) \\
& 2x_1 - x_2 \le & 9, & (4) \\
& x_1 + 2x_2 \le & 17, & (5) \\
& -x_1 + 2x_2 \le & 7, & (6) \\
& -x_1 + x_2 \le & 2, & (7) \\
& -2x_1 + x_2 \le & 1, & (8) \\
& -x_1 \le & 0, & (9) \\
& -x_2 \le & 0. & (10)
\end{array}
$$

(a) Solve the problem using Algorithm 3 and initial data

$$x_0 = \begin{bmatrix} 0 \\ 0 \end{bmatrix}, \quad D_0^{-1} = -I, \quad J_0 = \{9, 10\}.$$

(b) By introducing slack variables, transform the problem into the canonical form required by the simplex method. Solve using the revised simplex method with x_1 and x_2 nonbasic as initial data.

(c) Compare the computations for parts (a) and (b).

4.10. (a) State the dual of the problem of Exercise 4.9.

(b) Solve the dual problem using the revised simplex method beginning with

$$x_B = (2, 1)', \quad I_B = \{3, 6\}, \quad B^{-1} = \begin{bmatrix} 2 & 1 \\ 1 & 1 \end{bmatrix}.$$

(c) Solve the dual problem using Algorithm 3 and initial data analogous to that used in part (b). Note that the initial data for Algorithm 3 requires the inverse of a $(10, 10)$ matrix. This can be obtained by cleverly using the inverse of a $(2, 2)$ matrix and the results of Exercise 4.21, part (b).

(d) Compare the computations for parts (b) and (c).

(e) Compare the computations for Exercise 4.9, parts (a) and (b), with those for parts (b) and (c) of this exercise.

4.11. Apply the revised simplex method to the problem of Example 4.4. Compare the computations with those of the tableau form of the simplex method given in Section 4.3 for the problem of Example 4.4. Explain why the current basis inverse is always contained in columns 3, 4, and 5 of the current tableau. What is the generalization of this?

4.12. Suppose that the revised simplex method terminates at Step 1 with the optimal solution $x_B = B^{-1}b$. Assume that $x_B > 0$. Show that if

$$c_i - c_B' B^{-1} A_i > 0, \quad \text{for } i = 1, \dots, n \quad \text{and} \quad i \notin I_B,$$

the extreme point corresponding to x_B and I_B is the unique optimal solution to the problem

$$\min\{c'x \mid Ax = b, \ x \geq 0\}.$$

4.13. Suppose that the ith component of x becomes a nonbasic variable in the jth iteration of the revised simplex method. Show that this variable cannot become a basic variable in iteration $j + 1$.

4.14. Let x_B and $x_{\hat{B}}$ be the vectors of basic variables obtained in two consecutive iterations of the simplex method. Suppose that all components of x_B are strictly positive. Give a sufficient condition for the same to be true for $x_{\hat{B}}$.

4.15. Let B be the basis matrix associated with the optimal solution for the problem

$$\min\{c'x \mid Ax = b, \ x \geq 0\}.$$

Suppose that the column A_i of the initial simplex tableau contains the unit vector e_k. Show that the corresponding column of the final simplex tableau contains the kth column of B^{-1}.

4.16. Consider the model problem

$$\min\{c_1'x_1 + c_2'x_2 \mid A_1x_1 + A_2x_2 = b, \ x_2 \geq 0\}.$$

Note that this problem differs from the model problem required by the simplex method only in that the components of x_1 are unrestricted in sign.
(a) What are the optimality conditions for this problem?
(b) Modify the revised simplex method to solve this problem. *Hint:* The unrestricted variables are always basic.

4.17. Suppose that $(B^{-1}A_k)_l = 0$ in (4.25). Show that the lth equality constraint is redundant.

4.18. Verify (4.27).

4.19. Verify (4.28) and (4.29).

4.20. For any iteration of the revised simplex method, define

$$v_i = c_i + A_i'u, \quad \text{for all } i \in I_B.$$

Use the definition of u in Step 1 to show directly that

$$v_i = 0, \quad \text{for all } i \in I_B.$$

4.21. Let D be an (n, n) matrix partitioned as

$$D = \begin{bmatrix} B & N \\ 0 & \pm I \end{bmatrix},$$

where B is (m, m), N is $(m, n - m)$, 0 is $(n - m, m)$, and I is the $(n - m, n - m)$ identity matrix. (By using $\pm I$, we are handling two closely related problems simultaneously.)
(a) Show that B is nonsingular if and only if D is nonsingular.

(b) If D is nonsingular, show that

$$D^{-1} = \begin{bmatrix} B^{-1} & \mp B^{-1}N \\ 0 & \pm I \end{bmatrix}.$$

4.22. Show that if

$$\begin{bmatrix} x_B \\ x_{NB} \end{bmatrix}$$

is a nondegenerate extreme point, B is nonsingular.

PARAMETRIC
LINEAR PROGRAMMING

In many applications it is of interest to find out how the optimal solution for the LP

$$\min\{c'x \mid Ax \leq b\}$$

varies if some or all of the components of c and/or b are changed. More precisely we shall replace c with $c + tq$ and b with $b + tp$, where q is an n-vector, p is an m-vector, and t is a parameter, and try to express the optimal solution as an explicit function of t as the parameter increases from its initial value of 0 to some upper bound $\bar{t}$.

5.1 EXAMPLES

To illustrate the situation we first consider some examples with two variables and six inequality constraints.

Example 5.1

Determine an optimal solution for the problem

$$
\begin{array}{rrrrl}
\text{minimize:} & (7 - 2t)x_1 & + & (-10 + 2t)x_2 \\
\text{subject to:} & -4x_1 & + & 3x_2 & \leq & 9, & (1) \\
& -x_1 & + & 4x_2 & \leq & 25, & (2) \\
& 2x_1 & + & 5x_2 & \leq & 54, & (3) \\
& 2x_1 & + & x_2 & \leq & 30, & (4) \\
& -x_1 & & & \leq & 0, & (5) \\
& & - & x_2 & \leq & 0, & (6)
\end{array}
$$

for all values of t with $0 \leq t \leq 10$.

In this example $q = (-2, 2)'$ and $p = (0, 0, 0, 0, 0, 0)'$; that is, the constraints of the problem are independent of the parameter.

The feasible region is shown in Figure 5.1. It has the six extreme points $P_0 = (3, 7)'$, $P_1 = (7, 8)'$, $P_2 = (12, 6)'$, $P_3 = (15, 0)'$, $P_4 = (0, 0)'$, $P_5 = (0, 3)'$.

Because the feasible region is bounded we know from Section 2.4 that one of the extreme points is an optimal solution. Comparing the value of $c'x$ at the six extreme points, we find that

$$x_0 = \begin{bmatrix} 3 \\ 7 \end{bmatrix}$$

is an optimal solution for $t = 0$. Since the first and second constraints are active at x_0, $-c$ is in the cone spanned by the gradients of these constraints, that is, by the vectors $a_1 = (-4, 3)'$ and $a_2 = (-1, 4)'$ (see Figure 5.1).

It is clear that x_0 remains feasible for all values of t. Therefore, it follows from Theorem 3.1 that x_0 is an optimal solution for all t for which $-(c + tq) = (-7 + 2t, 10 - 2t)'$ is in the cone spanned by a_1 and a_2, that is, for all t for which

$$\begin{bmatrix} -4 & -1 \\ 3 & 4 \end{bmatrix} \begin{bmatrix} u_1 \\ u_2 \end{bmatrix} = \begin{bmatrix} -7 \\ 10 \end{bmatrix} + t \begin{bmatrix} 2 \\ -2 \end{bmatrix}$$

has a nonnegative solution. Since the inverse of the coefficient matrix of the equations above is

$$\begin{bmatrix} -4/13 & -1/13 \\ 3/13 & 4/13 \end{bmatrix}$$

Figure 5.1   Optimal extreme points for Example 5.1.

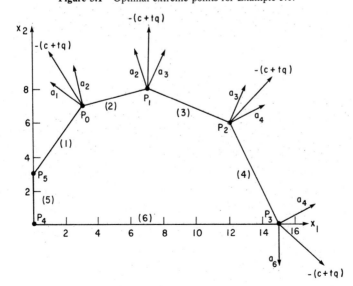

we obtain

$$
\begin{bmatrix} u_1(t) \\ u_2(t) \end{bmatrix} = \begin{bmatrix} -4/13 & -1/13 \\ 3/13 & 4/13 \end{bmatrix} \left(\begin{bmatrix} -7 \\ 10 \end{bmatrix} + t \begin{bmatrix} 2 \\ -2 \end{bmatrix} \right)
$$

$$
= \begin{bmatrix} 18/13 \\ 19/13 \end{bmatrix} + t \begin{bmatrix} -6/13 \\ -2/13 \end{bmatrix}.
$$

Observe that $u_1(t) \geq 0$ provided that $t \leq 3$ and $u_2(t) \geq 0$ provided that $t \leq 19/2$. Setting $t_0 = 0$ and $t_1 = 3$, we find that $x_0 = (3,7)'$ is an optimal solution for all t with $t_0 \leq t \leq t_1$. Since $u_1(t_1) = 0$, we have

$$
-(c + t_1 q) = a_2 u_2(t_1), \quad \text{with } u_2(t_1) = 1.
$$

Therefore, every point on the line segment connecting the extreme points $x_0 = (3,7)'$ and $x_1 = (7,8)'$ is an optimal solution for $t = t_1$. This suggests that x_1 is the optimal solution for $t > t_1$. To verify this, it suffices to show that $-(c + tq)$ is in the cone spanned by a_2 and $a_3 = (2,5)'$. This is the case for all t for which

$$
\begin{bmatrix} -1 & 2 \\ 4 & 5 \end{bmatrix} \begin{bmatrix} u_2 \\ u_3 \end{bmatrix} = \begin{bmatrix} -7 \\ 10 \end{bmatrix} + t \begin{bmatrix} 2 \\ -2 \end{bmatrix}
$$

has a nonnegative solution. The solution of these equations as a function of t is

$$
\begin{bmatrix} u_2(t) \\ u_3(t) \end{bmatrix} = \begin{bmatrix} -5/13 & 2/13 \\ 4/13 & 1/13 \end{bmatrix} \left(\begin{bmatrix} -7 \\ 10 \end{bmatrix} + t \begin{bmatrix} 2 \\ -2 \end{bmatrix} \right)
$$

$$
= \begin{bmatrix} 55/13 \\ -18/13 \end{bmatrix} + t \begin{bmatrix} -14/13 \\ 6/13 \end{bmatrix}
$$

$$
= \begin{bmatrix} u_2(t_1) \\ 0 \end{bmatrix} + \begin{bmatrix} -14/13 \\ 6/13 \end{bmatrix} (t - t_1).
$$

With $t_2 = 55/14$ it follows, therefore, that $x_1 = (7,8)'$ is an optimal solution for all t with $t_1 \leq t \leq t_2$.

Because $u_2(t_2) = 0$ and $u_3(t_2) = 3/7$ every point on the line joining the two extreme points $x_1 = (7,8)'$ and $x_2 = (12,6)'$ is an optimal solution for $t = t_2$. For $t > t_2$ we expect x_2 to be the optimal solution. Indeed, the equations

$$
\begin{bmatrix} 2 & 2 \\ 5 & 1 \end{bmatrix} \begin{bmatrix} u_3 \\ u_4 \end{bmatrix} = \begin{bmatrix} -7 \\ 10 \end{bmatrix} + t \begin{bmatrix} 2 \\ -2 \end{bmatrix}
$$

have the solution

$$\begin{bmatrix} u_3(t) \\ u_4(t) \end{bmatrix} = \begin{bmatrix} -1/8 & 1/4 \\ 5/8 & -1/4 \end{bmatrix} \left(\begin{bmatrix} -7 \\ 10 \end{bmatrix} + t \begin{bmatrix} 2 \\ -2 \end{bmatrix} \right)$$

$$= \begin{bmatrix} 27/8 \\ -55/8 \end{bmatrix} + t \begin{bmatrix} -3/4 \\ 7/4 \end{bmatrix}$$

$$= \begin{bmatrix} u_3(t_2) \\ 0 \end{bmatrix} + \begin{bmatrix} -3/4 \\ 7/4 \end{bmatrix} (t - t_2),$$

from which we conclude that $x_2 = (12, 6)'$ is an optimal solution for all t with $t_2 \leq t \leq t_3$, where $t_3 = 9/2$.

For $t = t_3$, $u_3(t) = 0$ and $u_4(t) = 1$. Thus all points on the line segment connecting $x_2 = (12, 6)'$ and $x_3 = (15, 0)'$ are optimal solutions. For $t > t_3$ the equations

$$\begin{bmatrix} 2 & 0 \\ 1 & -1 \end{bmatrix} \begin{bmatrix} u_4 \\ u_6 \end{bmatrix} = \begin{bmatrix} -7 \\ 10 \end{bmatrix} + t \begin{bmatrix} 2 \\ -2 \end{bmatrix}$$

have the solution

$$\begin{bmatrix} u_4(t) \\ u_6(t) \end{bmatrix} = \begin{bmatrix} 1/2 & 0 \\ 1/2 & -1 \end{bmatrix} \left(\begin{bmatrix} -7 \\ 10 \end{bmatrix} + t \begin{bmatrix} 2 \\ -2 \end{bmatrix} \right)$$

$$= \begin{bmatrix} -7/2 \\ -27/2 \end{bmatrix} + t \begin{bmatrix} 1 \\ 3 \end{bmatrix}$$

$$= \begin{bmatrix} u_4(t_3) \\ 0 \end{bmatrix} + \begin{bmatrix} 1 \\ 3 \end{bmatrix} (t - t_3).$$

Since $u_4(t)$ and $u_6(t)$ are nonnegative for all $t \geq t_3$, the extreme point $x_3 = (15, 0)'$ is an optimal solution for all $t \geq t_3$.

Summarizing the results, we see that there are values $t_1 < t_2 < t_3$ subdividing the given parameter interval $[0, 10]$ into four intervals,

$$[0, t_1], \quad [t_1, t_2], \quad [t_2, t_3], \quad [t_3, 10]$$

such that for all t in one of these intervals the same extreme point is an optimal solution for the given problem. The optimal solutions corresponding to two consecutive intervals are adjacent extreme points. For the value of t which is in both of these intervals, all points on the edge connecting these two extreme points are optimal solutions. The optimal value of $(c + tq)'x$ is a piecewise linear function of t given by

$$(c + tq)'x_0 = -49 + 8t, \quad \text{for} \quad 0 \le t \le t_1 = 3,$$
$$(c + tq)'x_1 = -31 + 2t, \quad \text{for} \quad t_1 \le t \le t_2 = 55/14,$$
$$(c + tq)'x_2 = 24 - 12t, \quad \text{for} \quad t_2 \le t \le t_3 = 9/2,$$
$$(c + tq)'x_3 = 105 - 30t, \quad \text{for} \quad t_3 \le t \le 10.$$

Since $(c + t_{j+1}q)'x_j = (c + t_{j+1}q)'x_{j+1}$ for $j = 0, 1, 2$, the optimal objective function value is continuous.

Example 5.2

Determine an optimal solution for the problem

$$\begin{array}{llr}
\text{minimize:} & -4x_1 - x_2 & \\
\text{subject to:} & -4x_1 + 3x_2 \le 9 - 5t, & (1) \\
& -x_1 + 4x_2 \le 25, & (2) \\
& 2x_1 + 5x_2 \le 54 - 4t, & (3) \\
& 2x_1 + x_2 \le 30, & (4) \\
& -x_1 \le 0, & (5) \\
& -x_2 \le -2t, & (6)
\end{array}$$

for all values of t with $0 \le t \le 10$. Here $q = (0,0)'$ and $p = (-5, 0, -4, 0, 0, -2)'$.

For $t = 0$ the feasible region is the same as in the preceding example and it can easily be verified that

$$x_1 = 15, \quad x_2 = 0$$

is the unique optimal solution for $t = 0$. The active constraints are

$$2x_1 + x_2 = 30 \quad \text{and} \quad -x_2 = -2t.$$

Obviously, the vector $(15, 0)'$ is not feasible for any $t > 0$. To obtain the optimal solution for $t > 0$ we write the solution of these equations as an explicit function of t:

$$\begin{bmatrix} x_1(t) \\ x_2(t) \end{bmatrix} = \begin{bmatrix} 1/2 & 1/2 \\ 0 & -1 \end{bmatrix} \left(\begin{bmatrix} 30 \\ 0 \end{bmatrix} + t \begin{bmatrix} 0 \\ -2 \end{bmatrix} \right)$$

$$= \begin{bmatrix} 15 \\ 0 \end{bmatrix} + t \begin{bmatrix} -1 \\ 2 \end{bmatrix}$$

$$= \begin{bmatrix} x_1(0) \\ x_2(0) \end{bmatrix} + \begin{bmatrix} -1 \\ 2 \end{bmatrix} (t - t_0),$$

where $t_0 = 0$. Because all gradients are independent of t, this linear function represents the optimal solution for all values of t for which the constraints are satisfied. To determine the maximal value of t for which this is true we substitute $x_1(t)$ and $x_2(t)$ into the constraints that are not active at $(x_1(0), x_2(0))'$. This gives the inequalities

$$
\left.
\begin{array}{rcccl}
-4x_1(t) + 3x_2(t) &\leq& 9 - 5t, & \text{or,} & 15t \leq 69, \\
- x_1(t) + 4x_2(t) &\leq& 25, & \text{or,} & 9t \leq 40, \\
2x_1(t) + 5x_2(t) &\leq& 54 - 4t, & \text{or,} & 12t \leq 24, \\
- x_1(t) &\leq& 0, & \text{or,} & t \leq 15.
\end{array}
\right\} \qquad (5.1)
$$

Therefore,

$$ x_1(t) = 15 - t, \quad x_2(t) = 2t $$

is an optimal solution for all t with $t_0 \leq t \leq t_1$ where

$$ t_1 = \min\left\{ \frac{69}{15}, \frac{40}{9}, \frac{24}{12}, \frac{15}{1} \right\} = 2. $$

For $t = t_1$, we have the active constraints

$$ 2x_1 + x_2 = 30, \quad -x_2 = -2t, \quad \text{and} \quad 2x_1 + 5x_2 = 54 - 4t, \qquad (5.2) $$

and the problem is now to select two constraints which will determine the optimal solution for $t > t_1$. In Figure 5.2 the feasible region for $t = t_1$, $R(t_1)$, is represented by the dashed lines. We expect that, for $t > t_1$, the constraints

$$ -x_2 = -2t \quad \text{and} \quad 2x_1 + 5x_2 = 54 - 4t $$

Figure 5.2   Feasible region for Example 5.2 for $t = 0, 1,$ and 2.

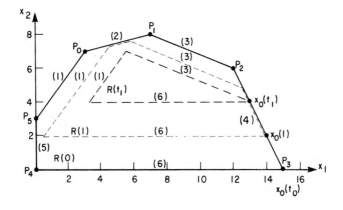

will determine the optimal solution. Indeed,

$$\begin{bmatrix} 0 & 2 \\ -1 & 5 \end{bmatrix} \begin{bmatrix} u_6 \\ u_3 \end{bmatrix} = \begin{bmatrix} 4 \\ 1 \end{bmatrix}$$

has the solution $u_3 = 2$, $u_6 = 9$. Thus

$$\begin{bmatrix} x_1(t) \\ x_2(t) \end{bmatrix} = \begin{bmatrix} 5/2 & 1/2 \\ -1 & 0 \end{bmatrix} \left(\begin{bmatrix} 0 \\ 54 \end{bmatrix} + t \begin{bmatrix} -2 \\ -4 \end{bmatrix} \right)$$

$$= \begin{bmatrix} 27 \\ 0 \end{bmatrix} + t \begin{bmatrix} -7 \\ 2 \end{bmatrix}$$

is the optimal solution for all $t \geq t_1$ with

$$
\begin{array}{llrl}
-4x_1(t) + 3x_2(t) \leq 9 - 5t, & \text{or,} & 39t \leq & 117, \\
-\ x_1(t) + 4x_2(t) \leq 25, & \text{or,} & 15t \leq & 52, \\
2x_1(t) + \ x_2(t) \leq 30, & \text{or,} & -12t \leq & -24, \\
-\ x_1(t) \qquad\qquad \leq 0, & \text{or,} & 7t \leq & 27.
\end{array}
$$

This gives

$$t_2 = \min \left\{ \frac{117}{39}, \frac{52}{15}, \frac{27}{7} \right\} = \frac{117}{39} = 3.$$

For $t = t_2$ we have the active constraints

$$-4x_1 + 3x_2 = 9 - 5t, \quad -x_2 = -2t, \quad \text{and} \quad 2x_1 + 5x_2 = 54 - 4t,$$

and it is not difficult to infer from Figure 5.2 that there is no feasible solution for $t > t_2$.

Summarizing these results we see that the optimal solution is given by

$$x_1(t) = 15 - t, \quad x_2(t) = 2t, \quad \text{for} \quad 0 \leq t \leq t_1 = 2,$$
$$x_1(t) = 27 - 7t, \quad x_2(t) = 2t, \quad \text{for} \quad t_1 \leq t \leq t_2 = 3.$$

No feasible solution exists for $t > t_2$. Because $15 - t_1 = 27 - 7t_1$, the components of the optimal solution are continuous piecewise linear functions of t.

Example 5.3

We consider the problem of Example 5.2 with $q = (0, 0)'$ replaced by $q = (1, -1/3)'$.

For $t = t_0 = 0$ the optimal solution is again $x_1 = 15$, $x_2 = 0$. Since both the objective function and the right-hand sides of the constraints depend on the parameter, we use the equations

$$2x_1 + x_2 = 30, \quad \text{and} \quad 2u_4 = 4 - t,$$
$$- x_2 = -2t, \qquad u_4 - u_6 = 1 + t/3,$$

to determine the optimal solution and the multipliers as functions of t. We have

$$\begin{bmatrix} x_1(t) \\ x_2(t) \end{bmatrix} = \begin{bmatrix} 1/2 & 1/2 \\ 0 & -1 \end{bmatrix} \left(\begin{bmatrix} 30 \\ 0 \end{bmatrix} + t \begin{bmatrix} 0 \\ -2 \end{bmatrix} \right)$$

$$= \begin{bmatrix} 15 \\ 0 \end{bmatrix} + t \begin{bmatrix} -1 \\ 2 \end{bmatrix},$$

and

$$\begin{bmatrix} u_4(t) \\ u_6(t) \end{bmatrix} = \begin{bmatrix} 1/2 & 0 \\ 1/2 & -1 \end{bmatrix} \left(\begin{bmatrix} 4 \\ 1 \end{bmatrix} + t \begin{bmatrix} -1 \\ 1/3 \end{bmatrix} \right)$$

$$= \begin{bmatrix} 2 \\ 1 \end{bmatrix} + t \begin{bmatrix} -1/2 \\ -5/6 \end{bmatrix}.$$

From Theorem 3.1 we conclude that

$$x_1(t) = 15 - t, \quad x_2(t) = 2t \tag{5.3}$$

is the optimal solution for all t for which the inequalities (5.1) and $u_4(t) \geq 0$, $u_6(t) \geq 0$; that is,

$$t \leq 4 \quad \text{and} \quad 5t \leq 6, \tag{5.4}$$

are satisfied. As in the preceding example the inequalities (5.1) are satisfied for all $t \leq t_1^*$ with

$$t_1^* = \min \left\{ \frac{69}{15}, \frac{40}{9}, \frac{24}{12}, \frac{15}{1} \right\} = 2.$$

The inequalities (5.4) imply that $t \leq \tilde{t}_1$, where

$$\tilde{t}_1 = \min \left\{ \frac{4}{1}, \frac{6}{5} \right\} = \frac{6}{5}.$$

Setting

$$t_1 = \min \{ \tilde{t}_1, t_1^* \} = \frac{6}{5}$$

we obtain the result that (5.3) is the optimal solution for all t with $t_0 \leq t \leq t_1$.

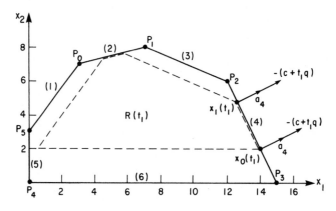

Figure 5.3   Feasible region for Example 5.3 for $t = 0$ and $t = 1$.

In Figure 5.3 the dashed lines again represent the feasible region $R(t_1)$ for $t = t_1$. We see that for $t > t_1$, the active constraint $-x_2 = -2t$ will be replaced by $2x_1 + 5x_2 = 54 - 4t$. For $t > t_1$ the optimal solution and the multipliers are, therefore, determined by

$$
\begin{aligned}
2x_1 + 5x_2 &= 54 - 4t, \\
2x_1 + x_2 &= 30,
\end{aligned}
\quad \text{and} \quad
\begin{aligned}
2u_3 + 2u_4 &= 4 - t, \\
5u_3 + u_4 &= 1 + t/3,
\end{aligned}
$$

which gives

$$
\begin{bmatrix} x_1(t) \\ x_2(t) \end{bmatrix} =
\begin{bmatrix} -1/8 & 5/8 \\ 1/4 & -1/4 \end{bmatrix}
\left[\begin{bmatrix} 54 \\ 30 \end{bmatrix} + t \begin{bmatrix} -4 \\ 0 \end{bmatrix} \right]
$$

$$
= \begin{bmatrix} 12 \\ 6 \end{bmatrix} + t \begin{bmatrix} 1/2 \\ -1 \end{bmatrix},
$$

and

$$
\begin{bmatrix} u_3(t) \\ u_4(t) \end{bmatrix} =
\begin{bmatrix} -1/8 & 1/4 \\ 5/8 & -1/4 \end{bmatrix}
\left[\begin{bmatrix} 4 \\ 1 \end{bmatrix} + t \begin{bmatrix} -1 \\ 1/3 \end{bmatrix} \right]
$$

$$
= \begin{bmatrix} -1/4 \\ 9/4 \end{bmatrix} + t \begin{bmatrix} 5/24 \\ -17/24 \end{bmatrix}.
$$

To compute t_2^* we substitute into the inactive constraints and obtain the inequalities

$$-4x_1(t) + 3x_2(t) \leq 9 - 5t, \quad \text{or,} \quad 0 \cdot t \leq 39,$$
$$- \ x_1(t) + 4x_2(t) \leq 25, \qquad \text{or,} \quad -9t \leq 26,$$
$$- \ x_1(t) \qquad\qquad \leq 0, \qquad\quad \text{or,} \quad -t \leq 24,$$
$$- \quad x_2(t) \leq -2t, \qquad \text{or,} \quad 3t \leq 6.$$

Thus

$$t_2^* = 2.$$

Similarly, $\tilde{t}_2$ is determined by the inequalities

$$u_3(t) \geq 0, \quad \text{or,} \quad -5t \leq -6,$$
$$u_4(t) \geq 0, \quad \text{or,} \quad 17t \leq 54,$$

which give

$$\tilde{t}_2 = \frac{54}{17}.$$

Therefore,

$$t_2 = \min\{t_2^*, \tilde{t}_2\} = 2.$$

As in Example 5.2 the constraints (5.2) are active for $t = 2$. Thus

$$x_1(t) = 27 - 7t, \quad x_2(t) = 2t \tag{5.5}$$

is feasible for $t_2 \leq t \leq t_3 = 3$, and there is no feasible solution for $t > t_3$. Because

$$2u_3 \qquad\quad = 4 - t,$$
$$5u_3 - u_6 = 1 + t/3,$$

has, for $t_2 \leq t \leq t_3$, the nonnegative solution

$$u_3(t) = 2 - \frac{t}{2}, \quad u_6(t) = 9 - \frac{17}{6}t$$

it follows that (5.5) is the optimal solution for $t_2 \leq t \leq t_3$.

Therefore, the optimal solution is given by

$$x_1(t) = 15 - t, \quad x_2(t) = 2t, \qquad \text{for} \quad 0 \leq t \leq t_1 = \frac{6}{5},$$
$$x_1(t) = 12 + \tfrac{1}{2}t, \ x_2(t) = 6 - t, \qquad \text{for} \quad t_1 \leq t \leq t_2 = 2,$$
$$x_1(t) = 27 - 7t, \quad x_2(t) = 2t, \qquad \text{for} \quad t_2 \leq t \leq t_3 = 3.$$

No feasible solution exists for $t > 3$. Because $15 - t_1 \neq 12 + (1/2)t_1$ and $2t_1 \neq 6 - t_1$ the components of the optimal solution are not continuous functions of t. Observe that $t_1 = \tilde{t}_1$; that is, the critical value t_1 is determined by a vanishing multiplier. The optimal solution "jumps" from the extreme point

$P_3(t_1)$ to the extreme point $P_2(t_1)$ (see Figure 5.3). In contrast, $t_2 = t_2^*$ is determined by a newly active constraint. In this case the components of the optimal solution are continuous for $t = t_2$ because $12 + (1/2)t_2 = 27 - t_2$ and $6 - t_2 = 2t_2$.

In the examples above the critical values t_j were always determined by the fact that either exactly one multiplier became zero or exactly one constraint became newly active. In either case it was not difficult to find an optimal solution for $t > t_j$. The situation is more complicated if several multipliers vanish and/or several constraints become newly active for $t = t_j$, as illustrated in the following examples.

Example 5.4

First we consider the problem of Example 5.1 with $q = (-2, 2)'$ replaced by $q = (-7, 10)'$.

For $t = 0$, $x_0 = (3, 7)'$ is the optimal solution and

$$\begin{bmatrix} u_1(t) \\ u_2(t) \end{bmatrix} = \begin{bmatrix} -4/13 & -1/13 \\ 3/13 & 4/13 \end{bmatrix} \left(\begin{bmatrix} -7 \\ 10 \end{bmatrix} + t \begin{bmatrix} 7 \\ -10 \end{bmatrix} \right)$$

$$= \begin{bmatrix} 18/13 \\ 19/13 \end{bmatrix} + t \begin{bmatrix} -18/13 \\ -19/13 \end{bmatrix}.$$

With $t_1 = 1$ we have $u_1(t_1) = u_2(t_1) = 0$. For $t > t_1$ the optimal solution is $x_1 = (15, 0)'$. This extreme point is not adjacent to $x_0 = (3, 7)'$.

Example 5.5

Next we consider the problem of Example 5.1 with $p = (-9, 1, -2, -10, 0, 0)'$ and $q = (0, 0)'$.

For $t = 0$ the optimal solution $x_0 = (3, 7)'$ is determined by the active constraints

$$-4x_1 + 3x_2 = 9 - 9t,$$
$$- x_1 + 4x_2 = 25 + t,$$

from which we obtain

$$\begin{bmatrix} x_1(t) \\ x_2(t) \end{bmatrix} = \begin{bmatrix} -4/13 & 3/13 \\ -1/13 & 4/13 \end{bmatrix} \left(\begin{bmatrix} 9 \\ 25 \end{bmatrix} + t \begin{bmatrix} -9 \\ 1 \end{bmatrix} \right)$$

$$= \begin{bmatrix} 3 \\ 7 \end{bmatrix} + t \begin{bmatrix} 3 \\ 1 \end{bmatrix}.$$

Substituting into the inactive constraints we have

$$2x_1(t) + 5x_2(t) \le 54 - 2t, \quad \text{or}, \quad 13t \le 13,$$
$$2x_1(t) + x_2(t) \le 30 - 10t, \quad \text{or}, \quad 17t \le 17,$$
$$- x_1(t) \qquad\qquad \le 0, \quad \text{or}, \quad -3t \le 3,$$
$$- x_2(t) \le 0, \qquad\qquad \text{or}, \quad -t \le 7.$$

This shows that for $t_1 = 1$ four constraints are active at the optimal solution

$$x_1(t_1) = 6, \quad x_2(t_1) = 8.$$

Figure 5.4 depicts the feasible region for $t = 1$. In order to find the constraints that determine the optimal solution for $t > 1$, the feasible region for $t = 3/2$ has been indicated by dashed lines. The figure shows that among the six possible choices of two of the four constraints active at $(6, 8)'$ only the equations

$$-4x_1 + 3x_2 = 9 - 9t \quad \text{and} \quad 2x_1 + x_2 = 30 - 10t \qquad (5.6)$$

determine an extreme point of the feasible region for $t > t_1$. Solving (5.6) we obtain

$$x_1(t) = \frac{81}{10} - \frac{21}{10}t, \quad x_2(t) = \frac{69}{5} - \frac{29}{5}t.$$

It is easy to verify that this is the optimal solution for all t with $t_1 \le t \le t_2$, where $t_2 = 69/29$ is determined by the condition $x_2(t) \ge 0$. There is no feasible solution for $t > t_2$.

Figure 5.4   Feasible region for Example 5.5 for $t = 1$ and $t > 1$.

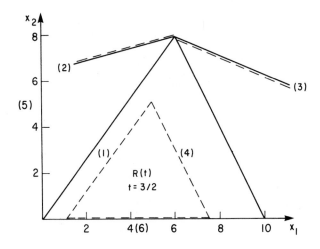

5.2 NUMERICAL METHODS FOR PARAMETRIC
LINEAR PROGRAMMING

We shall now show that the results obtained for the examples in Section 5.1 are typical for the general problem of determining an optimal solution for the LP

$$\min \{ (c + tq)'x \mid Ax \leq b + tp \} \tag{5.7}$$

for all t with $\underline{t} \leq t \leq \bar{t}$. Here A is an (m, n) matrix with rows a_1' , ..., a_m'; c, q, and x are n-vectors; b and p are m-vectors with components $b_1, \ldots, b_m$ and $p_1, \ldots, p_m$, respectively; and $\underline{t}$ and $\bar{t}$ are given scalars. Throughout this chapter we assume that rank$(A) = n$. From Section 2.4 we know that this is a necessary condition for the feasible region of (5.7) to have extreme points.

Assume that (5.7) has an optimal solution for $t = t_0 = \underline{t}$. Denote this optimal solution by $x_0 = x_0(t_0)$. From Section 2.4 we can assume that x_0 is an extreme point. Let $D_0^{-1} = \left[c_{10}, \ldots, c_{n0} \right]$ and $J_0 = \{ \alpha_{10}, \ldots, \alpha_{n0} \}$ be the associated matrix and index set, respectively. Without loss of generality, we may assume that $\alpha_{i0} = i$ for $i = 1, \ldots, n$. Setting

$$b_0 = (b_1, \ldots, b_n)' \quad \text{and} \quad p_0 = (p_1, \ldots, p_n)'$$

we can write the optimality conditions of Theorem 3.1 in the form

$$D_0' u_0(t_0) = -(c + t_0 q), \quad D_0 x_0(t_0) = b_0 + t_0 p_0, \tag{5.8}$$

$$u_0(t_0) \geq 0, \quad a_i' x_0(t_0) \leq b_i + t_0 p_i, \quad i = n + 1, \ldots, m. \tag{5.9}$$

Solving (5.8) for $u_0(t_0)$ and $x_0(t_0)$ we have

$$u_0(t_0) = g_{10} + t_0 g_{20}, \quad x_0(t_0) = h_{10} + t_0 h_{20},$$

where $g_{10} = -(D_0^{-1})'c$, $g_{20} = -(D_0^{-1})'q$, $h_{10} = D_0^{-1} b_0$, and $h_{20} = D_0^{-1} p_0$. Replacing t_0 with t and substituting into (5.9) we observe that

$$u_0(t) = g_{10} + t g_{20}, \quad x_0(t) = h_{10} + t h_{20} \tag{5.10}$$

satisfy the optimality conditions if and only if

$$g_{10} + t g_{20} \geq 0, \tag{5.11}$$

and

$$a_i'(h_{10} + t h_{20}) \leq b_i + t p_i, \quad \text{for } i = n + 1, \ldots, m. \tag{5.12}$$

Writing the inequalities (5.11) in the form

$$u_0(t) = g_{10} + t_0 g_{20} + (t - t_0) g_{20}$$

$$= u_0(t_0) + (t - t_0) g_{20} \geq 0,$$

we see that they are satisfied for every t with

$$-tg_{20} \leq u_0(t_0) - t_0 g_{20}.$$

If $g_{20} \geq 0$, then $u_0(t) \geq 0$ for all $t \geq t_0$. In this case we set $\tilde{t}_1 = \bar{t}$. Otherwise, we compute

$$\tilde{t}_1 = \min \left\{ \frac{(u_0)_i - t_0(g_{20})_i}{-(g_{20})_i} \mid \text{ all } i \text{ with } (g_{20})_i < 0 \right\} \qquad (5.13)$$

$$= t_0 + \min \left\{ \frac{(u_0)_i}{-(g_{20})_i} \mid \text{ all } i \text{ with } (g_{20})_i < 0 \right\},$$

where $(u_0)_i$ and $(g_{20})_i$ are the components of $u_0(t_0)$ and g_{20}, respectively. Because $u_0(t_0) \geq 0$, we have $\tilde{t}_1 \geq t_0$. Clearly, $\tilde{t}_1 > t_0$ if and only if $(u_0)_i > 0$ and/or $(g_{20})_i \geq 0$ for all $i = 1, \ldots, n$. Furthermore, $u_0(t) = g_{10} + tg_{20} \geq 0$ for $t_0 \leq t \leq \tilde{t}_1$.

Similarly, writing the inequalities (5.12) in the form

$$a_i'(h_{10} + t_0 h_{20} + (t - t_0)h_{20}) \leq b_i + t_0 p_i + (t - t_0)p_i, \quad i = n + 1, \ldots, m,$$

we observe that they are satisfied for every t with

$$(t - t_0)(a_i' h_{20} - p_i) \leq b_i + t_0 p_i - a_i' x_0(t_0), \quad i = n + 1, \ldots, m.$$

If $a_i' h_{20} - p_i \leq 0$ for all $i = n + 1, \ldots, m$, then $x_0(t) = h_{10} + th_{20}$ is feasible for every $t \geq t_0$. In this case we set $t_1^* = \bar{t}$. Otherwise, we compute

$$t_1^* = t_0 + \min \left\{ \frac{b_i + t_0 p_i - a_i' x_0(t_0)}{a_i' h_{20} - p_i} \mid \text{ all } i \text{ with } a_i' h_{20} - p_i > 0 \right\}. \qquad (5.14)$$

Since $x_0(t_0)$ is a feasible solution for $t = t_0$, we have $t_1^* \geq t_0$ and $t_1^* > t_0$ if and only if $a_i' x_0(t_0) < b_i + t_0 p_i$ and/or $a_i' h_{20} - p_i \leq 0$ for all $i = n + 1, \ldots, m$. Moreover, $x_0(t) = h_{10} + th_{20}$ is feasible for $t_0 \leq t \leq t_1^*$.

Finally, if we set

$$t_1 = \min\{\tilde{t}_1, t_1^*\},$$

then $t_1 \geq t_0$ and $x_0(t)$ as given by (5.10), is an optimal solution to (5.7) for every t with $t_0 \leq t \leq t_1$.

If $t_1 = \tilde{t}_1 \leq t_1^*$, then at least one component of $u_0(t)$ will be negative for $t > t_1$, indicating that the corresponding constraint will not be active at the optimal solution for $t > t_1$. On the other hand, if $t_1 = t_1^* \leq \tilde{t}_1$, $x_0(t)$ will not be feasible for $t > t_1$.

Depending on whether t_1 is equal to $\tilde{t}_1$ or t_1^*, we will have to consider two different cases.

Case 1: $t_1 = \tilde{t}_1 \leq t_1^{*}$

Let the minimum (5.13) be attained for $i = k$. Then the kth constraint will not be used to determine the optimal solution for $t > t_1$. In order to find a replacement for this constraint we substitute

$$x_0(t_1) - \sigma c_{k0}, \tag{5.15}$$

where $\sigma \geq 0$, into the constraints and try to determine the largest value of σ for which (5.15) is feasible. Because $a_k' c_{k0} = 1$ and $a_i' c_{k0} = 0$ for $i = 1, \ldots, n, i \neq k$, this gives the following relations:

$$a_i'(x_0(t_1) - \sigma c_{k0}) = a_i' x_0(t_1)$$

$$= b_i + t_1 p_i, \quad i = 1, \ldots, n, \ i \neq k,$$

$$a_k'(x_0(t_1) - \sigma c_{k0}) = a_k' x_0(t_1) - \sigma \leq b_k + t_1 p_k, \tag{5.16}$$

$$a_i'(x_0(t_1) - \sigma c_{k0}) = a_i' x_0(t_1) - \sigma a_i' c_{k0} \leq b_i + t_1 p_i, \quad i = n + 1, \ldots, m.$$

If $a_i' c_{k0} \geq 0$ for all $i = n + 1, \ldots, m$, then $x_0(t_1) - \sigma c_{k0}$ is feasible for all $\sigma \geq 0$. It will be shown in the proof of Theorem 5.1 that (5.7) then has no optimal solution for any $t > t_1$. If $a_i' c_{k0} < 0$ for at least one i, set

$$\sigma_0 = \min\left\{ \frac{a_i' x_0(t_0) - b_i - t_0 p_i}{a_i' c_{k0}} \ \Big|\ \text{all } i \text{ with } a_i' c_{k0} < 0 \right\}. \tag{5.17}$$

Let the minimum be attained for $i = l$ and define

$$x_1(t_1) = x_0(t_0) - \sigma_0 c_{k0}.$$

Then $x_1(t_1)$ is feasible and we infer from (5.16) and (5.17) that the constraints

$$a_i' x \leq b_i + t p_i, \quad i = 1, \ldots, n, \ i \neq k, \ i = l \tag{5.18}$$

are active at $x_1(t_1)$. Since it follows from Exercise 2.15 that $a_1, \ldots,$ $a_{k-1}, a_l, a_{k+1}, \ldots, a_n$ are linearly independent, $x_1(t_1)$ is an extreme point. Set

$$D_1' = \Big[a_1, \ldots, a_{k-1}, a_l, a_{k+1}, \ldots, a_n \Big], \quad u_1(t_1) = u_0(t_1), \tag{5.19}$$

$$b_1 = (b_1, \ldots, b_{k-1}, b_l, b_{k+1}, \ldots, b_n)', \tag{5.20}$$

and

$$p_1 = (p_1, \ldots, p_{k-1}, p_l, p_{k+1}, \ldots, p_n)'. \tag{5.21}$$

Because the kth component of $u_0(t_1)$ is zero we have $D_0' u_0(t_1) = D_1' u_1(t_1)$. Therefore, it follows from (5.8), (5.9), and (5.18) that

$$D_1' u_1(t_1) = -(c + t_1 q), \quad D_1 x_1(t_1) = b_1 + t_1 p_1,$$

$$u_1(t_1) \geq 0, \quad a_i' x_1(t_1) \leq b_i + t_1 p_i, \quad i = 1, \ldots, n, \ i \neq l, \ i = k.$$

By Theorem 3.1 this proves that $x_1(t_1)$ is an optimal extreme point for $t = t_1$. Furthermore, it will be shown in the proof of Theorem 5.1 that the kth component of $u_1(t)$ is positive for every $t > t_1$.

Case 2: $t_1 = t_1^* \leq \tilde{t}_1$

Let the minimum (5.14) be attained for $i = l$. Then the lth constraint is newly active at $x_0(t_1)$ and we have at least $n + 1$ active constraints. In order to find the constraint $a_i'x \leq b_i + tp_i$, $1 \leq i \leq n$, which will be replaced by this newly active constraint, we assume first that $a_l'c_{i0} > 0$ for at least one i, where $c_{10}, \ldots, c_{n0}$ denote again the columns of D_0^{-1}. Observe that

$$-(c + t_1 q) = (u_0(t_1))_1 a_1 + \cdots + (u_0(t_1))_n a_n \tag{5.22}$$

and

$$a_l = (a_l'c_1)a_1 + \cdots + (a_l'c_n)a_n.$$

Determine k such that

$$\omega_0 = \frac{(u_0(t_1))_k}{a_l'c_k} = \min\left\{\frac{(u_0(t_1))_i}{a_l'c_i} \mid \text{all } i \text{ with } a_l'c_i > 0\right\} \tag{5.23}$$

and set

$$(u_1(t_1))_k = \omega_0,$$

$$(u_1(t_1))_i = (u_0(t_1))_i - \omega_0 a_l'c_i, \quad i = 1, \ldots, n, \ i \neq k.$$

Then $u_1(t_1) \geq 0$. Adding

$$0 = \omega_0(a_l - (a_l'c_1)a_1 - \cdots - (a_l'c_n)a_n)$$

to (5.22) we obtain

$$-(c + t_1 q) = \sum_{\substack{i=1 \\ i \neq k}}^{n} (u_1(t_1))_i a_i + (u_1(t_1))_k a_l.$$

Therefore, setting

$$x_1(t_1) = x_0(t_1)$$

and defining D_1, b_1, and p_1 by (5.19), (5.20), and (5.21), respectively, we can write the optimality conditions for $x_1(t_1)$ in the form

$$D_1 u_1(t_1) = -(c + t_1 q), \quad D_1 x_1(t_1) = b_1 + t_1 p_1,$$

$$u_1(t_1) \geq 0, \quad a_i'x_1(t_1) \leq b_i + tp_i, \ i = n + 1, \ldots, m, \ i \neq l,$$

$$a_k'x_1(t_1) \leq b_k + t_1 p_k.$$

It will be shown in the proof of Theorem 5.1 that (5.7) has no feasible solution for any $t > t_1$ if $a_l'c_i \leq 0$ for $i = 1, \ldots, n$.

In both cases we can, therefore, replace $x_0(t_0)$ and D_0 with $x_1(t_1)$ and D_1, respectively, and repeat the described steps to obtain vectors h_{11}, h_{21}, and a critical value $t_2 \geq t_1$ such that

$$x_1(t) = h_{11} + th_{21}$$

is an optimal solution for all t with $t_1 \leq t \leq t_2$. It is important to observe that the two matrices D_0' and D_1' differ in exactly one column. Therefore, D_1^{-1} can easily be computed from D_0^{-1}.

In the proof of Theorem 5.1 it will be shown that:

1. If each critical value t_j determined in this way is strictly greater than its predecessor, the outlined method will terminate after a finite number of iterations with a critical value t_ν such that either $t_\nu = \bar{t}$ or there is no optimal solution for any $t > t_\nu$.
2. The following assumption is sufficient for the sequence of critical values to be strictly increasing.

Assumption.

1. $x_0(t_0)$ satisfies the strict complementary slackness condition.
2. For $j > 0$ each critical value t_j is determined by components of $u_{j-1}(t)$ which become zero for $t = t_j$ or by newly active constraints but not by both (i.e., $\tilde{t}_1 \neq t_1^*$).
3. For each critical value t_j the newly active constraint and the constraint to be replaced by it are both uniquely determined. This means that if $\tilde{t}_j < t_j^*$, the minima (5.13) and (5.17), and if $t_j^* < \tilde{t}_j$, the minima (5.14) and (5.23) are both attained for exactly one index.

A critical value t_j is said to be <u>nondegenerate</u> if the assumption above is satisfied.

We are now ready to describe the outlined algorithm in detail. In order to focus on the essential properties we assume that (5.7) has an optimal solution for $t = t_0 = \underline{t}$ and that every critical value is nondegenerate. The modifications of the algorithm required to handle degenerate critical values are described in the next section.

ALGORITHM 4

Model Problem:

$$\min\{(c + tq)'x \mid Ax \leq b + tp, \underline{t} \leq t \leq \bar{t}\} \qquad (5.24)$$

Initialization:

Set $t_0 = \underline{t}$. Use Algorithm 3 to determine an optimal extreme point $x_0(t_0)$ for the model problem with $t = t_0$. Let $D_0^{-1} = \left[c_{10}, \ldots, c_{n0} \right]$ and J_0

$= \{\alpha_{10}, \ldots, \alpha_{n0}\}$ be the corresponding matrix and index set, respectively. Set $j = 0$.

Step 1:   Computation of h_{1j} and h_{2j}.
Set

$$b_j = (b_{\alpha_{1j}}, \ldots, b_{\alpha_{nj}})', \quad p_j = (p_{\alpha_{1j}}, \ldots, p_{\alpha_{nj}})'.$$

Compute and print

$$h_{1j} = D_j^{-1}b_j, \quad h_{2j} = D_j^{-1}p_j.$$

Go to Step 2.

Step 2:   Computation of the Critical Value t_{j+1}.
Step 2.1:
Compute $g_{1j} = -(D_j^{-1})'c$, $g_{2j} = -(D_j^{-1})'q$, and set $u_j = g_{1j} + t_j g_{2j}$. If $g_{2j} \geq 0$, set $\tilde{t}_{j+1} = +\infty$ and go to Step 2.2. Otherwise, determine $\delta\tilde{t}_j$ and the smallest index k such that

$$\delta\tilde{t}_j = \frac{(u_j)_k}{-(g_{2j})_k} = \min\left\{ \frac{(u_j)_i}{-(g_{2j})_i} \mid \text{all } i \text{ with } (g_{2j})_i < 0 \right\}.$$

Set

$$\tilde{t}_{j+1} = t_j + \delta\tilde{t}_j.$$

Go to Step 2.2.

Step 2.2:
Compute $a_i'h_{2j} - p_i$ for all $i \notin J_j$. If $a_i'h_{2j} - p_i \leq 0$ for all $i \notin J_j$, set $t_{j+1}^* = +\infty$ and go to Step 2.3. Otherwise, determine δt_j^* and the smallest index l such that

$$\delta t_j^* = \frac{b_l + t_j p_l - a_l'x_j(t_j)}{a_l'h_{2j} - p_l}$$

$$= \min\left\{ \frac{b_i + t_j p_i - a_i'x_j(t_j)}{a_i'h_{2j} - p_i} \mid \text{all } i \notin J_j \text{ with } a_i'h_{2j} - p_i > 0 \right\}.$$

Set

$$t_{j+1}^* = t_j + \delta t_j^*.$$

Go to Step 2.3.

Step 2.3:
Set $t_{j+1} = \min\{\tilde{t}_{j+1}, t_{j+1}^*, \bar{t}\}$ and print t_{j+1}. If $t_{j+1} = \bar{t}$, stop. Otherwise, compute $x_j(t_{j+1}) = h_{1j} + t_{j+1}h_{2j}$. If $t_{j+1} = \tilde{t}_{j+1}$, go to Step 3; otherwise, go to Step 4.

Step 3:   Determination of a New Active Constraint.
Compute $a_i'c_{kj}$ for all $i \notin J_j$. If $a_i'c_{kj} \geq 0$ for all $i \notin J_j$, print the message
"problem is unbounded from below for $t > t_{j+1}$" and stop. Otherwise, deter-
mine σ_j and the smallest index l such that

$$\sigma_j = \frac{a_l'x_j(t_{j+1}) - b_l - t_{j+1}p_l}{a_l'c_{kj}}$$

$$= \min\left\{ \frac{a_i'x_j(t_{j+1}) - b_i - t_{j+1}p_i}{a_i'c_{kj}} \;\middle|\; \text{all } i \notin J_j \text{ with } a_i'c_{kj} < 0 \right\}.$$

Set

$$x_{j+1}(t_{j+1}) = x_j(t_{j+1}) - \sigma_j c_{kj}.$$

Go to Step 5.

Step 4:   Determination of a New Inactive Constraint.
Compute $a_i'c_{ij}$ for $i = 1, \ldots, n$. If $a_i'c_{ij} \leq 0$ for all i, print the message
"problem has no feasible solution for $t > t_{j+1}$" and stop. Otherwise, determine ω_j and the smallest index k such that

$$\omega_j = \frac{(g_{1j} + t_{j+1}g_{2j})_k}{a_l'c_{kj}}$$

$$= \min\left\{ \frac{(g_{1j} + t_{j+1}g_{2j})_i}{a_l'c_{ij}} \;\middle|\; \text{all } i \text{ with } a_l'c_{ij} > 0 \right\}.$$

Set

$$x_{j+1}(t_{j+1}) = x_j(t_{j+1})$$

and go to Step 5.

Step 5:   Update.
Set $D_{j+1}^{-1} = \Phi(D_j^{-1}, a_l, k)$ and $J_{j+1} = \{\alpha_{1,j+1}, \ldots, \alpha_{n,j+1}\}$, where

$$\alpha_{i,j+1} = \alpha_{ij}, \quad \text{for all } i \text{ with } i \neq k, \quad \text{and} \quad \alpha_{k,j+1} = l.$$

Replace j with $j + 1$ and go to Step 1.

      In order to illustrate Algorithm 4 we apply it to the problem of Example
5.3.

Example 5.6

$$\text{minimize:} \quad (-4 + t)x_1 + (-1 - \tfrac{1}{3}t)x_2$$

$$\text{subject to:} \qquad\qquad -4x_1 + 3x_2 \leq 9 - 5t, \qquad (1)$$

$$- x_1 + 4x_2 \leq 25, \qquad (2)$$

$$2x_1 + 5x_2 \leq 54 - 4t, \qquad (3)$$

$$2x_1 + x_2 \leq 30, \qquad (4)$$

$$- x_1 \qquad\quad \leq 0, \qquad (5)$$

$$- x_2 \leq -2t, \qquad (6)$$

for all t with $0 \leq t \leq 10$.

Initialization:

$$t_0 = 0, \quad x_0 = \begin{bmatrix} 15 \\ 0 \end{bmatrix}, \quad D_0^{-1} = \begin{bmatrix} 1/2 & 1/2 \\ 0 & -1 \end{bmatrix}, \quad J_0 = \{4,6\}, \quad j = 0.$$

Iteration 0

Step 1: $b_0 = (30,0)'$, $\quad p_0 = (0,-2)'$,

$$h_{10} = \begin{bmatrix} 1/2 & 1/2 \\ 0 & -1 \end{bmatrix} \begin{bmatrix} 30 \\ 0 \end{bmatrix} = \begin{bmatrix} 15 \\ 0 \end{bmatrix},$$

$$h_{20} = \begin{bmatrix} 1/2 & 1/2 \\ 0 & -1 \end{bmatrix} \begin{bmatrix} 0 \\ -2 \end{bmatrix} = \begin{bmatrix} -1 \\ 2 \end{bmatrix}.$$

Step 2.1: $g_{10} = - \begin{bmatrix} 1/2 & 0 \\ 1/2 & -1 \end{bmatrix} \begin{bmatrix} -4 \\ -1 \end{bmatrix} = \begin{bmatrix} 2 \\ 1 \end{bmatrix}$,

$$g_{20} = - \begin{bmatrix} 1/2 & 0 \\ 1/2 & -1 \end{bmatrix} \begin{bmatrix} 1 \\ -1/3 \end{bmatrix} = \begin{bmatrix} -1/2 \\ -5/6 \end{bmatrix},$$

$$u_0 = \begin{bmatrix} 2 \\ 1 \end{bmatrix} + 0 \cdot \begin{bmatrix} -1/2 \\ -5/6 \end{bmatrix} = \begin{bmatrix} 2 \\ 1 \end{bmatrix},$$

$$\delta \tilde{t}_0 = \min\left\{ \frac{2}{1/2}, \frac{1}{5/6} \right\} = \frac{6}{5}, \quad k = 2,$$

$$\tilde{t}_1 = 0 + \frac{6}{5} = \frac{6}{5}.$$

Step 2.2: $\delta t_0^* = \min\left\{\dfrac{69}{15}, \dfrac{40}{9}, \dfrac{24}{12}, -, \dfrac{15}{1}, -\right\} = 2, \quad l = 3,$

$t_1^* = 0 + 2 = 2.$

Step 2.3: $t_1 = \min\left\{\dfrac{6}{5}, 2, 10\right\} = \dfrac{6}{5},$

$x_0(t_1) = \begin{bmatrix} 15 \\ 0 \end{bmatrix} + \dfrac{6}{5}\begin{bmatrix} -1 \\ 2 \end{bmatrix} = \begin{bmatrix} 69/5 \\ 12/5 \end{bmatrix}.$ Transfer to Step 3.

Step 3: $\sigma_0 = \min\left\{\dfrac{-51}{-5}, \dfrac{-146/5}{-9/2}, \dfrac{-48/5}{-4}, -, \dfrac{-69/5}{-1/2}, -\right\} = \dfrac{12}{5},$

$l = 3,$

$x_1(t_1) = \begin{bmatrix} 69/5 \\ 12/5 \end{bmatrix} - \dfrac{12}{5}\begin{bmatrix} 1/2 \\ -1 \end{bmatrix} = \begin{bmatrix} 63/5 \\ 24/5 \end{bmatrix}.$

Step 5: $D_1^{-1} = \begin{bmatrix} 5/8 & -1/8 \\ -1/4 & 1/4 \end{bmatrix}, \quad J_1 = \{4, 3\}, \quad j = 1.$

Iteration 1

Step 1: $b_1 = (30, 54)', \quad p_1 = (0, -4)',$

$h_{11} = \begin{bmatrix} 5/8 & -1/8 \\ -1/4 & 1/4 \end{bmatrix}\begin{bmatrix} 30 \\ 54 \end{bmatrix} = \begin{bmatrix} 12 \\ 6 \end{bmatrix},$

$h_{21} = \begin{bmatrix} 5/8 & -1/8 \\ -1/4 & 1/4 \end{bmatrix}\begin{bmatrix} 0 \\ -4 \end{bmatrix} = \begin{bmatrix} 1/2 \\ -1 \end{bmatrix}.$

Step 2.1: $g_{11} = -\begin{bmatrix} 5/8 & -1/4 \\ -1/8 & 1/4 \end{bmatrix}\begin{bmatrix} -4 \\ -1 \end{bmatrix} = \begin{bmatrix} 9/4 \\ -1/4 \end{bmatrix},$

$g_{21} = -\begin{bmatrix} 5/8 & -1/4 \\ -1/8 & 1/4 \end{bmatrix}\begin{bmatrix} 1 \\ -1/3 \end{bmatrix} = \begin{bmatrix} -17/24 \\ 5/24 \end{bmatrix},$

$u_1 = \begin{bmatrix} 9/4 \\ -1/4 \end{bmatrix} + \dfrac{6}{5}\begin{bmatrix} -17/24 \\ 5/24 \end{bmatrix} = \begin{bmatrix} 7/5 \\ 0 \end{bmatrix},$

$\delta \tilde{t}_1 = \min\left\{\dfrac{7/5}{17/24}, -\right\} = \dfrac{168}{85}, \quad k = 1,$

$\tilde{t}_2 = \dfrac{6}{5} + \dfrac{168}{85} = \dfrac{54}{17}.$

Step 2.2: $\delta t_1^* = \min\left\{ -, -, -, -, -, \dfrac{12/5}{3} \right\} = \dfrac{4}{5}, \quad l = 6,$

$$t_2^* = \frac{6}{5} + \frac{4}{5} = 2.$$

Step 2.3: $t_2 = \min\left\{ \dfrac{54}{17}, 2, 10 \right\} = 2,$

$$x_1(t_2) = \begin{bmatrix} 12 \\ 6 \end{bmatrix} + 2 \begin{bmatrix} 1/2 \\ -1 \end{bmatrix} = \begin{bmatrix} 13 \\ 4 \end{bmatrix}. \quad \text{Transfer to Step 4.}$$

Step 4: $\omega_1 = \min\left\{ \dfrac{5/6}{1/4}, - \right\} = \dfrac{10}{3}, \quad k = 1,$

$$x_2(t_2) = x_1(t_2) = \begin{bmatrix} 13 \\ 4 \end{bmatrix}.$$

Step 5: $D_2^{-1} = \begin{bmatrix} 5/2 & 1/2 \\ -1 & 0 \end{bmatrix}, \quad J_2 = \{6, 3\}, \quad j = 2.$

Iteration 2

Step 1: $b_2 = (0, 54)', \quad p_2 = (-2, -4)',$

$$h_{12} = \begin{bmatrix} 5/2 & 1/2 \\ -1 & 0 \end{bmatrix} \begin{bmatrix} 0 \\ 54 \end{bmatrix} = \begin{bmatrix} 27 \\ 0 \end{bmatrix},$$

$$h_{22} = \begin{bmatrix} 5/2 & 1/2 \\ -1 & 0 \end{bmatrix} \begin{bmatrix} -2 \\ -4 \end{bmatrix} = \begin{bmatrix} -7 \\ 2 \end{bmatrix}.$$

Step 2.1: $g_{12} = -\begin{bmatrix} 5/2 & -1 \\ 1/2 & 0 \end{bmatrix} \begin{bmatrix} -4 \\ -1 \end{bmatrix} = \begin{bmatrix} 9 \\ 2 \end{bmatrix},$

$$g_{22} = -\begin{bmatrix} 5/2 & -1 \\ 1/2 & 0 \end{bmatrix} \begin{bmatrix} 1 \\ -1/3 \end{bmatrix} = \begin{bmatrix} -17/6 \\ -1/2 \end{bmatrix},$$

$$u_2 = \begin{bmatrix} 9 \\ 2 \end{bmatrix} + 2 \begin{bmatrix} -17/6 \\ -1/2 \end{bmatrix} = \begin{bmatrix} 10/3 \\ 1 \end{bmatrix},$$

$$\delta\tilde{t}_2 = \min\left\{ \frac{10/3}{17/6}, \frac{1}{1/2} \right\} = \frac{20}{17}, \quad k = 1,$$

$$\tilde{t}_3 = 2 + \frac{20}{17} = \frac{54}{17}.$$

Step 2.2: $\delta t_2^* = \min\{\dfrac{39}{39}, \dfrac{22}{15}, -, -, \dfrac{13}{7}, -\} = 1, \quad l = 1,$

$t_3^* = 2 + 1 = 3.$

Step 2.3: $t_3 = \min\left\{\dfrac{54}{17}, 3, 10\right\} = 3,$

$$x_2(t_3) = \begin{bmatrix} 27 \\ 0 \end{bmatrix} + 3\begin{bmatrix} -7 \\ 2 \end{bmatrix} = \begin{bmatrix} 6 \\ 6 \end{bmatrix}.$$ Transfer to Step 4.

Step 4: $a_i' c_{i2} \le 0, \ i = 1, 2$; stop, the problem has no feasible solution
for $t > 3$.

Summary of results:

$$x_0(t) = \begin{bmatrix} 15 \\ 0 \end{bmatrix} + t\begin{bmatrix} -1 \\ 2 \end{bmatrix}$$ is an optimal solution for $0 \le t \le \dfrac{6}{5}$,

$$x_1(t) = \begin{bmatrix} 12 \\ 6 \end{bmatrix} + t\begin{bmatrix} 1/2 \\ -1 \end{bmatrix}$$ is an optimal solution for $\dfrac{6}{5} \le t \le 2$,

$$x_2(t) = \begin{bmatrix} 27 \\ 0 \end{bmatrix} + t\begin{bmatrix} -7 \\ 2 \end{bmatrix}$$ is an optimal solution for $2 \le t \le 3$.

There is no feasible solution for $t > 3$.

A computer program which implements Algorithm 4 is given in Appendix B, Section B.3. The output from applying this program to the problem of Example 5.6 is shown in Figure B.57.

Theorem 5.1.

Assume that the model problem (5.24) has an optimal solution for $t = \underline{t}$ and that every critical value is nondegenerate. Then Algorithm 4 terminates after a finite number of iterations with critical values $\underline{t} = t_0 < t_1 < \cdots < t_\nu$ and vectors $h_{1j}, h_{2j}, \ j = 0, 1, \ldots, \nu - 1$, such that

(a) Either $t_\nu = \bar{t}$ or (5.24) has no optimal solution for any $t > t_\nu$.
(b) For $j = 0, 1, \ldots, \nu - 1$,

$$x_j(t) = h_{1j} + th_{2j}$$

is an optimal solution to (5.24) for all t with $t_j \le t \le t_{j+1}$.

Proof:

The initialization of Algorithm 4 and the assumption that t_0 is nondegenerate imply that

$$a_i' x_0(t_0) = b_i + t_0 p_i, \quad \text{for all } i \in J_0,$$

$$a_i' x_0(t_0) < b_i + t_0 p_i, \quad \text{for all } i \notin J_0, \tag{5.25}$$

and

$$u_0 = -(D_0^{-1})'(c + t_0 q) > 0. \tag{5.26}$$

By Step 1, $x_0(t_0) = h_{10} + t_0 h_{20}$. Thus it follows from (5.25) and Step 2.2 that $t_1^* > t_0$ and

$$x_0(t) = h_{10} + t h_{20} \tag{5.27}$$

is feasible for $t_0 \le t \le t_1^*$. Furthermore, we deduce from (5.26) and Step 2.1 that $u_0 = g_{10} + t_0 g_{20}$, $\tilde{t}_1 > t_0$ and $u_0(t) = g_{10} + t g_{20} \ge 0$ for $t_0 \le t \le \tilde{t}_1$. Therefore, $t_1 > t_0$ and, by Theorem 3.1, (5.27) is an optimal solution for $t_0 \le t \le t_1$.

Because t_1 is nondegenerate we have $\tilde{t}_1 \ne t_1^*$. Consider first the case that $\tilde{t}_1 \le t_1^*$ and suppose that $a_i' c_{k0} < 0$ for at least one $i \in J_0$. It follows from (5.25), the definition of $x_1(t_1)$ in Step 3, and the assumption that the minimum in Step 3 is unique that

$$a_i' x_1(t_1) = b_i + t_1 p_i, \quad \text{for all } i \in J_0, \ i \ne k,$$

$$a_k' x_1(t_1) < b_k + t_1 p_k,$$

$$a_l' x_1(t_1) = b_l + t_1 p_l,$$

$$a_i' x_1(t_1) < b_i + t_1 p_i, \quad \text{for all } i \notin J_0, \ i \ne l.$$

Since D_1' is obtained from D_0' by replacing the kth column with a_l, it follows that $x_1(t_1) = h_{11} + t_1 h_{21}$. Thus $t_2^* > t_1$ and

$$x_1(t) = h_{11} + t h_{21} \tag{5.28}$$

is feasible for $t_1 \le t \le t_2^*$. Let $u_0(t_1) = g_{10} + t_1 g_{20}$. Since the minimum in Step 2.1 is assumed to be unique we have

$$(u_0(t_1))_i > 0, \quad i = 1, \dots, n, \ i \ne k,$$

$$(u_0(t_1))_k = 0, \quad \text{and} \quad (g_{20})_k < 0.$$

Therefore,

$$-(c + t_1 q) = D_0' u_0(t_1) = D_1' u_1$$

with $u_1 = g_{11} + t_1 g_{21}$. Since $D_1^{-1} = \Phi(D_0^{-1}, a_l, k)$ we have $c_{k1} = (1/a_l' c_{k0}) c_{k0}$. Thus $(g_{20})_k < 0$ and $a_l' c_{k0} < 0$ imply that

$$(g_{21})_k = -c'_{k1}q = -\frac{c'_{k0}q}{a'_I c_{k0}} = \frac{(g_{20})_k}{a'_I c_{k0}} > 0.$$

This shows that $\tilde{t}_2 > t_1$ and $u_1(t) = g_{11} + t_1 g_{21} \geq 0$ for $t_1 \leq t \leq \tilde{t}_2$. Thus $t_2 > t_1$ and (5.28) is an optimal solution for $t_1 \leq t \leq t_2$.

Now let $t_1^* < \tilde{t}_1$ and suppose that $a'_I c_{i0} > 0$ for at least one i. By the definition of $x_0(t_1)$ and the uniqueness of the minimum in Step 2.2 we have

$$a'_i x_0(t_1) = b_i + t_1 p_i, \quad \text{for all } i \in J_0,$$

$$a'_I x_0(t_1) = b_I + t_1 p_I, \tag{5.29}$$

$$a'_i x_0(t_1) < b_i + t_1 p_i, \quad \text{for all } i \in J_0, \ i \neq l. \tag{5.30}$$

Let

$$u_0(t_1) = g_{10} + t_1 g_{20}, \quad w = (a'_I c_{10}, \ldots, a'_I c_{n0})',$$

and

$$\omega_0 = (u_0(t_1))_k / a'_I c_{k0}.$$

Then $\omega_0 > 0$ and

$$a_l = D'_0 w \quad \text{or} \quad 0 = \omega_0(a_l - D'_0 w).$$

Thus

$$-(c + t_1 q) = D'_0 u_0(t_1) = D'_0(u_0(t_1) - \omega_0 w) + \omega_0 a_l.$$

By the uniqueness of the minimum in Step 4 we have

$$(u_0(t_1) - \omega_0 w)_i > 0, \quad \text{for all } i \neq k,$$

and

$$(u_0(t_1) - \omega_0 w)_k = 0.$$

From this we conclude that

$$u_1 = -(D_1^{-1})'(c + t_1 q) = g_{11} + t_1 g_{21} > 0.$$

Thus $\tilde{t}_2 > t_1$ and $u_1(t) = g_{11} + t g_{21} \geq 0$ for all $t_1 \leq t \leq \tilde{t}_2$. Furthermore, it follows from (5.29) that

$$x_1(t_1) = h_{11} + t_1 h_{21} = x_0(t_1).$$

Let $x_0(t)$ be as defined by (5.27) and consider

$$x(t) = x_0(t) - \left[\frac{a'_I h_{20} - p_I}{a'_I c_{k0}}\right] c_{k0}(t - t_1).$$

Then

$$a'_i x(t) = a'_i x_0(t) = b_i + t p_i, \quad \text{for all } i \in J_0, \ i \neq k,$$

and

$$a_l'x(t) = b_l + tp_l.$$

By (5.29) this implies that

$$x(t) = x_1(t) = h_{11} + th_{21}.$$

Furthermore,

$$a_k'x(t) - b_k - tp_k = \frac{a_l'h_{20} - p_l}{a_l'c_{k0}}(t - t_1).$$

Since $a_l'h_{20} - p_l > 0$ and $a_l'c_{k0} > 0$ we have

$$a_k'x_1(t) < b_k + tp_k, \quad \text{for } t > t_1.$$

Therefore, it follows from (5.30) that $t_2^* > t_1$. Thus $t_2 > t_1$ and $x_1(t) = h_{11} + th_{21}$ is an optimal solution for $t_1 \leq t \leq t_2$.

Continuing this argument we see that Algorithm 4 generates critical values t_j and vectors h_{1j}, h_{2j} with the properties stated in Theorem 5.1. Since the definition of t_{j+1} implies that $h_{1j} + th_{2j}$ is not an optimal solution for any $t > t_{j+1}$, it follows that all matrices D_0', D_1', ... encountered in the algorithm are different. Thus the algorithm terminates after a finite number of iterations. It remains to show that (5.24) has no optimal solution for any $t > t_{j+1}$ if Algorithm 4 terminates at the jth iteration with Step 3 or Step 4. Suppose that termination occurs with Step 3. Let $t > t_{j+1}$ and let $\hat{x}$ be any feasible point. Then $a_i'c_{kj} \geq 0$ for all i implies that $\hat{x} - \sigma c_{kj}$ is feasible for all $\sigma \geq 0$. Moreover,

$$(c + tq)'(\hat{x} - \sigma c_{kj}) = (c + tq)'\hat{x} + \sigma(g_{1j} + tg_{2j})_k.$$

Since $(g_{1j} + tg_{2j})_k < 0$ for $t > t_{j+1}$ the problem is unbounded from below; that is, no optimal solution exists for $t > t_{j+1}$. If termination occurs with Step 4, then $a_l' = w'D_j'$ with $w \leq 0$ and $a_l'x_j(t) > b_l + tp_l$ for $t > t_{j+1}$.

Choose any x such that $a_i'x \leq b_i + tp_i$ for all $i \in J_j$. With $x = x_j(t) + (x - x_j(t))$ it follows that $a_i'(x - x_j(t)) \leq 0$ for all $i \in J_j$. Thus

$$a_l'x = a_l'x_j(t) + a_l'(x - x_j(t)) > b_l + w'D_j'(x - x_j(t)) > b_l$$

and (5.24) has no feasible solution for any $t > t_{j+1}$. ∎

Algorithm 4 can easily be modified to handle a parametric linear programming problem with both inequality and equality constraints. The only difference is that an equality constraint is never a candidate for a constraint that will become inactive. This means that the computation of the minimum in Step 2.1 is restricted to components of u_j that are associated with inequality constraints [i.e., $(u_j)_i$ with $\alpha_{ij} \leq m$], and the computation of the minimum in Step 4 is restricted to those c_{ij} of D_j^{-1} which correspond to gradients of inequality constraints (i.e., c_{ij} with $\alpha_{ij} \leq m$).

5.3 DEGENERATE CRITICAL VALUES

We consider again the model problem

$$\min\{(c + tq)'x \mid Ax \leq b + tp, \ \underline{t} \leq t \leq \bar{t}\}. \tag{5.31}$$

Let $x_j(t) = h_{1j} + th_{2j}$ be an optimal solution for $t_j \leq t \leq t_{j+1}$, where $h_{1j} = D_j^{-1}b_j$ and $h_{2j} = D_j^{-1}p_j$ are defined as in Step 1 of Algorithm 4. From Theorem 5.1, we know that either (5.31) has no optimal solution for any $t > t_{j+1}$ or there are $t_{j+2} > t_{j+1}$, D_{j+1}^{-1}, b_{j+1}, and p_{j+1} such that $x_{j+1}(t) = h_{1,j+1} + th_{2,j+1}$ is an optimal solution for $t_{j+1} \leq t \leq t_{j+2}$.

If t_{j+1} is a nondegenerate critical value, then D'_{j+1} and D'_j differ in exactly one column, as we have seen in the preceding section and Algorithm 4 is an efficient tool for determining D_{j+1}^{-1}, $h_{1,j+1}$, $h_{2,j+1}$, and t_{j+2}. If t_{j+1} is degenerate, then D'_{j+1} and D'_j may differ in several columns, as illustrated by Examples 5.4 and 5.5. The determination of D_{j+1}^{-1} is more difficult and requires the solution of an appropriate linear programming problem.

We will deal with degenerate critical values in two phases. Let D_j^{-1}, J_j, h_{1j}, h_{2j}, and t_{j+1} be defined as in Algorithm 4.

Phase 1:

Define the subset $I_1 \subset \{1, \ldots, m\}$ such that $i \in I_1$ if and only if

$$a_i'x_j(t_{j+1}) = b_i + t_{j+1}p_i \quad \text{and} \quad i \notin J_j,$$

where $x_j(t_{j+1}) = h_{1j} + t_{j+1}h_{2j}$. Consider the problem

$$\min\{(c + t_{j+1}q)'x \mid a_i'x \leq p_i \text{ for all } i \in J_j \cup I_1\}. \tag{5.32}$$

If (5.32) has no optimal solution it will be shown in Theorem 5.2 that (5.31) has no optimal solution for any $t > t_{j+1}$. If (5.32) has an optimal solution, let $\hat{x}_j$ be an optimal extreme point and $\hat{D}_j^{-1}$ and $\hat{J}_j$ the associated matrix and index set. Clearly, $\hat{D}_j^{-1} = D_j^{-1}$ if $I_1 = \emptyset$ (i.e., if there are no new active constraints).

Phase 2:

Set $\hat{u}_j = -(\hat{D}_j^{-1})'(c + t_{j+1}q)$ and define the subset $I_2 \subset \{1, \ldots, m\}$ such that $i \in I_2$ if and only if $i \in \hat{J}_j$ and $(\hat{u}_j)_k > 0$ where $\hat{\alpha}_{kj} = i$ (i.e., I_2 contains the indices of those constraints that are active at $\hat{x}_j$ and have strictly positive multipliers). Consider the problem

$$
\begin{aligned}
\text{minimize:} \quad & q'x \\
\text{subject to:} \quad & a_i'x \leq b_i + t_{j+1}p_i, \quad \text{for all } i \notin I_2, \\
& a_i'x = b_i + t_{j+1}p_i, \quad \text{for all } i \in I_2.
\end{aligned}
\right\} \tag{5.33}
$$

We can apply Algorithm 3 with the initial data $x_0 = \hat{x}_j$, $D_0^{-1} = \hat{D}_j^{-1}$, and $J_0 = \hat{J}_0$ to this problem with the following modification in the case that the minimum in Step 2 of Algorithm 3 is not unique. Let the set I_3 be defined such that $\nu \in I_3$ if and only if

$$\sigma_j = \frac{a_\nu x_j - b_\nu}{a_\nu' s_j}$$

$$= \min\left\{\frac{a_i' x_j - b_i}{a_i' s_j} \mid \text{all } i \notin J_j \text{ with } a_i' s_j < 0\right\} \tag{5.34}$$

and determine the smallest index l such that

$$\sigma_j^* = \frac{a_l' h_{2j} - p_l}{a_l' s_j} = \min\left\{\frac{a_i' h_{2j} - p_i}{a_i' s_j} \mid \text{all } i \in I_3\right\}. \tag{5.35}$$

If (5.33) has no optimal solution, it follows again from Theorem 5.2 that (5.31) has no optimal solution for any $t > t_{j+1}$. If (5.33) has an optimal solution, let x_{j+1} be an optimal extreme point and D_{j+1}^{-1} and J_{j+1} the associated matrix and index set. Set

$$b_{j+1} = (b_{\alpha_{1,j+1}}, \ldots, b_{\alpha_{n,j+1}})', \quad p_{j+1} = (p_{\alpha_{1,j+1}}, \ldots, p_{\alpha_{n,j+1}})',$$

and

$$h_{1,j+1} = D_{j+1}^{-1} b_{j+1}, \quad h_{2,j+1} = D_{j+1}^{-1} p_{j+1}. \tag{5.36}$$

We now have the following theorem.

Theorem 5.2.

  (a) If either of the problems (5.32) or (5.33) has no optimal solution, then (5.31) has no optimal solution for any $t > t_{j+1}$.
  (b) If both (5.32) and (5.33) have optimal solutions, then there is $t_{j+2} > t_{j+1}$ such that

$$x_{j+1}(t) = h_{1,j+1} + t h_{2,j+1}$$

  is an optimal solution to (5.31), where $h_{1,j+1}$ and $h_{2,j+1}$ are defined by (5.36).

Proof:
    (a) Suppose that there is $t^* > t_{j+1}$ and x^* such that x^* is an optimal solution for (5.31) for $t = t^*$. Then we have for all $i \in J_j \cup I_1$,

$$a_i' x^* = a_i' x_j(t_{j+1}) + a_i'(x^* - x_j(t_{j+1})) \le b_i + t^* p_i,$$

or

$$a_i'(x^* - x_j(t_{j+1})) \le b_i + t^* p_i - a_i' x_j(t_{j+1}) = (t^* - t_{j+1}) p_i.$$

Thus

$$\frac{a_i'(x^* - x_j(t_{j+1}))}{t^* - t_{j+1}} \leq p_i, \quad \text{for all } i \in J_j \cup I_1$$

which shows that

$$\frac{x^* - x_j(t_{j+1})}{t^* - t_{j+1}}$$

is a feasible solution for (5.32). Because $-(c + t_{j+1}q)'D_j^{-1} \geq 0$, $(c + t_{j+1}q)'x$ is bounded from below on the feasible region of (5.32). Thus it follows that (5.32) has an optimal solution.

Next assume that (5.33) has no optimal solution. Since $x_j(t_{j+1})$ is a feasible solution it follows again from Theorem 3.5 that $q'x$ is not bounded from below on the feasible region. There is a vector s such that $q's > 0$ and

$$a_i's \geq 0, \quad \text{for all } i \notin I_2 \quad \text{and} \quad a_i's = 0, \quad \text{for all } i \in I_2. \tag{5.37}$$

Since $(c + t_{j+1}q) \in \text{span}\{a_i \mid \text{for all } i \in I_2\}$, we have for any $t > t_{j+1}$ that

$$(c + tq)'s = (c + t_{j+1}q)'s + (t - t_{j+1})q's = (t - t_{j+1})q's > 0.$$

In conjunction with (5.37), this inequality shows that $(c + tq)'x$ is not bounded from below on the feasible region of (5.31).

(b) Observe that $\hat{\alpha}_{ij} = \alpha_{i,j+1}$ for all $i \in I_2$. Thus

$$-(c + t_{j+1}q) = \hat{D}_j'\hat{u}_j = \sum_{i \in I_2} (\hat{u}_j)_i a_{\alpha_{i,j+1}} \tag{5.38}$$

and with $v' = -q'D_{j+1}^{-1}$,

$$-q = D_{j+1}'v = \sum_{i \in I_2} v_i a_{\alpha_{i,j+1}} + \sum_{i \notin I_2} v_i a_{\alpha_{i,j+1}}.$$

Multiplying this equality by $(t - t_{j+1})$ and adding it to (5.38) we obtain

$$-(c + tq) = -(c + t_{j+1}q) - (t - t_{j+1})q$$

$$= \sum_{i \in I_2} \left[(\hat{u}_j)_i + (t - t_{j+1})v_i\right] a_{\alpha_{i,j+1}} + \sum_{i \notin I_2} (t - t_{j+1})v_i a_{\alpha_{i,j+1}}.$$

If $v_i \geq 0$ for all $i \notin I_2$, set $\tilde{t}_{j+2} = +\infty$, and otherwise set

$$\tilde{t}_{j+2} = t_{j+1} + \min\left\{\frac{(\hat{u}_j)_i}{-v_i} \mid \text{all } i \in I_2 \text{ with } v_i < 0\right\}.$$

Since $(\hat{u}_j)_i > 0$ for all $i \in I_2$ and $v_i \geq 0$ for all $i \notin I_2$, it follows that $\tilde{t}_{j+2} > t_{j+1}$ and

$$u_{j+1}(t) = -D_{j+1}^{-1}(c + tq) = -D_{j+1}^{-1}c - tD_{j+1}^{-1}q$$

$$= g_{1,j+1} + tg_{2,j+1} \geq 0 \qquad (5.39)$$

for all t with $t_{j+1} \leq t \leq \tilde{t}_{j+2}$. Next we set

$$\hat{b}_j = (b_{\hat{\alpha}_{1j}}, \ldots, b_{\hat{\alpha}_{nj}})', \quad \hat{p}_j = (p_{\hat{\alpha}_{1j}}, \ldots, p_{\hat{\alpha}_{nj}})',$$

and

$$\hat{x}_j(t) = \hat{D}_j^{-1}\hat{b}_j + t\hat{D}_j^{-1}\hat{p}_j$$

$$= \hat{h}_{1j} + t\hat{h}_{2j} \qquad (5.40)$$

$$= \hat{h}_{1j} + t_{j+1}\hat{h}_{2j} + (t - t_{j+1})\hat{h}_{2j}$$

$$= x_j(t_{j+1}) + (t - t_{j+1})\hat{x}_j.$$

Then we have

$$a_i'\hat{x}_j(t_{j+1}) < b_i + t_{j+1}p_i, \quad \text{for all } i \notin J_j \cup I_1$$

and for $t \geq t_{j+1}$,

$$a_i'\hat{x}_j(t) = a_i'x_j(t_{j+1}) + (t - t_{j+1})a_i'\hat{x}_j$$

$$\leq b_i + t_{j+1}p_i + (t - t_{j+1})p_i$$

$$= b_i + tp_i, \quad \text{for all } i \in J_j \cup I_1.$$

This shows that there is some $\hat{t}_{j+1} > t_{j+1}$ such that $\hat{x}_j(t)$ is a feasible solution for (5.31) for all t with $t_{j+1} \leq t \leq \hat{t}_{j+1}$.

Let σ_j, σ_j^* and l be determined by (5.34) and (5.35), respectively. Obtain $\hat{D}_{j+1}$ from $\hat{D}_j$ by replacing the kth column with a_l and let $\hat{J}_{j+1} = \{\hat{\alpha}_{1,j+1}, \ldots, \hat{\alpha}_{n,j+1}\}$ with $\hat{\alpha}_{k,j+1} = l$ and $\hat{\alpha}_{i,j+1} = \hat{\alpha}_{ij}$ for all $i \neq k$. Define

$$\hat{x}_{j+1}(t) = \hat{h}_{1,j+1} + t\hat{h}_{2,j+1}$$

by analogy to (5.40). We will show that then there is some $\hat{t}_{j+2} > t_{j+1}$ such that $\hat{x}_{j+1}(t)$ is a feasible solution for (5.31) for all t with $t_{j+1} \leq t \leq \hat{t}_{j+2}$. Repeating this argument until we obtain the optimal solution x_{j+1} for (5.33), we conclude from (5.39) and Theorem 3.1 that part (b) of the theorem holds.

Let

$$\tilde{x}_{j+1}(t) = \hat{x}_j(t) - \sigma_j\hat{c}_{kj} - (t - t_{j+1})\sigma_j^*\hat{c}_{kj},$$

where $\hat{c}_{kj}$ is the kth column of $\hat{D}_j^{-1}$. Then

$$a_l'\tilde{x}_{j+1}(t) = a_l'\hat{x}_j(t_{j+1}) + (t - t_{j+1})a_l'\hat{h}_{2j} - \sigma_j a_l'\hat{c}_{kj} - (t - t_{j+1})\sigma_j^* a_l'\hat{c}_{kj}$$

$$= b_l + t_{j+1}p_l + (t - t_{j+1})(a_l'\hat{h}_{2j} - a_l'\hat{c}_{kj}\sigma_j^*)$$

$$= b_l + t_{j+1}p_l + (t - t_{j+1})p_l$$

$$= b_l + tp_l, \quad \text{for all } t. \qquad (5.41)$$

Furthermore, for all $i = 1, \ldots, n$ with $i \neq k$ we have

$$a'_{\alpha_{ij}} \tilde{x}_{j+1}(t) = a'_{\alpha_{ij}} \hat{x}_j(t) = b_{\alpha_{ij}} + t p_{\alpha_{ij}}, \quad \text{for all } t. \tag{5.42}$$

It follows from (5.41) and (5.42) that $\hat{x}_{j+1}(t) = \tilde{x}_{j+1}(t)$ for all t. Define the set $\hat{I}$ such that $i \in \hat{I}$ if and only if $a'_i \tilde{x}_{j+1}(t_{j+1}) = b_i + \tilde{t}_{j+1} p_i$. Since by the definition of σ_j, $a'_i \tilde{x}_{j+1}(t_{j+1}) \leq b_i + \tilde{t}_{j+1} p_i$ for all i, it suffices to show that for every $i \in \hat{I}$, $a'_i \tilde{x}_{j+1}(t) \leq b_i + t p_i$ for $t \geq t_{j+1}$. Let $i \in \hat{I}$. Then

$$a'_i \tilde{x}_{j+1}(t) = b_i + \tilde{t}_{j+1} p_i - \sigma_j a'_i \hat{c}_{kj} + (t - t_{j+1})(a'_i \hat{h}_{2j} - \sigma^*_j a'_i \hat{c}_{kj}).$$

First assume that $\sigma_j > 0$. Then $a'_i \hat{c}_{kj} = 0$ and it follows that $i \in J_j \cup I_1$. Hence $a'_i \hat{h}_{2j} \leq p_i$ and $a'_i \tilde{x}_{j+1}(t) = b_i + \tilde{t}_{j+1} p_i - (t - t_{j+1}) a'_i \hat{h}_{2j} \leq b_i + t p_i$ for all $t \geq t_{j+1}$. Now let $\sigma_j = 0$. Then $\hat{I} = J_j \cup I_1$ and $l \in J_j \cup I_1$ which implies that $\sigma^*_j \geq 0$. Thus, if $a'_i \hat{c}_{kj} \geq 0$, then $a'_i \hat{h}_{2j} - \sigma^*_j a'_i \hat{c}_{kj} \leq a'_i \hat{h}_{2j} \leq p_i$ and, if $a'_i \hat{c}_{kj} < 0$, then it follows from the definition of σ^*_j that $a'_i \hat{h}_{2j} - \sigma^*_j a'_i \hat{c}_{kj} \leq p_i$. Therefore,

$$a'_i \tilde{x}_{j+1}(t) = b_i + \tilde{t}_{j+1} p_i + (t - t_{j+1})(a'_i \hat{h}_{2j} - \sigma^*_j a'_i \hat{c}_{kj})$$
$$\leq b_i + \tilde{t}_{j+1} p_i + (t - t_{j+1}) p_i$$
$$= b_i + t p_i, \quad \text{for all } t \geq t_{j+1}. \blacksquare$$

EXERCISES

5.1. Solve the following parametric linear programming problems for all $t \geq 0$. Graph the optimal objective function value as a function of t; that is, plot a graph of $(c + tq)'x(t)$ for $t \geq 0$, where $x(t)$ denotes the optimal solution as a function of t.

(a)   minimize: $-x_1 - (3 - t)x_2$

       subject to:
$$x_1 + 2x_2 \leq 8, \tag{1}$$
$$2x_1 - 3x_2 \leq 3, \tag{2}$$
$$-x_1 - x_2 \leq -1, \tag{3}$$
$$-x_1 + x_2 \leq 4. \tag{4}$$

(b)   minimize: $-tx_1 + (-10 + 4t)x_2$

       subject to:
$$-x_1 + x_2 \leq 2, \tag{1}$$
$$-x_1 - x_2 \leq -2, \tag{2}$$
$$x_1 + x_2 \leq 8, \tag{3}$$
$$x_1 - 2x_2 \leq 2, \tag{4}$$
$$-3x_1 + x_2 \leq -2. \tag{5}$$

5.2. Solve the following parametric linear programming problem for all $t \geq 0$. Sketch the path of the optimal solution and graph the optimal objective function value as a function of t (see Exercise 5.1).

$$
\begin{aligned}
\text{minimize:} \quad & -x_1 - 3x_2 \\
\text{subject to:} \quad & -x_1 + x_2 \leq 2, && (1) \\
& -x_1 - x_2 \leq -2, && (2) \\
& x_1 + x_2 \leq 8 - t, && (3) \\
& x_1 - 2x_2 \leq 2 - t/3, && (4) \\
& -3x_1 + x_2 \leq -2. && (5)
\end{aligned}
$$

5.3. Solve the following parametric linear programming problem for all $t \geq 0$.

$$
\begin{aligned}
\text{minimize:} \quad & (1 - t)x_1 + 2x_2 + 2x_3 \\
\text{subject to:} \quad & 3x_1 - x_2 + 2x_3 \leq 7, && (1) \\
& -4x_1 + 3x_2 + 8x_3 \leq 10, && (2) \\
& -2x_1 + 4x_2 \leq 12 - 2t, && (3) \\
& -x_1 \leq 0, && (4) \\
& -x_2 \leq 0, && (5) \\
& -x_3 \leq 0. && (6)
\end{aligned}
$$

5.4. Solve the parametric linear programming problems in Examples 5.1 through 5.3 for all $t \leq 0$.

5.5. Consider the problem

$$
\min\{c'x \mid a_i'x \leq b_i, \ i = 1, \ldots, m\}.
$$

The purpose of this exercise is to show how sensitivity analysis (see Section 3.4) can be performed using parametric linear programming.

(a) Suppose that an optimal solution has been obtained. Suppose also that just one b_i is now allowed to vary. Observe that the optimal solution, as it depends on b_i, may be obtained by solving the parametric problem (5.7) first with $q = 0$, $p = e_i$, ($e_i =$ the ith m-dimensional unit vector), $\underline{t} = 0$, and $\bar{t} = +\infty$, and then with $q = 0$, $p = -e_i$, $\underline{t} = 0$, and $\bar{t} = +\infty$. Use Algorithm 4 to perform this parametric analysis for constraints (1) and (2) for the problem of Example 3.6. Compare the results with those obtained in Example 3.6.

(b) Suppose that a c_i is now allowed to vary. Observe that the optimal solution, as it depends on c_i, may be obtained by solving the parametric problem (5.7) first with $q = e_i$ ($e_i =$ the ith n-dimensional unit vector), $p = 0$, $\underline{t} = 0$, and $\bar{t} = +\infty$, and then with $q = -e_i$, $p = 0$, $\underline{t} = 0$, and $t = +\infty$. Use Algorithm 4 to perform this parametric analysis for both components of the objective function for the problem of Example 3.6. Compare the results obtained with those obtained in Example 3.6.

5.6. Let $\underline{t} = t_0 < t_1 < \cdots < t_\nu = \bar{t}$ and $h_{1j}, h_{2j}, j = 0, \ldots, \nu - 1$ be such that

$$x_j(t) = h_{1j} + th_{2j}, \quad t_j \leq t \leq t_{j+1}, \, j = 0, \ldots, \nu - 1,$$

is an optimal solution for

$$\min \{ c'x \mid Ax \leq b + tp, \, \underline{t} \leq t \leq \bar{t} \}.$$

Show that

$$c'h_{2j} \leq c'h_{2,j+1}, \quad \text{for } j = 0, \ldots, \nu - 1.$$

5.7. Let $\underline{t} = t_0 \leq t_1 < \cdots < t_\nu = \bar{t}$ and $x_0, \ldots, x_{\nu-1}$ be such that, for $j = 0, \ldots, \nu - 1$, x_j is an optimal solution for

$$\min \{ (c + tq)'x \mid Ax \leq b \}$$

for all t with $t_j \leq t \leq t_{j+1}$. Show that

$$q'x_j \geq q'x_{j+1}, \quad \text{for } j = 0, \ldots, \nu - 1.$$

5.8. As an expert on linear programming you are consulted by someone who has solved the problem

$$\min \{ c'x \mid Ax \leq b + tp_j \}, \quad t \geq 0$$

for four different vectors p_1, p_2, p_3, p_4. The following table gives the optimal value of $c'x$ for each of the vectors p_j and $t = 0, 5, 10, 20$.

		p_j		
t	p_1	p_2	p_3	p_4
0	6	6	6	6
5	10	-2	14	6
10	18	3	4	9
20	30	12	1	16

Can all these results be correct? If not, which one, or ones, are not possible, and why?

5.9. Let x_0 be an optimal solution to the problem

$$\min \{ c'x \mid Ax \leq b \}.$$

Determine the maximal interval $[\gamma_1, \gamma_2]$ in which the kth component of c can vary without affecting the optimality of x_0.

5.10. Develop an algorithm based on the revised simplex method to solve the parametric linear programming problem

$$\min \{ (c + tq)'x \mid Ax = b + tp, \, x \geq 0 \}, \quad \underline{t} \leq t \leq \bar{t}.$$

5.11. Devise a parametric method to solve the problems

(a) $\min\{c'x + (d'x)(q'x) \mid Ax \le b\}$,

(b) $\min\left\{ \dfrac{c'x}{d'x} \mid Ax \le b \right\}$.

5.12. Suppose that the problem

$$\min\{(c + tq)'x \mid Ax \le b + tp\}$$

has an optimal solution for t_1 and $t_2 > t_1$. Show that it then has optimal solutions for every t with $t_1 \le t \le t_2$.

5.13. Suppose that the problem

$$\min\{(c + tq)'x \mid Ax \le b + tp\}$$

has no optimal solution for $t = \underline{t}$. If there are $t \ge \underline{t}$ for which optimal solutions exist, determine the smallest of these t.

5.14 Suppose that the problem

$$\min\{c'x \mid a_i'x \le b_i,\ i = 1,\ldots,m\}$$

has a unique optimal solution $\hat{x}$. Consider the augmented problem

$$\min\{c'x + t\lambda \mid a_i'x - \lambda \le b_i,\ \lambda \ge 0,\ i = 1,\ldots,m\}$$

for which $x_0 = 0$, $\lambda_0 = -\min\{0, b_1,\ldots,b_m\}$ is a feasible solution. Show that there is a value t_0 of the parameter t such that for all $t \ge t_0$ the augmented problem has the unique solution $(\hat{x}', 0)'$.

6

DUAL METHODS

In Chapter 3 we have seen that x_0 is an optimal solution for the model problem $\min\{c'x \mid Ax \le b\}$ if there is some $u \ge 0$ such that x_0 and u satisfy the primal feasibility, complementary slackness, and dual feasibility conditions. The algorithms described in Chapters 2 and 4 determine a sequence $\{x_j\}$ with decreasing values of the objective function for which the first two of these conditions are satisfied. Only the final optimal point also satisfies the dual feasibility condition. In this chapter we develop algorithms that generate sequences of points such that the complementary slackness and the dual feasibility conditions are satisfied. The corresponding sequence of objective function values is increasing. All but the final optimal point are primal infeasible.

6.1 A BASIC DUAL ALGORITHM

We consider again the model problem

$$\min\{c'x \mid Ax \le b\} \tag{6.1}$$

and its dual problem

$$\max\{-b'u \mid A'u = -c, \, u \ge 0\}. \tag{6.2}$$

The solution of this problem is particularly simple if the structure of the matrix A allows us to easily identify a nonsingular (n, n) submatrix of A,

$$D_0' = \begin{bmatrix} a_1, \ldots, a_n \end{bmatrix} \tag{6.3}$$

say, and the inverse matrix D_0^{-1} such that

$$x_0 = D_0^{-1} b_0, \quad \text{with } b_0 = (b_1, \ldots, b_n)' \tag{6.4}$$

is a feasible solution. Then x_0 is an extreme point and x_0, D_0^{-1}, and $J_0 = \{1, \ldots, n\}$ can be used as initial data for Algorithm 1.

As an example of such a situation consider

$$A = \begin{bmatrix} -I \\ A^* \end{bmatrix}, \quad b = \begin{bmatrix} 0 \\ b^* \end{bmatrix} \quad \text{with } b^* \geq 0,$$

where I is the (n, n) identity matrix. In this case, $D_0' = -I$ and $x_0 = 0$ is a feasible solution. If, on the other hand,

$$A = \begin{bmatrix} I \\ A^* \end{bmatrix} \quad \text{and} \quad c \leq 0,$$

then $D_0' = I$ has the property that $-c' D_0^{-1} \geq 0$. We observe that

$$u = (v', 0')' \quad \text{with } v' = -c' D_0^{-1}$$

is a feasible solution for the dual problem because $A'u = D_0'v = -c$ and $u \geq 0$. Thus the complementarity and the dual feasibility conditions are satisfied.

The algorithm to be developed in this section is a modification of Algorithm 1 designed to take advantage of the situation when an infeasible x_0 and a matrix D_0^{-1} are available with the property that $-c' D_0^{-1} \geq 0$ and the columns of D_0' are gradients of constraints active at x_0.

An important example of this arises in the following situation. Let x_0, D_0^{-1}, and J_0 be the final data obtained by applying Algorithm 1 to the model problem (6.1). Suppose that after solving the problem it is discovered that additional constraints have to be added which render x_0 infeasible. Then x_0 and D_0^{-1} have the properties described above. We illustrate this with the following example.

Example 6.1

$$\begin{array}{llll}
\text{minimize:} & -6x_1 - 8x_2 & & \\
\text{subject to:} & -3x_1 + 5x_2 \leq 25, & (1) \\
& x_1 + 3x_2 \leq 29, & (2) \\
& 3x_1 + 2x_2 \leq 45, & (3) \\
& -x_1 \qquad\quad \leq 0, & (4) \\
& \qquad -x_2 \leq 0, & (5) \\
& 5x_1 + 8x_2 \leq 82, & (6) \\
& x_1 \qquad\quad \leq 12. & (7)
\end{array}$$

This example is obtained from Example 2.5 by adding the constraints $x_1 \leq 12$ and $5x_1 + 8x_2 \leq 82$. The feasible regions for both problems are shown in Figure 6.1. The new problem has the extreme points P_0, P_1, $\tilde{P}_2$, $\tilde{P}_3$, and $\tilde{P}_4$. From Example 2.5, we know that $(11, 6)'$ is an optimal solution for the original problem. Thus

$$x_0 = \begin{bmatrix} 11 \\ 6 \end{bmatrix} \quad \text{and} \quad D_0^{-1} = \begin{bmatrix} -2/7 & 3/7 \\ 3/7 & -1/7 \end{bmatrix} \quad \text{with } D_0' = \begin{bmatrix} 1 & 3 \\ 3 & 2 \end{bmatrix}$$

have the required properties. The new constraint (6) renders x_0 infeasible. To recover a feasible point we replace one of the constraints active at x_0 with this constraint. Replacing (3) by (6) gives the extreme point $\hat{P}_2$ as the solution of

$$x_1 + 3x_2 = 29 \quad \text{and} \quad 5x_1 + 8x_2 = 82.$$

Since

$$u_2 \begin{bmatrix} 1 \\ 3 \end{bmatrix} + u_6 \begin{bmatrix} 5 \\ 8 \end{bmatrix} = -\begin{bmatrix} -6 \\ -8 \end{bmatrix}$$

implies that $u_2 = -8/7$ and $u_6 = 10/7$, the dual feasibility condition is not satisfied. Replacing (2) by (6) we obtain $\hat{P}_3$ from the equations

$$3x_1 + 2x_2 = 45 \quad \text{and} \quad 5x_1 + 8x_2 = 82.$$

Furthermore, we have

$$u_3 \begin{bmatrix} 3 \\ 2 \end{bmatrix} + u_6 \begin{bmatrix} 5 \\ 8 \end{bmatrix} = -\begin{bmatrix} -6 \\ -8 \end{bmatrix}$$

with $u_3 = 4/7$ and $u_6 = 6/7$. Because $\hat{P}_3$ is not feasible we replace (3) by (7). The equations

$$x_1 = 12 \quad \text{and} \quad 5x_1 + 8x_2 = 82$$

determine the feasible extreme point $\tilde{P}_3$. The dual feasibility condition is satisfied because

$$u_7 \begin{bmatrix} 1 \\ 0 \end{bmatrix} + u_6 \begin{bmatrix} 5 \\ 8 \end{bmatrix} = -\begin{bmatrix} -6 \\ -8 \end{bmatrix}$$

has solution $u_6 = 1$ and $u_7 = 1$. Thus $\tilde{P}_3$ is an optimal extreme point for the problem of Example 6.1.

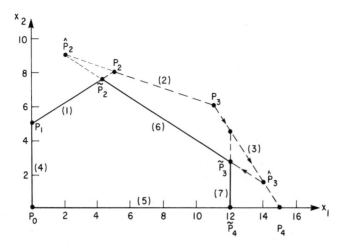

Figure 6.1 Graphical illustration of Example 6.1.

Geometrically, we can interpret the foregoing procedure as follows. We start with an infeasible extreme point P_3 and choose a constraint l, say, which is violated at P_3. Then we select an edge emanating from P_3 and move along this edge to an adjacent extreme point $\hat{P}_3$ with the following properties:

1. All constraints active at P_3 are satisfied at $\hat{P}_3$.
2. The lth constraint is active at $\hat{P}_3$.
3. The dual feasibility condition is satisfied at $\hat{P}_3$.

The procedure terminates if an extreme point is obtained which is either feasible or has no adjacent extreme point with the properties above. In the first case we have an optimal solution, in the second case the problem has no feasible solution. This is illustrated in the following example.

Example 6.2

$$
\begin{aligned}
\text{minimize:} \quad & x_1 + x_2 \\
\text{subject to:} \quad & x_1 + 2x_2 \le 12, && (1) \\
& -x_1 \le -8, && (2) \\
& -x_2 \le -5. && (3)
\end{aligned}
$$

Figure 6.2 shows that the problem of Example 6.2 has no feasible solution. At the extreme point P_1 the dual feasibility condition is satisfied. However, neither of the adjacent extreme points P_2 nor P_3 satisfy both constraints (2) and (3) which are active at P_1.

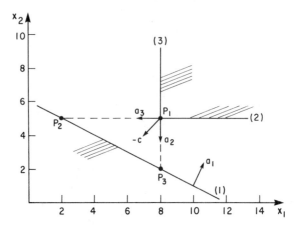

Figure 6.2 Graphical illustration of Example 6.2.

As in Algorithm 1, the dual algorithm uses an extreme point x_0 and the associated matrix D_0^{-1} as initial data and determines a sequence of points $x_0, x_1, \ldots, x_p$ and matrices $D_0^{-1}, D_1^{-1}, \ldots, D_p^{-1}$, say. For each j, D_j' and D_{j+1}' differ by exactly one column and

$$c'x_j \leq c'x_{j+1}, \quad -c'D_j^{-1} \geq 0.$$

Furthermore, x_p is either an optimal solution for (6.1) or we obtain the information that no optimal solution exists.

To describe this basic idea in more detail we assume that the matrix D_0' as defined by (6.3) has the property $-c'D_0^{-1} \geq 0$. Let x_0 be as in (6.4). The first step is to determine whether x_0 is feasible. For this purpose let l be the smallest index such that

$$a_l'x_0 - b_l = \max\{a_i'x_0 - b_i, \quad i = 1, \ldots, m\}.$$

If $a_l'x_0 - b_l \leq 0$, then x_0 is a feasible, and hence optimal, solution for (6.1). If $a_l'x_0 - b_l > 0$, we try to find a new extreme point x_1 for which the lth constraint is active and the dual feasibility condition is satisfied.

Thus in contrast to Algorithm 1 we first determine the constraint that will be newly active at the new extreme point x_1. To select the constraint that becomes inactive let

$$D_0^{-1} = \left[c_{10}, \ldots, c_{n0} \right]$$

and consider the effect of setting the search direction s_0 equal to c_{k0}. Then $x_1 = x_0 - \sigma_0 s_0$ for some $\sigma_0 > 0$. Because $a_l'x_0 - b_l > 0$ it follows that

$$a_l'x_1 = a_l'x_0 - \sigma_0 a_l's_0 = b_l, \quad \text{for some } \sigma_0 > 0$$

if and only if $a_i' s_0 > 0$. Thus only columns c_{i0} with $a_i' c_{i0} > 0$ are possible candidates for the search direction s_0. If $a_i' c_{i0} \leq 0$ for all $i = 1, \ldots, n$, then (6.1) has no feasible solution (see the proof of Theorem 6.1).

Suppose that $a_l' c_{k0} > 0$. Let

$$v' = -c' D_0^{-1} \quad \text{and} \quad w' = a_l' D_0^{-1}.$$

Then $v \geq 0$ and

$$-c = D_0' v = v_1 a_1 + \cdots + v_k a_k + \cdots + v_n a_n, \tag{6.5}$$

$$a_l = D_0' w = w_1 a_1 + \cdots + w_k a_k + \cdots + w_n a_n.$$

Adding the expression

$$0 = \omega(a_l - w_1 a_1 - \cdots - w_n a_n)$$

to (6.5) we obtain

$$-c = \omega a_l + (v_1 - \omega w_1) a_1 + \cdots + (v_k - \omega w_k) a_k + \cdots + (v_n - \omega w_n) a_n.$$

Because the kth constraint is not active at x_1, the dual feasibility condition is satisfied if

$$\omega \geq 0, \quad v_k - \omega w_k = 0, \quad v_i - \omega w_i \geq 0, \quad i = 1, \ldots, n.$$

We have $w_k = a_l' c_{k0} > 0$ and $u_i \geq 0$ for all $i = 1, \ldots, n$. Thus these conditions are satisfied if

$$\omega = \frac{v_k}{w_k} \leq \frac{v_i}{w_i}, \quad \text{for all } i \text{ with } w_i > 0.$$

Therefore, the index of the constraint which becomes inactive is equal to the smallest index k such that

$$\frac{v_k}{w_k} = \min \left\{ \frac{v_i}{w_i} \mid \text{all } i \text{ with } w_i > 0 \right\}.$$

Because the lth constraint is newly active at x_1, the step size is determined by the equation

$$a_l'(x_0 - \sigma s_0) = b_l;$$

i.e.,

$$\sigma_0 = \frac{a_l' x_0 - b_l}{a_l' s_0} > 0.$$

Finally, $-c' s_0 = v' D_0 c_{k0} = v' e_k = v_k \geq 0$ implies that

$$c' x_1 = c'(x_0 - \sigma_0 s_0) = c' x_0 - \sigma_0 c' s_0 \geq c' x_0.$$

We now formulate the procedure above as

ALGORITHM 5

Model Problem: $\min \{ c'x \mid a_i'x \leq b_i , \ i = 1, \ldots, m \}$

Initialization:
Start with any point x_0, $D_0' = \left[a_{\alpha_{10}}, \ldots, a_{\alpha_{n0}} \right]$, $J_0 = \{ \alpha_{10}, \ldots, \alpha_{n0} \}$, and D_0^{-1}, where the columns of D_0' are gradients of constraints which are active at x_0, and, $-c'D_0^{-1} \geq 0$. Compute $c'x_0$ and set $j = 0$.

Step 1:   Computation of Search Direction s_j.
Let $D_j^{-1} = \left[c_{1j}, \ldots, c_{nj} \right]$ and $J_j = \{ \alpha_{1j}, \ldots, \alpha_{nj} \}$. Determine the smallest index l such that

$$a_l'x_j - b_l = \max \{ a_i'x_j - b_i \mid i = 1, \ldots, m \text{ with } i \notin J_j \}.$$

If $a_l'x_j - b_l \leq 0$, stop with optimal solution x_j. Otherwise, compute

$$w_i = a_l'c_{ij}, \quad \text{for } i = 1, \ldots, n.$$

If $w_i \leq 0$ for all i, print the message "problem has no feasible solution" and stop. Otherwise, compute the multipliers

$$v_i = -c'c_{ij}, \quad \text{for } i = 1, \ldots, n,$$

and determine the smallest index k such that

$$\frac{v_k}{w_k} = \min \left\{ \frac{v_i}{w_i} \mid \text{all } i \text{ with } w_i > 0 \right\}.$$

Set $s_j = c_{kj}$ and go to Step 2.

Step 2:   Computation of Maximum Feasible Step Size σ_j.
Set

$$\sigma_j = \frac{a_l'x_j - b_l}{a_l's_j}$$

and go to Step 3.

Step 3:   Update.
Set $x_{j+1} = x_j - \sigma_j s_j$ and compute $c'x_{j+1}$. Set $D_{j+1}^{-1} = \Phi(D_j^{-1}, a_l, k)$ and $J_{j+1} = \{ \alpha_{1,j+1}, \ldots, \alpha_{n,j+1} \}$, where

$$\alpha_{i,j+1} = \alpha_{ij}, \quad \text{for all } i \text{ with } i \neq k, \quad \text{and} \quad \alpha_{k,j+1} = l.$$

Replace j with $j + 1$ and go to Step 1.

We illustrate Algorithm 5 by applying it to the problem of

Example 6.3

$$
\begin{aligned}
\text{minimize:} \quad & -16x_1 - 11x_2 \\
\text{subject to:} \quad -2x_1 - \quad & x_2 \le -4, & (1) \\
-x_1 \quad & \le \quad 0, & (2) \\
& x_2 \le \quad 8, & (3) \\
x_1 + 2x_2 & \le 17, & (4) \\
5x_1 + 3x_2 & \le 43, & (5) \\
x_1 \quad & \le \quad 8, & (6) \\
-x_2 & \le \quad 0. & (7)
\end{aligned}
$$

The feasible region for this problem is shown in Figure 6.3. It has extreme points $P_0, \ldots, P_6$.

As the initial point we choose Q_0 with active constraints $x_1 \le 8$ and $x_2 \le 8$.

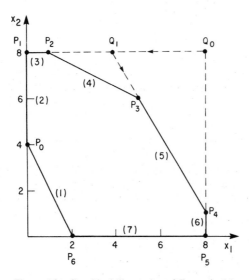

Figure 6.3   Graphical illustration of Example 6.3.

Initialization:

$$
x_0 = \begin{bmatrix} 8 \\ 8 \end{bmatrix}, \quad D_0' = D_0^{-1} = \begin{bmatrix} 1 & 0 \\ 0 & 1 \end{bmatrix}, \quad J_0 = \{6, 3\},
$$

$$
-c'D_0^{-1} = (16, 11), \quad c'x_0 = -216, \quad j = 0.
$$

Iteration 0

Step 1: $a_5'x_0 - b_5 = \max\{-20, -8, -, 7, 21, -, -8\} = 21,\quad l = 5,$

$w_1 = 5,\quad w_2 = 3,\quad v_1 = 16,\quad v_2 = 11,$

$$\frac{v_1}{w_1} = \min\left\{\frac{16}{5}, \frac{11}{3}\right\} = \frac{16}{5},\quad k = 1,$$

$$s_0 = \begin{bmatrix} 1 \\ 0 \end{bmatrix}.$$

Step 2: $\sigma_0 = \dfrac{21}{5}.$

Step 3: $x_1 = \begin{bmatrix} 8 \\ 8 \end{bmatrix} - \dfrac{21}{5}\begin{bmatrix} 1 \\ 0 \end{bmatrix} = \begin{bmatrix} 19/5 \\ 8 \end{bmatrix},\quad c'x_1 = \dfrac{-744}{5},$

$$D_1^{-1} = \begin{bmatrix} 1/5 & -3/5 \\ 0 & 1 \end{bmatrix},\quad J_1 = \{5, 3\},\quad j = 1.$$

Iteration 1

Step 1: $a_4'x_1 - b_4 = \max\left\{\dfrac{-58}{5}, \dfrac{-19}{5}, -, \dfrac{14}{5}, -, \dfrac{-21}{5}, -8\right\} = \dfrac{14}{5},$

$l = 4,$

$$w_1 = \frac{1}{5},\quad w_2 = \frac{7}{5},\quad v_1 = \frac{16}{5},\quad v_2 = \frac{7}{5},$$

$$\frac{v_2}{w_2} = \min\left\{\frac{16/5}{1/5}, \frac{7/5}{7/5}\right\} = 1,\quad k = 2,$$

$$s_1 = \begin{bmatrix} -3/5 \\ 1 \end{bmatrix}.$$

Step 2: $\sigma_1 = \dfrac{14/5}{7/5} = 2.$

Step 3: $x_2 = \begin{bmatrix} 19/5 \\ 8 \end{bmatrix} - 2\begin{bmatrix} -3/5 \\ 1 \end{bmatrix} = \begin{bmatrix} 5 \\ 6 \end{bmatrix},\quad c'x_2 = -146,$

$$D_2^{-1} = \begin{bmatrix} 2/7 & -3/7 \\ -1/7 & 5/7 \end{bmatrix},\quad J_2 = \{5, 4\},\quad j = 2.$$

Iteration 2

Step 1: $a_3'x_2 - b_3 = \max\{-12, -5, -2, -, -, -3, -6\} = -2$,

$l = 3$.

$a_3'x_2 - b_3 \leq 0$; stop with optimal solution x_2.

A computer program which implements Algorithm 5 is given in Appendix B, Section B.4.1. The output from applying this program to the problem of Example 6.3 is shown in Figure B.64.

The main properties of Algorithm 5 are established in the following theorem.

Theorem 6.1.

Let $x_0, x_1, \ldots, x_j, \ldots$ be the sequence of iterates obtained by applying Algorithm 5 to the model problem

$$\min\{c'x \mid Ax \leq b\}.$$

If $c'x_j < c'x_{j+1}$ for every j, Algorithm 5 terminates after a finite number of steps with either an optimal solution or the information that the problem has no feasible solution.

Proof:

By construction, at every iteration j,

$$a_i'x_j = b_i, \quad \text{for all } i \in J_j$$

and

$$v_j' = -c'D_j^{-1} \geq 0.$$

Therefore, setting $u = (u_1, \ldots, u_m)'$ with $u_i = 0$ for all $i \notin J_j$ and $u_i = v_k$ if $i = \alpha_{kj}$, we have $u \geq 0$,

$$-c = D_j'v_j = A'u, \quad \text{and} \quad u'(Ax_j - b) = 0;$$

that is, the dual feasibility condition and the complementary slackness condition are satisfied for every j. If Algorithm 5 terminates with "optimal solution x_j," then x_j is feasible and, by Theorem 3.1, indeed optimal.

Suppose that Algorithm 5 terminates with "problem has no feasible solution." Then

$$a_l'x_j > b_l \quad \text{and} \quad a_l' = w'D_j \quad \text{with } w \leq 0. \tag{6.6}$$

Suppose to the contrary that there exists an x with $a_i'x \leq b_i$ for all $i \in J_j$. With $x = x_j + (x - x_j)$ we have

$$b_i \geq a_i'x = a_i'x_j + a_i'(x - x_j) = b_i + a_i'(x - x_j), \quad \text{for all } i \in J_j.$$

Thus $a_i'(x - x_j) \le 0$ for all $i \in J_j$ and it follows from (6.6) that

$$a_l'x = a_l'x_j + a_l'(x - x_j) > b_l + w'D_j(x - x_j) > b_l.$$

The assumption that there exists an x with $Ax \le b$ leads to an contradiction and is therefore false.

If $c'x_j < c'x_{j+1}$ for every j, all iterates x_j are different. Since each x_j is uniquely determined by the equations

$$a_i'x_j = b_i, \quad \text{for } i \in J_j,$$

the possible number of iterations is bounded by the number of different subsets of $\{a_1, \ldots, a_m\}$ containing n elements each. Hence Algorithm 5 terminates after a finite number of steps. ■

The assumption $c'x_j < c'x_{j+1}$ is critical for the proof of the finite termination property. If $c'x_j = c'x_{j+1}$, the possibility of cycling exists. As usual, cycling can be prevented by using an appropriate version of Bland's rules.

First we observe that $c'x_j = c'x_{j+1}$ if and only if v_k determined in Step 1 of Algorithm 5 is equal to zero. Indeed, since $\sigma_j > 0$ we have

$$c'x_{j+1} = c'x_j - \sigma_j c's_j = c'x_j + \sigma_j v'D_j'c_{kj} = c'x_j + \sigma_j v_k.$$

Thus if $v_k = 0$, we modify the selection of l in Step 1 by choosing l to be the smallest index such that $a_l'x_j - b_l > 0$. Then Algorithm 5 is guaranteed to terminate in a finite number of steps. To see this we apply the revised simplex method to the problem

$$\min\{b'u \mid A'u = -c, u \ge 0\} \tag{6.7}$$

which is equivalent to the dual problem (6.2).

We claim that applying Algorithm 5 to (6.1) is equivalent to applying the revised simplex method to (6.7). At each iteration the basis matrix B of the revised simplex method corresponds to D_j'. Suppose that for some j we have $B = D_j'$. The reader will be asked in Exercise 6.4 to show that the following statements are true.

1. The index l of the variable to become basic, determined in Step 1 of the revised simplex method, is the same as the index of the constraint to become active, selected in Step 1 of Algorithm 5. (A variable of the dual problem which becomes positive corresponds to a primal constraint which becomes active.)

2. The index of the variable to become nonbasic in Step 2 of the revised simplex method is the same as the index of the constraint which becomes inactive in Algorithm 5. (A zero dual variable corresponds to an inactive primal constraint.)

3. The revised simplex method invokes Bland's rules if and only if Algorithm 5 does so. Again both methods determine the same index as in case 1.

If we set the initial basis matrix B equal to D_0', it follows, therefore, that each iteration of Algorithm 5 corresponds to an iteration of the revised simplex method applied to (6.7). Thus Theorem 4.1 implies that Algorithm 5 with Bland's rules terminates after a finite number of steps.

6.2 THE DUAL SIMPLEX METHOD

In this section we develop a dual method for the revised simplex method analogous to that of Section 6.1. Consider the model problem

$$\min \{ c'x \mid Ax = b, x \geq 0 \}. \tag{6.8}$$

Let B be a known basis matrix for (6.8) for which dual feasibility is satisfied but primal feasibility is not. We develop a variation of the revised simplex method which solves (6.8) using B as initial data.

The dual of (6.8) is

$$\max \{ -b'u \mid -A'u \leq c \}.$$

Letting A be of dimension (m, n) and denoting the ith column of A by A_i, the equivalent minimization problem is

$$\min \{ b'u \mid -A_i'u \leq c_i, i = 1, \ldots, m \}. \tag{6.9}$$

This is precisely the model problem required by Algorithm 1. Suppose that we solve (6.9) using Algorithm 1. Then we also obtain an optimal solution for the dual of (6.9). Since the dual of (6.9) is (6.8), we will have also found an optimal solution for (6.8).

Let u_0 be the dual feasible solution associated with B. Using the notation of the revised simplex method, we have

$$u_0 = -(B^{-1})'c_B.$$

Letting $I_B = \{ \beta_1, \ldots, \beta_m \}$ denote the basic index set, we can write

$$B = \left[A_{\beta_1}, \ldots, A_{\beta_m} \right].$$

With x_0, D_0^{-1}, and J_0 replaced with u_0, $-(B^{-1})'$, and I_B, respectively, we see that the latter three quantities are appropriate initial data for the application of Algorithm 1 to (6.9). We proceed to derive the dual simplex method by formulating each step of Algorithm 1 in terms of the quantities used in the revised simplex method.

The vector of multipliers for the active constraints of (6.9) is computed in Step 1 of Algorithm 1. Since $(D_0^{-1})' = -B^{-1}$, this vector is

$$-(-B^{-1})b = B^{-1}b = x_B.$$

The smallest multiplier is

$$(x_B)_k = \min \{ (x_B)_i \mid i = 1, \ldots, m \}$$

and corresponds to the β_kth constraint of (6.9). If this multiplier is nonnegative, u_0 is optimal for (6.9) and x_B together with $x_{NB} = 0$ is optimal for (6.8). Otherwise, Algorithm 1 proceeds by using the kth column of $-(B^{-1})'$ as a search direction. Let s_0 denote this m-vector. In Step 2 of Algorithm 1, if $A_i's_0 \geq 0$ for $i = 1, \ldots, n$, then (6.9) is unbounded from below and consequently the primal problem (6.8) has no feasible solution. Otherwise, Step 2 proceeds with the computation of the maximum feasible step size and its associated index l:

$$\sigma_0 = \frac{-A_l'u_0 - c_l}{-A_l's_0}$$

$$= \min\left\{\frac{-A_i'u_0 - c_i}{-A_i's_0} \mid \text{all } i = 1, \ldots, n \text{ with } -A_i's_0 < 0\right\}.$$

Simplifying gives

$$\sigma_0 = \frac{A_l'u_0 + c_l}{A_l's_0}$$

$$= \min\left\{\frac{A_i'u_0 + c_i}{A_i's_0} \mid \text{all } i = 1, \ldots, n \text{ with } A_i's_0 > 0\right\}.$$

Step 3 replaces B^{-1} with

$$\left[\Phi((B^{-1})', A_l, k)\right]'$$

and updates I_B by replacing β_k with l. Step 3 also obtains the next iterate for (6.9) as

$$u_1 = u_0 - \sigma_0 s_0.$$

In terms of the new basis inverse, u_1 is the solution of the linear equations $-B'u_1 = c_B$ and may be computed equivalently as

$$u_1 = -(B^{-1})'c_B.$$

In the revised simplex method, k denotes the nonbasic variable which is to become basic, and β_l refers to the basic variable which is to become nonbasic. Because we based the analysis above on Algorithm 1, the rôles of k and l have been interchanged. To emphasize the similar features of the two methods, we reverse the interchange in the formal statement of the algorithm. We also replace s_0 with s_B. Of course, this does not affect the computations.

Summarizing these steps gives the dual simplex method.

DUAL SIMPLEX METHOD

Model Problem: $\min\{c'x \mid Ax = b, x \geq 0\}$

Initialization:
Start with basic index set $I_B = \{\beta_1, \ldots, \beta_m\}$, basis matrix

$$B = \left[A_{\beta_1}, \ldots, A_{\beta_m}\right],$$

B^{-1}, dual feasible solution $u = -(B^{-1})'c_B$, and basic solution $x_B = B^{-1}b$.
Compute $c_B'x_B$.

Step 1:   Computation of Search Direction s_B.
Compute the smallest index l such that

$$(x_B)_l = \min\{(x_B)_i \mid i = 1, \ldots, m\}.$$

If $(x_B)_l \geq 0$, stop with optimal solution $x_B = B^{-1}b$ and $x_{NB} = 0$.
If $(x_B)_l < 0$, set s_B equal to the lth column of $-(B^{-1})'$ and go to Step 2.

Step 2:   Computation of Maximum Feasible Step Size.
If $A_i's_B \leq 0$ for $i = 1, \ldots, n$ and $i \notin I_B$, the problem has no feasible solution: stop.
Otherwise, compute the smallest index k such that

$$\frac{A_k'u + c_k}{A_k's_B} = \min\left\{\frac{A_i'u + c_i}{A_i's_B} \mid \text{all } i \notin I_B \text{ with } A_i's_B > 0\right\}$$

and go to Step 3.

Step 3:   Update.
Replace B^{-1} with $\left[\Phi((B^{-1})', A_k, l)\right]'$, β_l with k, u with $-(B^{-1})'c_B$, and x_B with $B^{-1}b$. Compute $c_B'x_B$ and go to Step 1.

We illustrate the computations for the dual simplex method in

Example 6.4

> minimize: $2x_1 + 14x_2 + 36x_3$
>
> subject to: $-2x_1 + x_2 + 4x_3 - x_4 \quad\quad = 5,$ (1)
>
> $\quad\quad -x_1 - 2x_2 - 3x_3 \quad\quad + x_5 = 2,$ (2)
>
> $x_i \geq 0, \ i = 1, \ldots, 5.$

Note that this problem was solved in Example 4.1 using the revised simplex method. A dual feasible solution is obtained by taking $I_B = \{1, 4\}$. This gives

$$B^{-1} = \begin{bmatrix} 0 & -1 \\ -1 & 2 \end{bmatrix} \quad \text{and} \quad u = -\begin{bmatrix} 0 & -1 \\ -1 & 2 \end{bmatrix}\begin{bmatrix} 2 \\ 0 \end{bmatrix} = \begin{bmatrix} 0 \\ 2 \end{bmatrix}.$$

It is straightforward to verify that u satisfies dual feasibility. The steps of the dual simplex method are as follows.

Initialization:

$$I_B = \{1,4\}, \quad B^{-1} = \begin{bmatrix} 0 & -1 \\ -1 & 2 \end{bmatrix}, \quad u = \begin{bmatrix} 0 \\ 2 \end{bmatrix}, \quad x_B = \begin{bmatrix} -2 \\ -1 \end{bmatrix},$$

$$c_B' x_B = -4.$$

Iteration 0

Step 1: $\quad (x_B)_1 = \min\{-2, -1\} = -2, \quad l = 1,$

$$s_B = \begin{bmatrix} 0 \\ 1 \end{bmatrix}.$$

Step 2: $\quad \dfrac{A_5' u + c_5}{A_5' s_B} = \min\left\{-,-,-,-,\dfrac{2}{1}\right\} = 2, \quad k = 5.$

Step 3: $\quad B^{-1} \leftarrow \begin{bmatrix} 0 & 1 \\ -1 & 0 \end{bmatrix}, \quad I_B \leftarrow \{5,4\}, \quad u \leftarrow \begin{bmatrix} 0 \\ 0 \end{bmatrix}, \quad x_B \leftarrow \begin{bmatrix} 2 \\ -5 \end{bmatrix},$

$$c_B' x_B = 0.$$

Iteration 1

Step 1: $\quad (x_B)_2 = \min\{2, -5\} = -5, \quad l = 2,$

$$s_B = \begin{bmatrix} 1 \\ 0 \end{bmatrix}.$$

Step 2: $\quad \dfrac{A_3' u + c_3}{A_3' s_B} = \min\left\{-,\dfrac{14}{1},\dfrac{36}{4},-,-\right\} = 9, \quad k = 3.$

Step 3: $\quad B^{-1} \leftarrow \begin{bmatrix} 3/4 & 1 \\ 1/4 & 0 \end{bmatrix}, \quad I_B \leftarrow \{5,3\}, \quad u \leftarrow \begin{bmatrix} -9 \\ 0 \end{bmatrix},$

$$x_B \leftarrow \begin{bmatrix} 23/4 \\ 5/4 \end{bmatrix}, \quad c_B' x_B = 45.$$

Iteration 2

Step 1: $\quad (x_B)_2 = \min\left\{\dfrac{23}{4}, \dfrac{5}{4}\right\} = \dfrac{5}{4}, \quad l = 2.$

$(x_B)_2 \geq 0$; stop with optimal solution $x = \left(0, 0, \dfrac{5}{4}, 0, \dfrac{23}{4} \right)'$.

A computer program which implements the dual simplex method is given in Appendix B, Section B.4.2. The output from applying this program to the problem of Example 6.4 is shown in Figure B.71.

Comparison of the revised and dual simplex methods show that they differ in the order in which the new basic and nonbasic variables are chosen. The revised simplex method maintains primal feasibility at each step and iteratively proceeds to satisfy dual feasibility. The new basic variable (index k) is chosen first and corresponds to the most violated dual constraint. The new nonbasic variable (index β_l) is then chosen to maintain primal feasibility. The dual simplex method maintains dual feasibility at each step and iteratively proceeds to satisfy primal feasibility. The new nonbasic variable (index β_l) is chosen first and corresponds to the most violated primal constraint. The new basic variable (index k) is then chosen to maintain dual feasibility.

Since the dual simplex method for (6.8) is precisely Algorithm 1 applied to (6.9), termination of the former in a finite number of steps with either an optimal solution or the information that the problem is unbounded from below follows directly from Theorem 2.2 provided that no degenerate dual extreme point is encountered.

Bland's rules can be used to prevent cycling when a degenerate extreme point is encountered by the dual simplex method. Degeneracy is recognized by a zero step size in Step 2, and we proceed by using Bland's rules until a strictly positive step is taken or an optimal solution is obtained. Since the dual simplex method for (6.8) is precisely Algorithm 1 applied to (6.9), we can use Bland's rules as formulated for the latter algorithm and express them in the notation of the former algorithm. This gives

Step 1: Let l be such that β_l is the smallest index with $(x_B)_l < 0$.

Step 2: Let k be the smallest index such that

$$\frac{A_k' u + c_k}{A_k' s_B} = \min \left\{ \frac{A_i' u + c_i}{A_i' s_B} \mid \text{all } i \text{ with } A_i' s_B > 0 \right\}.$$

With Theorem A.1, we now have

Theorem 6.2.

Let the dual simplex method be applied to

$$\min \{ c'x \mid Ax = b, \, x \geq 0 \}$$

with Bland's rules being used whenever a degenerate dual basic feasible solution is encountered. Then in a finite number of steps, either an optimal solution is obtained or it is determined that the problem has no feasible solution.

EXERCISES

6.1. Use Algorithm 5 to solve the following problems.

(a)   minimize: $- x_2$

     subject to: $3x_1 + x_2 \leq 2,$ (1)

                           $-3x_1 + x_2 \leq 3,$ (2)

                           $4x_1 + x_2 \leq 2,$ (3)

                           $2x_1 + 3x_2 \leq -1,$ (4)

                           $- x_1 + x_2 \leq 1,$ (5)

                           $x_1 \qquad \leq 1,$ (6)

                                 $x_2 \leq 2.$ (7)

    Initial point: $x_0 = (1, 2)'$.

(b)   minimize: $-2x_1 - 7x_2 - 6x_3$

     subject to: $x_1 + x_2 + 2x_3 \leq 2,$ (1)

                         $- x_1 + x_2 + x_3 \leq 1,$ (2)

                       $2x_1 + x_2 - 3x_3 \leq -1,$ (3)

                       $- x_1 + x_2 + x_3 \leq -1,$ (4)

                       $x_1 \qquad \leq 2,$ (5)

                         $x_2 \qquad \leq 1,$ (6)

                            $x_3 \leq 3.$ (7)

    Initial point: $x_0 = (2, 1, 3)'$.

6.2. Solve each of the following problems using the dual simplex method and the initial data indicated.

(a)   minimize: $10x_1 \qquad + 3x_3 + x_4$

     subject to: $x_1 + 2x_2 - x_3 - x_4 = -2,$ (1)

                      $3x_1 - 3x_2 + x_3 + 2x_4 = 5,$ (2)

                   $x_i \geq 0, \ i = 1, \ldots, 4.$

$$I_B = \{2, 3\}, \quad B^{-1} = \begin{bmatrix} -1 & -1 \\ -3 & -2 \end{bmatrix}, \quad u = (9, 6)'.$$

(b)   minimize: $3x_1 + 2x_2 + 3x_3 + 8x_4$

     subject to: $x_1 + x_2 + 2x_3 + 3x_4 - x_5 = 4,$ (1)

                      $2x_1 - x_2 + x_3 + 4x_4 \qquad = 5,$ (2)

                   $x_i \geq 0, \ i = 1, \ldots, 5$

$$I_B = \{2, 5\}, \quad B^{-1} = \begin{bmatrix} 0 & -1 \\ -1 & -1 \end{bmatrix}, \quad u = (0, 2)'.$$

(c)   minimize: $-2x_1 + 5x_2 - x_3 + x_4$

subject to: $-4x_1 + x_2 - x_3 - 3x_4 - x_5 = -23,$ (1)

$x_1 + x_2 - x_3 + 2x_4 - x_6 = 7,$ (2)

$x_i \geq 0, \quad i = 1, \ldots, 6.$

$$I_B = \{2, 6\}, \quad B^{-1} = \begin{bmatrix} 1 & 0 \\ 1 & -1 \end{bmatrix}, \quad u = (-5, 0)'.$$

(d)   minimize: $10x_1 + 8x_2 + 3x_3 - 4x_4$

subject to: $6x_1 + 3x_2 + 2x_3 - x_4 = 9,$ (1)

$5x_1 - 2x_2 + x_3 + 4x_4 = 1,$ (2)

$x_i \geq 0, \quad i = 1, \ldots, 4.$

$$I_B = \{3, 4\}, \quad B^{-1} = \begin{bmatrix} 4/9 & 1/9 \\ -1/9 & 2/9 \end{bmatrix}, \quad u = \left(-\frac{16}{9}, \frac{5}{9}\right)'.$$

(e)   minimize: $6x_1 + 4x_2 + x_3$

subject to: $-x_1 + x_2 - x_3 + x_4 = 1,$ (1)

$-2x_1 - 2x_2 + x_3 + x_5 = -1,$ (2)

$x_i \geq 0, \quad i = 1, \ldots, 5.$

$$I_B = \{1, 3\}, \quad B^{-1} = \begin{bmatrix} -1/3 & -1/3 \\ -2/3 & 1/3 \end{bmatrix}, \quad u = \left(\frac{8}{3}, \frac{5}{3}\right)'.$$

(f)   minimize: $3x_1 + 4x_2 + x_3 + 5x_4 + 5x_5$

subject to: $3x_1 + x_2 + x_3 + x_4 = -1,$ (1)

$x_1 + 2x_2 - x_3 + x_5 = 1,$ (2)

$x_i \geq 0, \quad i = 1, \ldots, 5.$

$$I_B = \{1, 5\}, \quad B^{-1} = \begin{bmatrix} 1/3 & 0 \\ -1/3 & 1 \end{bmatrix}, \quad u = \left(\frac{2}{3}, -5\right)'.$$

6.3. Develop a generalization of Algorithm 5 which can handle problems of the form

$$\min\{c'x \mid A_1x \leq b_1, \quad A_2x = b_2\}.$$

6.4. Prove statements 1 to 3 following the formulation of the model problem (6.7).

7

UPPER BOUNDING
TECHNIQUES

The algorithms presented in Chapter 2 were for a model problem having general inequality and equality constraints. Nonnegativity constraints, or lower bounds of zero, occur naturally in some applications. The revised simplex method takes advantage of the simple algebraic structure of these lower bounds. Some problems may possess upper bounds as well as lower bounds. Section 7.1 presents a modification of the revised simplex method which implicitly accounts for both upper and lower bounds. Section 7.2 deals with problems for which subsets of problem variables are required to sum to unity. This is called generalized upper bounding.

7.1 THE UPPER-BOUNDED SIMPLEX METHOD

The model problem for the simplex method has equality constraints and lower bounds of zero on all variables. Many linear programming problems have upper bounds on some or all of the variables as well as lower bounds which may be nonzero. In this section we modify the revised simplex method to account for lower and upper bounds on all the variables. We use the model problem

$$\min\{c'x \mid Ax = b, \ d \le x \le e\}, \qquad (7.1)$$

where x is an n-vector, A is an (m, n) matrix and d and e are n-vectors of lower and upper bounds, respectively. We allow the possibility of $d_i = -\infty$ or $e_i = +\infty$ and interpret this to mean that either variable i has no lower bound or no upper bound, respectively. Thus the model problem has lower and upper bounds on some, but not necessarily all variables. This problem could be converted to the form of the model problem required by the revised simplex method by first replacing x with $x - d$ (giving lower bounds of zero on all transformed variables) and then adding slack variables to each of the upper bound constraints. The second transformation doubles the number of variables and increases the number of rows of A from m to $m + n$. This would increase the size of the basis inverse from (m, m) to $(m + n, m + n)$. By accounting for the upper bounds directly, we give a variation of the revised simplex method which requires a basis inverse of only (m, m).

The optimality conditions for (7.1) are

$$\left.\begin{aligned}
Ax = b, \quad d \le x \le e, \\
A'u - v + w = -c, \quad v \ge 0, \quad w \ge 0, \\
v'(-x + d) = 0, \quad w'(x - e) = 0.
\end{aligned}\right\} \qquad (7.2)$$

Let x be any extreme point for (7.1). Assuming that rank$(A) = m$, $n - m$ components of x must be at their lower or upper bounds. Let

$$L = \{i \mid x_i = d_i\} \quad \text{and} \quad U = \{i \mid x_i = e_i\}.$$

Then L and U are unordered index sets which contain the indices of all variables at their lower and upper bounds, respectively. These variables are called <u>nonbasic</u> and the remainder are called <u>basic</u>. Let x_B denote the m-vector of basic variables, and let B denote the associated submatrix of columns of A. Then $Ax = b$ implies that

$$Bx_B + \sum_{i \in L} d_i A_i + \sum_{i \in U} e_i A_i = b,$$

where A_i denotes the ith column of A and the notation

$$\sum_{i \in L}$$

means the sum over all indices i in the set L. The above may be solved for the basic variables in terms of the nonbasic:

$$x_B = B^{-1}\left[b - \sum_{i \in L} d_i A_i - \sum_{i \in U} e_i A_i\right].$$

The corresponding objective function value is

$$c_B'x_B + \sum_{i \in L} c_i d_i + \sum_{i \in U} c_i e_i.$$

The derivation of a solution procedure for (7.1) follows our derivation of the revised simplex method from Algorithm 1. We first require the multipliers for all active constraints and obtain them as follows. In agreement with the complementary slackness part of (7.2), all multipliers associated with inactive constraints have value zero. Letting c_B denote the subvector of c associated with basic variables, u is the solution of the m simultaneous linear equations

$$B'u = -c_B$$

so that

$$u = -(B^{-1})'c_B.$$

Having determined u, the multipliers for variables at their bounds are found from (7.2):

$$v_i = A_i'u + c_i, \quad \text{all } i \in L,$$
$$w_i = -A_i'u - c_i, \quad \text{all } i \in U.$$

The smallest such multiplier is found by determining k_1 and k_2 with

$$v_{k_1} = \min\{v_i \mid \text{all } i \in L\},$$
$$w_{k_2} = \min\{w_i \mid \text{all } i \in U\}.$$

If $v_{k_1} \geq 0$ and $w_{k_2} \geq 0$, all the optimality conditions for (7.1) are satisfied and the current extreme point is optimal. Otherwise, an active bound constraint associated with the smallest multiplier is to become inactive. Let

$$k = \begin{cases} k_1, & \text{if } v_{k_1} \leq w_{k_2}, \\ k_2, & \text{if } v_{k_1} > w_{k_2}. \end{cases}$$

There are two cases to be considered.

Case 1: $k = k_1$

Variable x_k is to be increased from its lower bound of d_k. We require a search direction which allows the basic variables to change, the kth variable to increase, and which ensures that the remaining nonbasic variables are unchanged. A suitable choice is

$$s_j = \begin{bmatrix} s_B \\ -1 \\ 0 \end{bmatrix},$$

where s_B denotes the portion of the search direction associated with the current basic variables and the -1 corresponds to the variable k. Letting x_{NB} denote the nonbasic variables other than the kth, we choose s_B so that the new point

$$\begin{bmatrix} x_B \\ d_k \\ x_{NB} \end{bmatrix} - \sigma \begin{bmatrix} s_B \\ -1 \\ 0 \end{bmatrix}$$

satisfies $Ax = b$. Substitution gives

$$Bx_B + \sum_{i \in L} d_i A_i + \sum_{i \in U} e_i A_i - \sigma B s_B + \sigma A_k = b.$$

The current extreme point also satisfies $Ax = b$ so that

$$Bs_B = A_k,$$

or,

$$s_B = B^{-1} A_k. \tag{7.3}$$

Case 2: $k = k_2$

Here variable x_k is to be decreased from its upper bound of e_k. Arguing as in Case 1 gives

$$s_j = \begin{bmatrix} s_B \\ 1 \\ 0 \end{bmatrix}.$$

The requirement that

$$\begin{bmatrix} x_B \\ e_k \\ 0 \end{bmatrix} - \sigma \begin{bmatrix} s_B \\ 1 \\ 0 \end{bmatrix}$$

satisfy $Ax = b$ leads to

$$s_B = -B^{-1} A_k. \tag{7.4}$$

Having determined s_B from either (7.3) or (7.4), we next determine the maximum feasible step size σ_B. Let d_B denote the m-vector of lower bounds ordered according to x_B, and let e_B denote the analogous vector of upper bounds. As σ is varied, the basic variables change to $x_B - \sigma s_B$. Requiring these components to lie within their respective bounds imposes restrictions on σ:

$$(d_B)_i \le (x_B)_i - \sigma (s_B)_i \le (e_B)_i, \quad i = 1, \ldots, m.$$

These imply that

$$\sigma \le ((x_B)_i - (d_B)_i)/(s_B)_i, \quad \text{all } i \text{ with } (s_B)_i > 0,$$

$$\sigma \le ((x_B)_i - (e_B)_i)/(s_B)_i, \quad \text{all } i \text{ with } (s_B)_i < 0.$$

Let σ_1, l_1, σ_2, and l_2 be such that

$$\sigma_1 = \frac{(x_B)_{l_1} - (d_B)_{l_1}}{(s_B)_{l_1}}$$

$$= \min\left\{\frac{(x_B)_i - (d_B)_i}{(s_B)_i} \;\middle|\; \text{all } i \text{ with } (s_B)_i > 0\right\},$$

$$\sigma_2 = \frac{(x_B)_{l_2} - (e_B)_{l_2}}{(s_B)_{l_2}}$$

$$= \min\left\{\frac{(x_B)_i - (e_B)_i}{(s_B)_i} \;\middle|\; \text{all } i \text{ with } (s_B)_i < 0\right\}.$$

Then the constraints above are satisfied for all σ with

$$0 \le \sigma \le \min\{\sigma_1, \sigma_2\}.$$

Restricting σ as above ensures that the current basic variables remain within their bounds. The new basic variable must also satisfy its bounds. In Case 1 this implies that

$$d_k + \sigma \le e_k, \quad \text{or,} \quad \sigma \le e_k - d_k,$$

and in Case 2,

$$e_k - \sigma \ge d_k, \quad \text{or,} \quad \sigma \le e_k - d_k.$$

Defining $\sigma_3 = e_k - d_k$ and $l_3 = k$, the maximum feasible step size is

$$\sigma_B = \min\{\sigma_1, \sigma_2, \sigma_3\}.$$

Let

$$l = \begin{cases} l_1, & \text{if } \sigma_B = \sigma_1, \\ l_2, & \text{if } \sigma_B = \sigma_2, \\ l_3, & \text{if } \sigma_B = \sigma_3. \end{cases}$$

Because we have allowed the possibility that $d_i = -\infty$ or $e_i = +\infty$, it may occur that $\sigma_B = +\infty$. Since the search direction is always chosen so that the objective function is a strictly decreasing function of the step size, $\sigma_B = +\infty$ implies that (7.1) is unbounded from below and the algorithm terminates with a message to that effect.

Assuming that $\sigma_B < +\infty$, a new basic feasible solution is obtained. To formulate the corresponding change in basic/nonbasic variables, we require an index set similar to that used in the revised simplex method. For the current basic feasible solution, let

$$I_B = \{\beta_1, \ldots, \beta_m\},$$

where

$$B = \left[A_{\beta_1}, \ldots, A_{\beta_m}\right].$$

The index sets L and U must be modified according to k and l as follows. If $k = k_1$, variable k has been increased from its lower bound: k is deleted from L. If $k = k_2$, variable k has been decreased from its upper bound: k is deleted from U. If $\sigma_B < \sigma_3$ and $l = l_1$, variable β_l has been reduced to its lower bound: β_l is added to L. If $\sigma_B < \sigma_3$ and $l = l_2$, variable β_l has been increased to its upper bound: β_l is added to U. If $\sigma_B = \sigma_3$ and $k = k_1$, variable k has been increased from its lower bound to its upper bound: k is added to U. If $\sigma_B = \sigma_3$ and $k = k_2$, variable k has been decreased from its upper bound to its lower bound: k is added to L.

If $\sigma_B = \sigma_3$, both I_B and B^{-1} are unchanged. If $\sigma_B < \sigma_3$, variable k has become basic and variable β_l has become nonbasic: β_l is replaced with k and B^{-1} is replaced with

$$\left[\Phi((B^{-1})', A_k, l)\right]'.$$

Summarizing our analysis gives

UPPER-BOUNDED SIMPLEX METHOD

Model Problem: $\min\{c'x \mid Ax = b, \quad d \leq x \leq e\}$

Initialization:
Start with basis index set $I_B = \{\beta_1, \ldots, \beta_m\}$, basis matrix

$$B = \left[A_{\beta_1}, \ldots, A_{\beta_m}\right],$$

B^{-1}, index sets L and U, and basic feasible solution x_B, x_{NB}. Compute

$$c'x = c_B'x_B + \sum_{i \in L} c_i d_i + \sum_{i \in U} c_i e_i.$$

Step 1:   **Computation of Search Direction s_B.**
Compute

$$u = -(B^{-1})'c_B,$$
$$v_i = A_i'u + c_i, \quad \text{all } i \in L,$$
$$w_i = -A_i'u - c_i, \quad \text{all } i \in U,$$

k_1 and k_2 such that

$$v_{k_1} = \min\{v_i \mid \text{all } i \in L\},$$
$$w_{k_2} = \min\{w_i \mid \text{all } i \in U\}.$$

If $v_{k_1} \geq 0$ and $w_{k_2} \geq 0$, stop with optimal solution x_B, x_{NB}. Otherwise, compute

$$k = \begin{cases} k_1, & \text{if } v_{k_1} \leq w_{k_2}, \\ k_2, & \text{if } v_{k_1} > w_{k_2}, \end{cases}$$

$$s_B = \begin{cases} B^{-1}A_k, & \text{if } k = k_1, \\ -B^{-1}A_k, & \text{if } k = k_2, \end{cases}$$

and go to Step 2.

Step 2:   Computation of Maximum Feasible Step Size σ_B.
Compute

$$\sigma_1 = \frac{(x_B)_{l_1} - (d_B)_{l_1}}{(s_B)_{l_1}}$$

$$= \min\left\{ \frac{(x_B)_i - (d_B)_i}{(s_B)_i} \;\middle|\; \text{all } i \text{ with } (s_B)_i > 0 \right\},$$

$$\sigma_2 = \frac{(x_B)_{l_2} - (e_B)_{l_2}}{(s_B)_{l_2}}$$

$$= \min\left\{ \frac{(x_B)_i - (e_B)_i}{(s_B)_i} \;\middle|\; \text{all } i \text{ with } (s_B)_i < 0 \right\},$$

$$\sigma_3 = e_k - d_k, \quad l_3 = k,$$

$$\sigma_B = \min\{\sigma_1, \sigma_2, \sigma_3\},$$

$$l = \begin{cases} l_1, & \text{if } \sigma_B = \sigma_1, \\ l_2, & \text{if } \sigma_B = \sigma_2, \\ l_3, & \text{if } \sigma_B = \sigma_3. \end{cases}$$

If $\sigma_B = +\infty$, print "the problem is unbounded from below" and stop. Otherwise, go to Step 3.

Step 3:   Update.
If $k = k_1$, replace L with $L - \{k\}$.[1]
If $k = k_2$, replace U with $U - \{k\}$.

---

[1] $L - \{k\}$ means the set obtained from L by deleting k. Similarly, $L + \{\beta_l\}$ means the set obtained from L by adding β_l.

If $\sigma_B < \sigma_3$ and $l = l_1$, replace L with $L + \{\beta_l\}$.

If $\sigma_B < \sigma_3$ and $l = l_2$, replace U with $U + \{\beta_l\}$.

If $\sigma_B = \sigma_3$ and $k = k_1$, replace U with $U + \{k\}$.

If $\sigma_B = \sigma_3$ and $k = k_2$, replace L with $L + \{k\}$.

If $\sigma_B < \sigma_3$, replace β_l with k and B^{-1} with

$$\left[\Phi((B^{-1})', A_k, l)\right]'.$$

If $\sigma_B = \sigma_3$, B^{-1} and I_B remain unchanged.

Replace x_B with

$$B^{-1}\left[b - \sum_{i \in L} d_i A_i - \sum_{i \in U} e_i A_i\right],$$

compute $c'x = c_B'x_B + \sum_{i \in L} c_i d_i + \sum_{i \in U} c_i e_i$, and go to Step 1.

We illustrate the upper-bounded simplex method in the following example.

Example 7.1

$$\begin{aligned}
\text{minimize:} \quad & -4x_1 - 3x_2 \\
\text{subject to:} \quad & x_1 + x_2 - x_3 = 0, \quad (1) \\
& 2x_1 + 4x_2 - x_4 = 9, \quad (2)
\end{aligned}$$

$$1 \le x_1 \le 4, \; 1 \le x_2 \le 3, \; 3 \le x_3 \le 6, \; 0 \le x_4 \le +\infty.$$

Initialization:

$$I_B = \{2, 4\}, \quad B^{-1} = \begin{bmatrix} 1 & 0 \\ 4 & -1 \end{bmatrix}, \quad L = \{1, 3\}, \quad U = \emptyset,$$

$$x_B = \begin{bmatrix} 2 \\ 1 \end{bmatrix}, \quad c'x = -10.$$

Iteration 0

Step 1: $\quad u = -\begin{bmatrix} 1 & 4 \\ 0 & -1 \end{bmatrix}\begin{bmatrix} -3 \\ 0 \end{bmatrix} = \begin{bmatrix} 3 \\ 0 \end{bmatrix},$

$$v_3 = \min\{-1, -, -3, -\} = -3, \quad k_1 = 3,$$

$$w_- = \min\{-, -, -, -\} = -, \quad k_2 = -,$$

$$k = k_1 = 3,$$

$$s_B = B^{-1}A_3 = \begin{bmatrix} -1 \\ -4 \end{bmatrix}.$$

Step 2: $\quad \sigma_1 = \min\{-,-\} = +\infty, \quad l_1 = -,$

$$\sigma_2 = \min\left\{\frac{2-3}{-1}, \frac{1-\infty}{-4}\right\} = 1, \quad l_2 = 1,$$

$$\sigma_3 = 3, \quad l_3 = 3,$$

$$\sigma_B = \min\{+\infty, 1, 3\} = 1, \quad l = l_2 = 1.$$

Step 3: $\quad L \leftarrow \{1\}, \quad U \leftarrow \{2\}, \quad I_B \leftarrow \{3, 4\},$

$$B^{-1} \leftarrow \begin{bmatrix} -1 & 0 \\ 0 & -1 \end{bmatrix}, \quad x_B \leftarrow \begin{bmatrix} 4 \\ 5 \end{bmatrix}, \quad c'x = -13.$$

Iteration 1

Step 1: $\quad u = -\begin{bmatrix} -1 & 0 \\ 0 & -1 \end{bmatrix}\begin{bmatrix} 0 \\ 0 \end{bmatrix} = \begin{bmatrix} 0 \\ 0 \end{bmatrix},$

$$v_1 = \min\{-4, -, -, -\} = -4, \quad k_1 = 1,$$

$$w_2 = \min\{-, 3, -, -\} = 3, \quad k_2 = 2,$$

$$k = k_1 = 1,$$

$$s_B = B^{-1}A_1 = \begin{bmatrix} -1 \\ -2 \end{bmatrix}.$$

Step 2: $\quad \sigma_1 = \min\{-,-\} = +\infty, \quad l_1 = -,$

$$\sigma_2 = \min\left\{\frac{4-6}{-1}, \frac{5-\infty}{-2}\right\} = 2, \quad l_2 = 1,$$

$$\sigma_3 = 3, \quad l_3 = 1,$$

$$\sigma_B = \min\{+\infty, 2, 3\} = 2, \quad l = l_2 = 1.$$

Step 3: $\quad L \leftarrow \varnothing, \quad U \leftarrow \{2, 3\}, \quad I_B \leftarrow \{1, 4\},$

$$B^{-1} \leftarrow \begin{bmatrix} 1 & 0 \\ 2 & -1 \end{bmatrix}, \quad x_B \leftarrow \begin{bmatrix} 3 \\ 9 \end{bmatrix}, \quad c'x = -21.$$

Iteration 2

Step 1: $u = - \begin{bmatrix} 1 & 2 \\ 0 & -1 \end{bmatrix} \begin{bmatrix} -4 \\ 0 \end{bmatrix} = \begin{bmatrix} 4 \\ 0 \end{bmatrix}$,

$v_- = \min\{-, -, -, -\} = -,\quad k_1 = -,$

$w_2 = \min\{-, -1, 4, -\} = -1,\quad k_2 = 2,$

$k = k_2 = 2,$

$s_B = -B^{-1}A_2 = \begin{bmatrix} -1 \\ 2 \end{bmatrix}.$

Step 2: $\sigma_1 = \min\left\{-, \dfrac{9-0}{2}\right\} = \dfrac{9}{2},\quad l_1 = 2,$

$\sigma_2 = \min\left\{\dfrac{3-4}{-1}, -\right\} = 1,\quad l_2 = 1,$

$\sigma_3 = 2,\quad l_3 = 2,$

$\sigma_B = \min\left\{\dfrac{9}{2}, 1, 2\right\} = 1,\quad l = l_2 = 1.$

Step 3: $U \leftarrow \{3\},$

$U \leftarrow \{1, 3\},\quad L \leftarrow \varnothing,\quad I_B \leftarrow \{2, 4\},$

$B^{-1} \leftarrow \begin{bmatrix} 1 & 0 \\ 4 & -1 \end{bmatrix},\quad x_B \leftarrow \begin{bmatrix} 2 \\ 7 \end{bmatrix},\quad c'x = -22.$

Iteration 3

Step 1: $u = - \begin{bmatrix} 1 & 4 \\ 0 & -1 \end{bmatrix} \begin{bmatrix} -3 \\ 0 \end{bmatrix} = \begin{bmatrix} 3 \\ 0 \end{bmatrix}$,

$v_- = \min\{-, -, -, -\} = -,\quad k_1 = -,$

$w_1 = \min\{1, -, 3, -\} = 1,\quad k_2 = 1,$

$k = k_2 = 1.$

$w_1 \geq 0$; stop with optimal solution $x = (4, 2, 6, 7)'.$

A computer program which implements the upper-bounded simplex method is given in Appendix B, Section B.5.1. The output from applying this program to the problem of Example 7.1 is shown in Figure B.79.

During one iteration of the upper-bounded simplex method, the objective function is changed by $-\sigma_B c'_B s_B + \sigma_B c_k$ (if $k = k_1$) or $-\sigma_B c'_B s_B - \sigma_B c_k$ (if $k = k_2$). Using the definitions of s_B and u, the two possible changes may be expressed as $\sigma_B v_{k_1}$ (if $k = k_1$) and $\sigma_B w_{k_2}$ (if $k = k_2$). In either case, the objective function is strictly reduced provided that $\sigma_B > 0$. In the absence of degeneracy, no extreme point can be repeated. The model problem has finitely many extreme points, so termination with an optimal solution or the information that the problem is unbounded from below must occur in a finite number of steps.

If $\sigma_B = 0$, then at least $n + 1$ constraints are active. The current extreme point is thus degenerate and Bland's rules may be used to ensure that cycling does not occur. It is shown in Appendix A that the use of Bland's rules implies that no active set will ever be repeated. For the upper-bounded simplex method, this means that no basis will ever be repeated.

Theorem 7.1.

Let the upper-bounded simplex method be applied to (7.1) with Bland's rules being used whenever a degenerate basic feasible solution is encountered. Then the method will terminate in a finite number of steps with either an optimal solution or the information that the problem is unbounded from below.

7.2 GENERALIZED UPPER BOUNDING

In this section we consider linear programming problems of the form

$$\left.\begin{array}{l}
\text{minimize:} \quad \displaystyle\sum_{i=1}^{n_0} c_i x_i \;+\; \sum_{i=n_0+1}^{n_1} c_i x_i \;+\cdots+\; \sum_{i=n_{r-1}+1}^{n_r} c_i x_i \\[3ex]
\text{subject to:} \quad \displaystyle\sum_{i=1}^{n_0} A_i x_i \;+\; \sum_{i=n_0+1}^{n_1} A_i x_i \;+\cdots+\; \sum_{i=n_{r-1}+1}^{n_r} A_i x_i \;=\; b, \\[3ex]
\displaystyle\sum_{i=n_0+1}^{n_1} x_i \;=\; 1, \\[3ex]
\cdots \\[3ex]
\displaystyle\sum_{i=n_{r-1}+1}^{n_r} x_i \;=\; 1, \\[3ex]
x_i \geq 0, \qquad \text{for all } i.
\end{array}\right\} \quad (7.5)$$

Here, c_i and x_i are scalars and b and A_i are m-vectors. Let n denote the total number of problem variables in (7.5). Problem (7.5) has $m + r + n$ constraints of which the first m are general equality constraints, the next r restrict r subsets of the variables to sum to unity, and the last n are nonnegativity constraints.

This problem can, of course, be solved by the revised simplex method. Because it has $m + r$ equality constraints every basis matrix is of dimension $(m + r, m + r)$. Taking advantage of the special structure of the equations of (7.5) we can, however, obtain a solution by using a "working basis" of dimension (m, m). All the quantities needed to carry out an iteration of the revised simplex method can be computed with this working basis. This is a substantial computational saving if r is large compared to m.

Define

$$S_0 = \{j \mid 1 \leq j \leq n_0\},$$

and for $i = 1, \ldots, r,$

$$S_i = \{j \mid n_{i-1} + 1 \leq j \leq n_i\}.$$

Thus S_i with $i > 0$ contains the indices of the variables which occur in the $(m + i)$th equation of the constraints of (7.5).

Note that $n = \sum_{i=0}^{r} n_i$. Define the r-vectors $C_1, \ldots, C_n$ such that

$$
\left.
\begin{aligned}
C_i &= 0, && \text{for all } i \in S_0, \\
C_i &= e_\nu, && \text{for all } i \in S_\nu, \quad \nu = 1, \ldots, r,
\end{aligned}
\right\}
\tag{7.6}
$$

where e_ν is the r-dimensional unit vector (i.e., the νth component of e_ν is one and all other components are zero). Then the constraints of the minimization problem above can be written in the form

$$\sum_{i=1}^{n} \begin{bmatrix} A_i \\ C_i \end{bmatrix} x_i = \begin{bmatrix} b \\ e \end{bmatrix} \tag{7.7}$$

$$x_i \geq 0, \quad \text{for } i = 1, \ldots, n,$$

where $e = (1, \ldots, 1)'$ is an r-vector.

Assume that

$$\hat{B} = \begin{bmatrix} A_{\beta_1}, \ldots, A_{\beta_{m+r}} \\ C_{\beta_1}, \ldots, C_{\beta_{m+r}} \end{bmatrix}$$

is any basis matrix for (7.7) and let

$$I_{\hat{B}} = \{\beta_1, \ldots, \beta_{m+r}\}$$

be the corresponding index set. Denote the $(m + r)$-vector of basic variables by $x_{\hat{B}}$. Because all nonbasic variables are zero, it follows from (7.7) that, for all $i = 1, \ldots, r$, there is at least one index k_i such that

$$k_i \in S_i \quad \text{and} \quad x_{k_i} \text{ is a basic variable.}$$

These variables $x_{k_1}, x_{k_2}, \ldots, x_{k_r}$ are called <u>key variables</u> and the associated columns A_{k_i} are called <u>key columns</u>.

Rearranging the columns of $\hat{B}$ if necessary, we may assume that the last r columns of $\hat{B}$ are the key columns. More precisely, we assume that

$$A_{\beta_{m+i}} = A_{k_i}, \quad \text{for } i = 1, \ldots, r. \tag{7.8}$$

Then $C_{\beta_{m+i}} = e_i$ for $i = 1, \ldots, r$ and it follows that

$$\hat{B} = \begin{bmatrix} A_{\beta_1}, \ldots, A_{\beta_m}, & A_{\beta_{m+1}}, \ldots, A_{\beta_{m+r}} \\ C_{\beta_1}, \ldots, C_{\beta_m}, & I \end{bmatrix},$$

where I is the (r, r) identity matrix. Setting

$$A = \begin{bmatrix} A_{\beta_1}, \ldots, A_{\beta_m} \end{bmatrix}, \quad K = \begin{bmatrix} A_{\beta_{m+1}}, \ldots, A_{\beta_{m+r}} \end{bmatrix},$$

and

$$C = \begin{bmatrix} C_{\beta_1}, \ldots, C_{\beta_m} \end{bmatrix},$$

we have

$$\hat{B} = \begin{bmatrix} A & K \\ C & I \end{bmatrix}. \tag{7.9}$$

Finally, let

$$x_{\hat{B}} = \begin{bmatrix} x_B \\ x_K \end{bmatrix}, \tag{7.10}$$

where x_B is the m-vector of nonkey basic variables and x_K denotes the r-vector of key basic variables.

By definition $x_{\hat{B}}$ is the uniquely determined solution of the equations

$$\hat{B} x_{\hat{B}} = \begin{bmatrix} b \\ e \end{bmatrix}.$$

Using (7.9) and (7.10), we can write these equations in the form

$$Ax_B + Kx_K = b, \tag{7.11}$$

$$Cx_B + \quad x_K = e. \tag{7.12}$$

Solving (7.12) for x_K and substituting into (7.11), we obtain

$$(A - KC)x_B = b - Ke \quad \text{and} \quad x_K = e - Cx_B. \tag{7.13}$$

Since x_B is uniquely determined it follows that the (m, m) matrix $A - KC$ is nonsingular. It is the working basis associated with $\hat{B}$. We denote it by B and its columns by $B_1, \ldots, B_m$. Thus

$$B = \left[B_1, \ldots, B_m\right] = A - KC. \tag{7.14}$$

Since the ith column of KC is KC_{β_i} it follows from the definition of the vectors C_i that

$$B_i = A_{\beta_i} - KC_{\beta_i} = \begin{cases} A_{\beta_i}, & \text{if } \beta_i \in S_0, \\ A_{\beta_i} - A_{\beta_{m+\nu}}, & \text{if } \beta_i \in S_\nu \text{ and } 1 \le \nu \le r. \end{cases} \tag{7.15}$$

Observe that by (7.8), $A_{\beta_{m+\nu}}$ is the key column associated with the set S_ν.

We are now ready to show how all quantities computed in an iteration of the revised simplex method with the aid of $\hat{B}$ can be derived from the working basis B.

In order to compute the vector $x_{\hat{B}} = (x_B', x_K')'$ of basic variables, let

$$d = b - Ke = b - \sum_{i=1}^{r} A_{\beta_{m+i}}.$$

Then it follows from (7.13) and (7.14) that

$$x_B = B^{-1}d \tag{7.16}$$

and for $i = 1, \ldots, r$,

$$(x_K)_i = x_{\beta_{m+i}} = 1 - \sum_{\nu} (x_B)_\nu, \tag{7.17}$$

where the summation is over all $\nu \le m$ such that $\beta_\nu \in S_i$; that is, the ith key basic variable is 1 minus the sum of all the other basic variables that belong to the set S_i.

Step 1 of the revised simplex method requires the computation of the multipliers v_i for all active inequality constraints (i.e., for $i \notin I_{\hat{B}}$). Using the notation of Chapter 4 we have for all $i \notin I_{\hat{B}}$,

$$v_i = c_i + (A_i', C_i')u \tag{7.18}$$

where $u'\hat{B} = -c_{\hat{B}}'$ and $(c_{\hat{B}})_i = c_{\beta_i}, i = 1, \ldots, m + r$.

Let u_1 consist of the first m components of u, denote the remaining vector by u_2, and partition $c_{\hat{B}}$ in accordance with (7.10) into c_B and c_K. Using (7.9), we can then write $u'\hat{B} = -c_{\hat{B}}'$ as follows:

$$(u_1', u_2') \begin{bmatrix} A & K \\ C & I \end{bmatrix} = -(c_B', c_K')$$

or

$$u_1'A + u_2'C = -c_B' \quad \text{and} \quad u_1'K + u_2' = -c_K'.$$

This gives

$$u_2 = -c_K - K'u_1, \tag{7.19}$$

$$(A - KC)'u_1 = -c_B + C'c_K. \tag{7.20}$$

The ith component of $C'c_K$ is $C_{\beta_i}'c_K = (c_K)_\nu = c_{\beta_{m+\nu}}$ if $\beta_i \in S_\nu$ for some $1 \leq \nu \leq r$ and zero if $\beta_i \in S_0$. Thus if we set, for all $i = 1, \dots, m$,

$$d_i = \begin{cases} (c_B)_i, & \text{if } \beta_i \in S_0, \\ (c_B)_i - c_{\beta_{m+\nu}}, & \text{if } \beta_i \in S_\nu \text{ for some } 1 \leq \nu \leq r, \end{cases}$$

and $d_B = (d_1, \dots, d_m)'$, it follows from (7.14) and (7.20) that

$$u_1 = -(B^{-1})'d_B$$

and from (7.19) that

$$(u_2)_i = -(c_K)_i - A_{\beta_{m+i}}'u_1$$
$$= -c_{\beta_{m+i}} - A_{\beta_{m+i}}'u_1.$$

Substituting into (7.18) we have

$$v_i = c_i + A_i'u_1 + C_i'u_2$$
$$= c_i - A_i'(B^{-1})'d_B, \quad \text{if } i \in S_0$$

and

$$v_i = c_i + A_i'u_1 + (u_2)_\nu$$
$$= c_i - c_{\beta_{m+\nu}} + (A_i - A_{\beta_{m+\nu}})'u_1, \quad \text{if } i \in S_\nu, \ 1 \leq \nu \leq r.$$

Determine the smallest k such that

$$v_k = \min \{ v_i \mid \text{for all } i \notin I_{\hat{\beta}} \}.$$

If $v_k \geq 0$, the current basic solution is optimal. If $v_k < 0$, x_k becomes a basic variable.

To determine the basic variable which becomes nonbasic, we first compute the vector $s_{\hat{\beta}}$ as the solution of the equations

$$\hat{B}s_{\hat{\beta}} = \begin{bmatrix} A_k \\ C_k \end{bmatrix}.$$

Partitioning $s_{\hat{B}}$ into $(s_B', s_K')'$ and using (7.9) we obtain the relations

$$As_B + Ks_K = A_k,$$

$$Cs_B + \quad s_K = C_k,$$

which give

$$s_K = C_k - Cs_B,$$

$$s_B = B^{-1}(A_k - KC_k).$$

If $k \in S_0$, then $C_k = 0$ and

$$s_B = B^{-1}A_k,$$

$$(s_K)_i = -\sum_{\nu} (s_B)_\nu,$$

where the summation is over all $\nu \le m$ such that $\beta_\nu \in S_i$. If $k \in S_\rho$ with $\rho \ge 1$, then $C_k = e_\rho$ and

$$s_B = B^{-1}(A_k - A_{\beta_{m+\rho}}),$$

$$(s_K)_i = -\sum_{\nu} (s_B)_\nu, \quad \text{for } i \ne \rho,$$

$$(s_K)_\rho = 1 - \sum_{\nu} (s_B)_\nu,$$

where both summations are over all $\nu \le m$ such that $\beta_\nu \in S_i$ and $\beta_\nu \in S_\rho$, respectively. The problem is unbounded from below if $s_B \le 0$ and $s_K \le 0$. If at least one component of $s_{\hat{B}} = (s_B', s_K')'$ is positive, determine σ_B and the smallest index l such that

$$\sigma_B = \frac{(x_{\hat{B}})_l}{(s_{\hat{B}})_l} = \min\left\{ \frac{(x_{\hat{B}})_i}{(s_{\hat{B}})_i} \mid \text{all } i \text{ with } (s_{\hat{B}})_i > 0 \right\}.$$

Now the new basis matrix is obtained from $\hat{B}$ by replacing the lth column with

$$\begin{bmatrix} A_k \\ C_k \end{bmatrix}.$$

Denote the working basis associated with the new basis matrix by $\tilde{B}$. In order to determine $\tilde{B}$ and its inverse we have to consider three cases.

Case 1: $l \le m$
    Here the column of $\hat{B}$ which is replaced is not a key column. The definition of B_i in (7.15) shows that only the lth column of B is affected by the change in $\hat{B}$. Denoting the lth column of $\tilde{B}$ by $\tilde{B}_l$ we have

$$\tilde{B}_l = \begin{cases} A_k, & \text{if } k \in S_0, \\ A_k - A_{\beta_{m+\nu}}, & \text{if } k \in S_\nu, \quad 1 \le \nu \le r. \end{cases}$$

Thus $\tilde{B}^{-1} = \left[\Phi((B^{-1})', \tilde{B}_l, k) \right]'$. The new index set is obtained by setting $\beta_l = k$.

Case 2: $l > m$, $\beta_l \in S_\nu$, and $\beta_i \notin S_\nu$ for $i = 1, \ldots, m$

In this case a key column of $\hat{B}$ is replaced and the corresponding key variable is the only basic variable that belongs to the set S_ν. Since at least one variable of S_ν is an element of the new vector of basic variables, it follows that $k \in S_\nu$. Furthermore, A_{β_l} does not occur in the definition of $B_1, \ldots, B_m$ in (7.15). Thus $\tilde{B} = B$. The new index set is obtained by setting $\beta_l = k$.

Case 3: $l > m$, $\beta_l \in S_\nu$, and $\beta_\rho \in S_\nu$ for some $\rho \le m$

We can reduce this situation to Case 1 by interchanging columns l and ρ of the basis matrix $\hat{B}$. Let us first investigate the effect of this change on the definition of the working matrix B. Denote the modified working basis by B^* and its columns by $B_1^*, \ldots, B_m^*$. It follows from (7.15) that $B_i^* = B_i$ for all i with $\beta_i \notin S_\nu$. Since A_{β_ρ} and A_{β_l} have been interchanged we have for all i with $\beta_i \in S_\nu$,

$$B_i^* = A_{\beta_i} - A_{\beta_\rho} = B_i - (A_{\beta_\rho} - A_{\beta_l}) = B_i - B_\rho, \quad \text{if } i \ne \rho,$$

$$B_\rho^* = A_{\beta_l} - A_{\beta_\rho} = -B_\rho.$$

Denoting the columns of $(B^{-1})'$ and $(B^{*-1})'$ by $b_1, \ldots, b_m$ and $b_1^*, \ldots, b_m^*$, respectively, we conclude from Exercise 7.7 that

$$b_i^* = b_i, \quad \text{for all } i \ne \rho,$$

$$b_\rho^* = -\sum_i b_i,$$

where the summation is over all i such that $\beta_i \in S_\nu$. Thus B^{*-1} can easily be obtained from B^{-1}. Finally, as in Case 1,

$$\tilde{B}^{-1} = \left[\Phi((B^{*-1})', \tilde{B}_\rho, \rho) \right]',$$

where $\tilde{B}_\rho = A_k - A_{\beta_\rho}$. The new index set is determined by setting $\beta_l = \beta_\rho$ and then $\beta_\rho = k$.

Based on the results above, we can now describe an algorithm for solving the given minimization problem. The algorithm determines exactly the same sequence of basic solutions as the revised simplex method. The difference is that instead of using an $(m + r, m + r)$ basis matrix we derive all quantities from an (m, m) working basis.

GENERALIZED UPPER-BOUNDED SIMPLEX METHOD

Model Problem:

$$\min\left\{ \sum_{i=1}^{n} c_i x_i \;\middle|\; \sum_{i=1}^{n_0} \begin{bmatrix} A_i \\ C_i \end{bmatrix} x_i + \cdots + \sum_{i=n_{r-1}+1}^{n_r} \begin{bmatrix} A_i \\ C_i \end{bmatrix} x_i = \begin{bmatrix} b \\ e \end{bmatrix}, \; x_i \geq 0 \right\},(7.21)$$

where $C_1, \ldots, C_n$ are defined by (7.6).

Initialization:
Determine a basis matrix $\hat{B}$ and the corresponding index set $I_{\hat{B}} = \{\beta_1, \ldots, \beta_{m+r}\}$ such that for $i = 1, \ldots, r$ column $m + i$ of $\hat{B}$ is a key column and $\beta_{m+i} \in S_i$. Define the working basis matrix $B = \left[B_1, \ldots, B_m \right]$ with $B_i = A_{\beta_i}$ if $\beta_i \in S_0$ and $B_i = A_{\beta_i} - A_{\beta_{m+\nu}}$ if $\beta_i \in S_\nu$ for some $1 \leq \nu \leq r$. Compute B^{-1}, x_B and x_K according to (7.16) and (7.17), respectively, and $c'x = c_B' x_B + c_K' x_K$.

Step 1:   Computation of Search Direction s_B.
For $i = 1, \ldots, m$ compute

$$d_i = \begin{cases} (c_B)_i, & \text{if } \beta_i \in S_0, \\ (c_B)_i - c_{\beta_{m+\nu}}, & \text{if } \beta_i \in S_\nu \text{ for some } 1 \leq \nu \leq r. \end{cases}$$

Set $d_B = (d_1, \ldots, d_m)'$ and compute $u_1 = -(B^{-1})' d_B$.
For every $i \notin I_{\hat{B}}$ set

$$v_i = c_i + A_i' u_1, \qquad\qquad\qquad \text{if } i \in S_0,$$

$$v_i = c_i - c_{\beta_{m+\nu}} + (A_i - A_{\beta_{m+\nu}})' u_1, \quad \text{if } i \in S_\nu, \; 1 \leq \nu \leq r.$$

Determine the smallest index k such that

$$v_k = \min\{v_i \mid \text{all } i \notin I_{\hat{B}}\}.$$

If $v_k \geq 0$, stop with optimal solution $(x_B', x_K')'$.
If $v_k < 0$, go to Step 2.

Step 2:   Computation of Maximum Feasible Step Size σ_B.
If $k \in S_0$, set

$$s_B = B^{-1} A_k$$

$$(s_K)_i = -\sum_\nu (s_B)_\nu, \quad i = 1, \ldots, r,$$

where the summation is over all $\nu \leq m$ such that $\beta_\nu \in S_i$. If $k \in S_\rho$ with $\rho \geq 1$, set

$$s_B = B^{-1}(A_k - A_{\beta_{m+\rho}}),$$

$$(s_K)_i = -\sum_\nu (s_B)_\nu, \quad \text{for } i = 1, \ldots, r, \ i \neq \rho,$$

$$(s_K)_\rho = 1 - \sum_\nu (s_B)_\nu,$$

where both summations are over all $\nu \leq m$ such that $\beta_\nu \in S_i$ and $\beta_\nu \in S_\rho$, respectively. If $s_B \leq 0$ and $s_K \leq 0$ print "the problem is unbounded from below" and stop. If at least one component of $s_{\tilde{B}} = (s_B', s_K')'$ is positive determine σ_B and the smallest index l such that

$$\sigma_B = \frac{(x_{\tilde{B}})_l}{(s_{\tilde{B}})_l} = \min\left\{ \frac{(x_{\tilde{B}})_i}{(s_{\tilde{B}})_i} \mid \text{all } i \text{ with } (s_{\tilde{B}})_i > 0 \right\}.$$

Go to Step 3.

Step 3:   Update.
If $l \leq m$, go to Step 3.1. If $l > m$, $\beta_l \in S_\nu$ for some $1 \leq \nu \leq r$, and $\beta_i \notin S_\nu$ for $i = 1, \ldots, m$, go to Step 3.2. If $l > m$, $\beta_l \in S_\nu$ for some $1 \leq \nu \leq r$, and $\beta_\rho \in S_\nu$ for at least one $1 \leq \rho \leq m$, go to Step 3.3.

Step 3.1:
Set

$$\tilde{B}_l = \begin{cases} A_k, & \text{if } k \in S_0, \\ A_k - A_{\beta_{m+\nu}}, & \text{if } k \in S_\nu \text{ for some } 1 \leq \nu \leq r. \end{cases}$$

Replace B^{-1} with $\left[\Phi((B^{-1})', \tilde{B}_l, k) \right]'$ and obtain the new index set by setting $\beta_l = k$. Go to Step 3.4.

Step 3.2:
Leave B^{-1} unchanged, obtain the new index set by setting $\beta_l = k$ and go to Step 3.4.

Step 3.3:
Set $(B^{*^{-1}})' = \left[b_1^*, \ldots, b_m^* \right]$ with

$$b_i^* = b_i, \quad \text{for all } i \neq \rho,$$

$$b_\rho^* = -\sum_i b_i,$$

where the summation is over all i such that $\beta_i \in S_\nu$ and $b_1, \ldots, b_m$ are the columns of $(B^{-1})'$. Set

$$\tilde{B}_\rho = \begin{cases} A_k, & \text{if } k \in S_0, \\ A_k - A_{\beta_{m+\nu}}, & \text{if } k \in S_\nu \text{ for some } 1 \leq \nu \leq r. \end{cases}$$

Replace B^{-1} with $\left[\Phi((B^{*-1})', \tilde{B}_\rho, \rho) \right]'$ and obtain the new index set by setting $\beta_l = \beta_\rho$ and then $\beta_\rho = k$. Go to Step 3.4.

Step 3.4:
Compute

$$x_B = B^{-1}\left(b - \sum_{i=1}^{r} A_{\beta_{m+i}}\right),$$

$$(x_K)_i = 1 - \sum_{\nu} (x_B)_\nu, \quad \text{for } i = 1, \ldots, r,$$

where the last summation is over all $\nu \leq m$ such that $\beta_\nu \in S_i$. Compute $c'x = c'_B x_B + c'_K x_K$ and go to Step 1.

We illustrate the generalized upper-bounded simplex method by applying it to the following example.

Example 7.2

minimize: $-7x_1 - 4x_2 + 5x_3 - 3x_4 - 2x_5 - 2x_6 - 4x_7$

subject to: $3x_1 + 2x_2 + 18x_3 + 14x_4 + 8x_5 + 2x_6 - 4x_7 = 16, (1)$

$2x_1 + x_2 + 12x_3 + 8x_4 + 4x_5 + 2x_6 + 2x_7 = 13, (2)$

$x_3 + x_4 + x_5 = 1, (3)$

$x_6 + x_7 = 1, (4)$

$x_i \geq 0, \ i = 1, \ldots, 7.$

Observe that $m = r = 2$, $S_0 = \{1, 2\}$, $S_1 = \{3, 4, 5\}$, and $S_2 = \{6, 7\}$.

Initialization:

$$\hat{B} = \begin{bmatrix} 14 & 2 & 18 & -4 \\ 8 & 2 & 12 & 2 \\ 1 & 0 & 1 & 0 \\ 0 & 1 & 0 & 1 \end{bmatrix}, \quad I_{\hat{B}} = \{4, 6, 3, 7\}, \quad \text{key variables: } x_3 \text{ and } x_7,$$

$$B_1 = A_{\beta_1} - A_{\beta_3} = \begin{bmatrix} -4 \\ -4 \end{bmatrix},$$

$$B_2 = A_{\beta_2} - A_{\beta_4} = \begin{bmatrix} 6 \\ 0 \end{bmatrix}, \quad B = \begin{bmatrix} -4 & 6 \\ -4 & 0 \end{bmatrix}, \quad B^{-1} = \begin{bmatrix} 0 & -1/4 \\ 1/6 & -1/6 \end{bmatrix},$$

$$x_B = B^{-1}(b - A_{\beta_3} - A_{\beta_4}) = \begin{bmatrix} 1/4 \\ 1/2 \end{bmatrix},$$

$$(x_K)_1 = 1 - (x_B)_1 = \frac{3}{4}, \quad (x_K)_2 = 1 - (x_B)_2 = \frac{1}{2},$$

$$c'x = 0.$$

Iteration 0

Step 1: $d_1 = c_4 - c_{\beta_3} = -8$, $d_2 = c_6 - c_{\beta_4} = 2$,

$$u_1 = -(B^{-1})' \begin{bmatrix} -8 \\ 2 \end{bmatrix} = \begin{bmatrix} -1/3 \\ -5/3 \end{bmatrix},$$

$$v_1 = c_1 + A_1' u_1 = \frac{-34}{3},$$

$$v_2 = c_2 + A_2' u_1 = \frac{-19}{3},$$

$$v_5 = c_5 - c_{\beta_3} + (A_5 - A_{\beta_3})' u_1 = \frac{29}{3},$$

$$v_1 = \min \left\{ \frac{-34}{3}, \frac{-19}{3}, -, -, \frac{29}{3}, -, - \right\} = \frac{-34}{3}, \quad k = 1.$$

Step 2: $k \in S_0$,

$$s_B = B^{-1} A_1 = \begin{bmatrix} -1/2 \\ 1/6 \end{bmatrix},$$

$$(s_K)_1 = -(s_B)_1 = \frac{1}{2}, \quad (s_K)_2 = -(s_B)_2 = \frac{-1}{6},$$

$$\sigma_B = \min \left\{ -, \frac{1/2}{1/6}, \frac{3/4}{1/2}, - \right\} = \frac{3}{2}, \quad l = 3.$$

Step 3: $l > 2$, $\beta_3 \in S_1$,   and $\beta_1 \in S_1$, $\rho = 1$. Transfer to Step 3.3.

Step 3.3: $b_1^* = -b_1 = \begin{bmatrix} 0 \\ 1/4 \end{bmatrix}$, $b_2^* = b_2$,

$$(B^{*-1})' = \begin{bmatrix} 0 & 1/6 \\ 1/4 & -1/6 \end{bmatrix}, \quad \tilde{B}_1 = A_1 = \begin{bmatrix} 3 \\ 2 \end{bmatrix},$$

$$B^{-1} \leftarrow \left[\Phi((B^{*-1})', \tilde{B}_1, 1) \right]' = \begin{bmatrix} 0 & 1/2 \\ 1/6 & -1/4 \end{bmatrix},$$

$$I_{\hat{B}} \leftarrow \{1, 6, 4, 7\}.$$

Step 3.4: $x_B = B^{-1}(b - A_{\beta_3} - A_{\beta_4}) = \begin{bmatrix} 0 & 1/2 \\ 1/6 & -1/4 \end{bmatrix} \begin{bmatrix} 6 \\ 3 \end{bmatrix} = \begin{bmatrix} 3/2 \\ 1/4 \end{bmatrix}$,

$(x_K)_1 = 1$, $(x_K)_2 = 1 - (x_B)_2 = \dfrac{3}{4}$,

$c'x = -17$.

Iteration 1

Step 1: $d_1 = c_1 = -7$, $d_2 = c_6 - c_{\beta_4} = 2$,

$u_1 = -(B^{-1})' \begin{bmatrix} -7 \\ 2 \end{bmatrix} = \begin{bmatrix} -1/3 \\ 4 \end{bmatrix}$,

$v_2 = c_2 + A_2'u_1 = \dfrac{-2}{3}$,

$v_3 = c_3 - c_{\beta_3} + (A_3 - A_{\beta_3})'u_1 = \dfrac{68}{3}$,

$v_5 = c_5 - c_{\beta_3} + (A_5 - A_{\beta_3})'u_1 = -13$,

$v_5 = \min\left\{ -, \dfrac{-2}{3}, \dfrac{68}{3}, -, -13, -, - \right\} = -13$, $k = 5$.

Step 2: $k \in S_1$, $\rho = 1$,

$s_B = B^{-1}(A_5 - A_{\beta_3}) = \begin{bmatrix} -2 \\ 0 \end{bmatrix}$,

$(s_K)_1 = 1$, $(s_K)_2 = -(s_B)_2 = 0$,

$\sigma_B = \left\{ -, -, \dfrac{1}{1}, - \right\} = 1$, $l = 3$.

Step 3: $l > 2$, $\beta_3 \in S_1$, $\beta_1 \notin S_1$, and $\beta_2 \notin S_1$.
Transfer to Step 3.2.

Step 3.2: B^{-1} remains unchanged, $I_{\dot{B}} \leftarrow \{1, 6, 5, 7\}$.

Step 3.4: $x_B = B^{-1}(b - A_{\beta_3} - A_{\beta_4}) = \begin{bmatrix} 0 & 1/2 \\ 1/6 & -1/4 \end{bmatrix} \begin{bmatrix} 12 \\ 7 \end{bmatrix} = \begin{bmatrix} 7/2 \\ 1/4 \end{bmatrix}$,

$(x_K)_1 = 1$, $(x_K)_2 = 1 - (x_B)_2 = \dfrac{3}{4}$,

$c'x = -30$.

Iteration 2

Step 1: $d_1 = c_1 = -7, \quad d_2 = c_6 - c_{\beta_4} = 2,$

$$u_1 = -(B^{-1})' \begin{bmatrix} -7 \\ 2 \end{bmatrix} = \begin{bmatrix} -1/3 \\ 4 \end{bmatrix},$$

$$v_2 = c_2 + A_2'u_1 = \frac{-2}{3},$$

$$v_3 = c_3 - c_{\beta_3} + (A_3 - A_{\beta_3})'u_1 = \frac{107}{3},$$

$$v_4 = c_4 - c_{\beta_3} + (A_4 - A_{\beta_3})'u_1 = 13,$$

$$v_2 = \min\left\{ -, \frac{-2}{3}, \frac{107}{3}, 13, -, -, - \right\} = \frac{-2}{3}, \quad k = 2.$$

Step 2: $k \in S_0,$

$$s_B = B^{-1}A_2 = \begin{bmatrix} 1/2 \\ 1/12 \end{bmatrix},$$

$$(s_K)_1 = 0, \quad (s_K)_2 = -(s_B)_2 = \frac{-1}{12},$$

$$\sigma_B = \min\left\{ \frac{7/2}{1/2}, \frac{1/4}{1/12}, -, - \right\} = 3, \quad l = 2.$$

Step 3: $l \leq 2.$ Transfer to Step 3.1.

Step 3.1: $\tilde{B}_2 = A_2 = \begin{bmatrix} 2 \\ 1 \end{bmatrix},$

$$B^{-1} \leftarrow \left[\Phi((B^{-1})', \tilde{B}_2, 2) \right]' = \begin{bmatrix} -1 & 2 \\ 2 & -3 \end{bmatrix},$$

$$I_{\hat{B}} \leftarrow \{1, 2, 5, 7\}.$$

Step 3.4: $x_B = B^{-1}(b - A_{\beta_3} - A_{\beta_4}) = \begin{bmatrix} -1 & 2 \\ 2 & -3 \end{bmatrix} \begin{bmatrix} 12 \\ 7 \end{bmatrix} = \begin{bmatrix} 2 \\ 3 \end{bmatrix},$

$$(x_K)_1 = 1, \quad (x_K)_2 = 1,$$

$$c'x = -32.$$

Iteration 3

Step 1: $d_1 = c_1 = -7, \quad d_2 = c_2 = -4,$

$$u_1 = -(B^{-1})' \begin{bmatrix} -7 \\ -4 \end{bmatrix} = \begin{bmatrix} 1 \\ 2 \end{bmatrix},$$

$$v_3 = c_3 - c_{\beta_3} + (A_3 - A_{\beta_3})'u_1 = 33,$$

$$v_4 = c_4 - c_{\beta_3} + (A_4 - A_{\beta_3})'u_1 = 13,$$

$$v_6 = c_6 - c_{\beta_4} + (A_6 - A_{\beta_4})'u_1 = 8,$$

$$v_6 = \min\{ -, -, 33, 13, -, 8, - \} = 8, \quad k = 6.$$

$v_6 \geq 0$; stop with optimal solution $x = (2, 3, 0, 0, 1, 0, 1)'$.

A computer program which implements the generalized upper-bounded simplex method is given in Appendix B, Section B.5.2. The output from applying this program to the problem of Example 7.2 is shown in Figure B.88.

If the algorithm encounters a degenerate basic feasible solution, cycling is a theoretical possibility. As in the revised simplex method, this problem can be overcome by using Bland's rules.

Theorem 7.2.

Let the generalized upper-bounded simplex method be applied to the model problem (7.21) with Bland's rules being used whenever a degenerate basic feasible solution is encountered. Then in a finite number of iterations either an optimal solution is obtained or it is determined that the problem is unbounded from below.

Proof:

By construction, the generalized upper-bounded simplex method generates exactly the same sequence of basic solutions as the revised simplex method. Thus the statements of the theorem follow from Theorem 4.1. ■

E X E R C I S E S

7.1. Solve the following problem using the upper-bounded simplex method and the initial data indicated.

$$\begin{array}{llll} \text{minimize:} & -10x_1 + & x_2 \\ \text{subject to:} & x_1 + 3x_2 + x_3 & = 9, & (1) \\ & 2x_1 + & x_2 & + x_4 = 8, & (2) \end{array}$$

$$1 \leq x_1 \leq \tfrac{13}{4}, \quad 1 \leq x_2 \leq \tfrac{5}{2}, \quad 0 \leq x_3 \leq +\infty, \quad 0 \leq x_4 \leq +\infty.$$

$$I_B = \{1, 4\}, \quad B^{-1} = \begin{bmatrix} 1 & 0 \\ -2 & 1 \end{bmatrix}, \quad L = \{3\}, \quad U = \{2\}, \quad x_B = \begin{bmatrix} 3/2 \\ 5/2 \end{bmatrix}.$$

7.2. Solve the following problem using the upper-bounded simplex method and the initial data indicated.

$$\text{minimize:} \quad 2x_1 - x_2$$

subject to:

$$-x_1 - 2x_2 + x_3 \qquad\qquad = 6, \qquad (1)$$
$$x_1 - 2x_2 \qquad + x_4 \qquad = 4, \qquad (2)$$
$$-x_1 + x_2 \qquad\qquad + x_5 = 1, \qquad (3)$$

$$-2 \le x_1 \le -\tfrac{1}{2}, \quad -2 \le x_2 \le -\tfrac{1}{2}, \quad 0 \le x_i \le +\infty, \quad i = 3, 4, 5.$$

$$I_B = \{3, 4, 5\}, \quad B^{-1} = \begin{bmatrix} 1 & 0 & 0 \\ 0 & 1 & 0 \\ 0 & 0 & 1 \end{bmatrix}, \quad L = \{2\}, \quad U = \{1\}, \quad x_B = \begin{bmatrix} 3/2 \\ 1/2 \\ 5/2 \end{bmatrix}.$$

7.3. Solve the following problem using the upper-bounded simplex method and the initial data indicated.

$$\text{minimize:} \quad -x_1 - 10x_2$$

subject to:

$$x_1 + x_2 + x_3 \qquad\qquad\qquad = 11, \qquad (1)$$
$$-x_1 - x_2 \qquad + x_4 \qquad\qquad = -7, \qquad (2)$$
$$x_1 - x_2 \qquad\qquad + x_5 \qquad = 7, \qquad (3)$$
$$-x_1 + x_2 \qquad\qquad\qquad + x_6 = -3, \qquad (4)$$

$$\tfrac{11}{2} \le x_1 \le \tfrac{17}{2}, \quad \tfrac{1}{2} \le x_2 \le \tfrac{7}{2}, \quad 0 \le x_i \le +\infty, \quad i = 3, 4, 5, 6.$$

$$I_B = \{1, 3, 4, 6\}, \quad B^{-1} = \begin{bmatrix} 0 & 0 & 1 & 0 \\ 1 & 0 & -1 & 0 \\ 0 & 1 & 1 & 0 \\ 0 & 0 & 1 & 1 \end{bmatrix}, \quad L = \{2, 5\}, \quad U = \varnothing,$$

$$x_B = \begin{bmatrix} 15/2 \\ 3 \\ 1 \\ 4 \end{bmatrix}.$$

7.4. Solve the following problem using the generalized upper-bounded simplex method and the initial data indicated.

$$\text{minimize:} \quad -x_1$$

subject to:

$$x_1 \qquad + 2x_3 + 5x_4 + x_5 - x_6 - 12x_7 = 8, \qquad (1)$$
$$x_1 + x_2 - x_3 + 4x_4 + 2x_5 - 3x_6 + 6x_7 = 4, \qquad (2)$$
$$x_2 + x_3 \qquad\qquad\qquad = 1, \qquad (3)$$
$$x_4 + x_5 \qquad\qquad = 1, \qquad (4)$$
$$x_6 + x_7 = 1, \qquad (5)$$

$$x_i \ge 0, \quad i = 1, \ldots, 7.$$

$$\hat{B} = \begin{bmatrix} 1 & -1 & 2 & 5 & -12 \\ 1 & -3 & -1 & 4 & 6 \\ 0 & 0 & 1 & 0 & 0 \\ 0 & 0 & 0 & 1 & 0 \\ 0 & 1 & 0 & 0 & 1 \end{bmatrix}, \ I_{\hat{B}} = \{1, 6, 3, 4, 7\}, \text{ key variables } x_3, x_4, \text{ and}$$

x_7.

7.5. For the model problem of Section 7.2 show that the number of sets S_i containing two or more basic variables is at most $m - 1$.

7.6. Let the matrices $\hat{B}$ and B be defined by (7.9) and (7.14), respectively. Show that $\hat{B}$ is singular if and only if B is singular.

7.7. Let the matrices $B = \begin{bmatrix} B_1, \ldots, B_m \end{bmatrix}$, $(B^{-1})' = \begin{bmatrix} b_1, \ldots, b_m \end{bmatrix}$, the set $I \subset \{1, \ldots, m\}$, and the integer $\rho \in I$ be given. Set $B^* = \begin{bmatrix} B_1^*, \ldots, B_m^* \end{bmatrix}$ with $B_\rho^* = -B_\rho$, $B_i^* = B_i$ for all $i \notin I$, $B_i^* = B_i - B_\rho$ for all $i \neq \rho$ and $i \in I$. Show that then $(B^{*-1})' = \begin{bmatrix} b_1^*, \ldots, b_m^* \end{bmatrix}$ with $b_i^* = b_i$ for all $i \neq \rho$ and $b_\rho^* = -\sum_{i \in I} b_i$.

APPLICATIONS

Linear programming may be used to solve a wide variety of problems. In this chapter we give the reader an introduction to the applications of linear programming, as well as mathematical modeling, by considering five application areas.

8.1 THE DIET PROBLEM

A diet for a certain type of livestock is to be constructed. There are n types of food available and each contains varying amounts of m kinds of nutrients (e.g., phosphorus and iron), necessary for the animal's health and well-being. The known data are:

$$a_{ij} = \text{quantity of nutrient } i \text{ per one unit of food } j, \quad i = 1, \ldots, m,$$
$$j = 1, \ldots, n,$$
$$r_i = \text{livestock requirement for nutrient } i, \quad i = 1, \ldots, m,$$
$$c_j = \text{cost of one unit of food } j, \quad j = 1, \ldots, n.$$

The problem is to construct a diet of least cost which meets the m nutritional requirements. Letting x_j denote the amount of the jth food in the diet, the jth nutritional requirement is

$$\sum_{j=1}^{n} a_{ij}x_j \geq r_i,$$

and the cost of the diet is

$$\sum_{j=1}^{n} c_j x_j.$$

The minimum cost of the diet is obtained by solving the linear programming problem

$$\text{minimize:} \quad \sum_{j=1}^{n} c_j x_j$$

$$\text{subject to:} \quad \sum_{j=1}^{n} a_{ij} x_j \geq r_i, \quad i = 1, \ldots, m,$$

$$x_j \geq 0, \quad j = 1, \ldots, n.$$

The nonnegativity restrictions express the conditions that the diet cannot contain a negative amount of any food.

A simple numerical example of this problem is as follows.

Example 8.1

We have five foods and three requirements. The data are summarized in Table 8.1.

TABLE 8.1   Diet Problem Data

		Food				
	1	2	3	4	5	
Cost	2.0	1.75	1.9	2.8	2.1	
Nutrient		Quantity				Requirement
1	0.15	0.08	0.17	0.20	0.11	1.3
2	0.75	0.83	0.78	0.95	0.70	8.0
3	1.20	1.10	1.15	1.30	1.06	12.1

The linear programming problem to be solved is thus

minimize: $2.0x_1 + 1.75x_2 + 1.9x_3 + 2.8x_4 + 2.1x_5$

subject to: $0.15x_1 + 0.08x_2 + 0.17x_3 + 0.2x_4 + 0.11x_5 \geq 1.3,$ (1)

$0.75x_1 + 0.83x_2 + 0.78x_3 + 0.95x_4 + 0.70x_5 \geq 8.0,$ (2)

$1.2x_1 + 1.1x_2 + 1.15x_3 + 1.3x_4 + 1.06x_5 \geq 12.1,$ (3)

$$x_i \geq 0, \quad i = 1, \ldots, 5.$$

The optimal solution is $x_0 = (0, 5.9158, 4.8632, 0, 0)'$, and the optimal solution for the dual is $u_0 = (0.81579, 0, 1.53158, 0.03974, 0, 0, 0.64579, 0.38679)'$. The minimum cost is 19.59263 units.

8.2 PATTERN RECOGNITION

As a simple example of a problem in pattern recognition, we consider the classification of m objects as members of either one of two given sets S_1 and S_2. We assume that all information available about the ith object is contained in an n-vector x_i of measurements. In order to assign each object to one of the two sets we use a decision function $d(x)$. The ith object is declared to be a member of S_1 or S_2 depending on whether $d(x_i)$ is positive or negative. It is assumed that the decision function is of the form

$$d(x) = \sum_{i=1}^{k} \alpha_i f_i(x) + \alpha_0, \tag{8.1}$$

where $f_1(x), \ldots, f_k(x)$ are given functions and the parameters $\alpha_0, \ldots, \alpha_k$ are to be determined. Let

$$y_1, \ldots, y_p \quad \text{and} \quad z_1, \ldots, z_q \tag{8.2}$$

be the measurements of p objects known to belong to S_1 and q objects known to belong to S_2, respectively. Then we try to choose the parameters in (8.1) in such a way that the difference

$$\max_{j=1,\ldots,p} \left\{ \sum_{i=1}^{k} \alpha_i f_i(y_j) + \alpha_0 \right\} - \max_{j=1,\ldots,q} \left\{ \sum_{i=1}^{k} \alpha_i f_i(z_j) + \alpha_0 \right\} \tag{8.3}$$

becomes as large as possible subject to the normalization constraints

$$-1 \le \alpha_i \le 1, \quad i = 1, \ldots, k. \tag{8.4}$$

It is not difficult to see that maximizing (8.3) subject to (8.4) is equivalent to solving the linear programming problem

$$\text{maximize:} \qquad \lambda - \tau$$

$$\text{subject to:} \qquad \sum_{i=1}^{k} \alpha_i f_i(y_i) \ge \lambda, \quad j = 1, \ldots, p,$$

$$\sum_{i=1}^{k} \alpha_i f_i(z_j) \le \tau, \quad j = 1, \ldots, q,$$

$$-1 \le \alpha_i \le 1, \quad i = 1, \ldots, k,$$

with the $k + 2$ variables $\lambda, \tau, \alpha_1, \ldots, \alpha_k$.

Since $\lambda = \tau = \alpha_1 = \cdots = \alpha_k = 0$ is a feasible solution and the objective function $\tau - \lambda$ is bounded from above on the feasible region, it follows from Theorem 3.5 that the LP has an optimal solution. Let $\bar{\lambda}$, $\bar{\tau}$, $\bar{\alpha}_1, \ldots, \bar{\alpha}_k$ be an optimal solution. Suppose first that $\bar{\lambda} - \bar{\tau} > 0$. Setting

$$d(x) = \sum_{i=1}^{k} \bar{\alpha}_i f_i(x) + \bar{\alpha}_0, \quad \bar{\alpha}_0 = -\frac{1}{2}(\bar{\lambda} + \bar{\tau}), \tag{8.5}$$

we have

$$d(y_j) \geq \bar{\lambda} + \bar{\alpha}_0 = \frac{1}{2}(\bar{\lambda} - \bar{\tau}) > 0, \quad \text{for } j = 1, \ldots, p,$$

and

$$d(z_j) \leq \bar{\tau} + \bar{\alpha}_0 = \frac{1}{2}(\bar{\tau} - \bar{\lambda}) < 0, \quad \text{for } j = 1, \ldots, q;$$

that is, the decision function (8.5) classifies the $p + q$ objects with measurements (8.2) correctly. If, on the other hand, $\bar{\lambda} - \bar{\tau} = 0$, then there is no decision function of the form (8.1) which gives a correct classification of these $p + q$ objects. The set of functions $f_1(x), \ldots, f_k(x)$ is not adequate for the given problem.

Figure 8.1 illustrates the procedure above for $n = 2$ and the case of linear decision functions [i.e., $f_1(x) = x_1$, $f_2(x) = x_2$]. The vectors of measurements for the $p = 10$ objects in S_1 and the $q = 8$ objects in S_2 are indicated by squares and circles, respectively.

Figure 8.1 Pattern recognition: linear decision function.

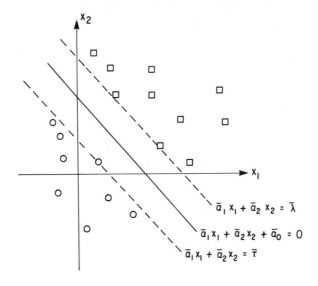

8.3   GRADUATION OF ACTUARIAL DATA

Let $x_1, \ldots, x_n$ be a sequence of numbers. The sequence of first-order differences $\Delta x_1, \ldots, \Delta x_{n-1}$ is defined by

$$\Delta x_i = x_{i+1} - x_i, \quad i = 1, \ldots, n - 1.$$

Taking differences of the new sequence we obtain second-order differences $\Delta^2 x_1, \ldots, \Delta^2 x_{n-2}$, where

$$
\begin{aligned}
\Delta^2 x_i = \Delta(\Delta x_i) &= \Delta x_{i+1} - \Delta x_i \\
&= x_{i+2} - x_{i+1} - (x_{i+1} - x_i) \\
&= x_{i+2} - 2x_{i+1} + x_i.
\end{aligned}
\tag{8.6}
$$

Repeating this procedure we can define the kth-order differences $\Delta^k x_1, \ldots, \Delta^k x_{n-k}$ for any positive integer $k < n$. For $k = 3$ we obtain from (8.6)

$$
\begin{aligned}
\Delta^3 x_i &= \Delta^2 x_{i+1} - \Delta^2 x_i \\
&= x_{i+3} - 2x_{i+2} + x_{i+1} - (x_{i+2} - 2x_{i+1} + x_i) \\
&= x_{i+3} - 3x_{i+2} + 3x_{i+1} - x_i.
\end{aligned}
\tag{8.7}
$$

In actuarial work it is often required to replace a given sequence of numbers $y_1, \ldots, y_n$ by a sequence of graduated values $x_1, \ldots, x_n$ which are smooth in the sense that the kth order differences of the numbers $x_1, \ldots, x_n$ are small. The values of k commonly used are $k = 2$ and $k = 3$.

To be meaningful the graduated data should not deviate too much from the original data. If the kth-order differences of $y_1, \ldots, y_n$ are not small, the two requirements of fit and smoothness are contradictory. Various graduation methods have been proposed [Miller, 1946] which try to overcome this problem. A well-known example is the Whittaker-Henderson Type B graduation formula, which determines the graduated values $x_1, \ldots, x_n$ by minimizing the function

$$\sum_{i=1}^{n} \omega_i (y_i - x_i)^2 + t \sum_{i=1}^{n-k} (\Delta^k x_i)^2. \tag{8.8}$$

Here ω_i, $i = 1, \ldots, n$, are positive weights and $t \geq 0$ is a parameter that indicates the relative emphasis we are placing on fit,

$$F = \sum_{i=1}^{n} \omega_i (y_i - x_i)^2,$$

and smoothness,

$$S = \sum_{i=1}^{n-k} (\Delta^k x_i)^2. \tag{8.9}$$

From (8.7) it follows easily that $\Delta^k x_i$ is a linear function of $x_{i+k}, \ldots, x_i$. Therefore, (8.8) is a quadratic function in the unknowns $x_1, \ldots, x_n$ (i.e., we have an unconstrained quadratic minimization problem).

In many applications it may be be desirable to control the deviation of the graduated values x_i from the given data y_i more precisely than is possible by the use of Whittaker-Henderson formulas. For example, we may want to be sure that for all i

$$-\alpha_i \leq y_i - x_i \leq \beta_i, \tag{8.10}$$

where $\alpha_i \geq 0$ and $\beta_i \geq 0$ are suitably chosen constants. In this case we could replace (8.8) with

$$\min \left\{ \sum_{i=1}^{n-k} (\Delta^k x_i)^2 \mid -\alpha_i \leq y_i - x_i \leq \beta_i, \ i = 1, \ldots, n \right\}$$

which has variables $x_1, \ldots, x_n$. This problem is one of minimizing a quadratic function subject to linear constraints and is called a quadratic programming problem. Furthermore, instead of trying to make (8.9) as small as possible subject to (8.10), we could minimize

$$\max_{i=1,\ldots,n-k} |\Delta^k x_i|. \tag{8.11}$$

Introducing a new variable λ we can replace (8.10) with

$$-\lambda \leq \Delta^k x_i \leq \lambda, \ i = 1, \ldots, n-k.$$

Therefore, minimizing (8.11) subject to the constraints (8.10) is equivalent to solving the linear programming problem

minimize: λ

subject to: $-\lambda \leq \Delta^k x_i \leq \lambda, \ i = 1, \ldots, n-k,$

$\qquad\qquad -\alpha_i \leq y_i - x_i \leq \beta_i, \ i = 1, \ldots, n,$

with variables $x_1, \ldots, x_n$, and λ.

8.4 ENGINEERING PLASTICITY

An important area of structural mechanics is the analysis of structures under various types of forces. A typical problem is to discover the effect of wind on a tall building. Another is to analyze the effect of snow loading on a roof structure. Under generally low loading levels, the response of a structure is typically elastic. That is, as an external force is increased, the structure deforms to accommodate the force, but when the force is reduced or removed, elastic behavior dictates that the structure return to its original undeformed position. An increased external force results in larger deformations until the elastic limit of the structure is reached.

When the elastic limit is reached, the structure begins to behave plasti-cally, in the sense that permanent deformations occur that are not recoverable upon removing the load. At a certain loading level, one, or more generally, several individual elements of the structure will have reached their elastic lim-its such that the structure is at the point of incipient plastic collapse.

The goal of plastic analysis is to determine the limit load of a structure and its associated elastic-plastic deformation profile at incipient collapse. Rigid-plastic analysis to determine the plastic collapse limit load may be for-mulated as a linear programming problem [Cohn and Maier, 1979; Maier, Grierson, and Best, 1977]. Elastic-plastic analysis to determine the deforma-tion profile at incipient collapse may be formulated as a quadratic program-ming problem.

Consider the pin-jointed truss shown in Figure 8.2. The truss consists of

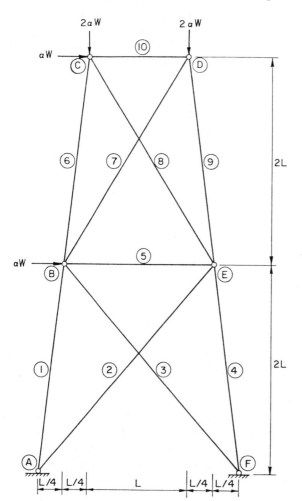

Figure 8.2   Pin-jointed truss.

10 elements and is subjected to the indicated external forces (loads) at nodes *B*, *C*, and *D*. Nodes *A* and *F* are fixed to a foundation and may not move. Let *W* be a fixed constant and α a scalar variable. Forces of αW are applied horizontally to nodes *B* and *C* and vertical forces of $2\alpha W$ are applied downward at nodes *C* and *D*. This is called proportional loading. The loading parameter α is to be increased from zero (no external loading) to the value at which the truss is at incipient collapse.

Let Q_i denote the axial force in element *i*. We use the sign convention that an axial force is positive when the member is in compression and negative when in tension. Furthermore, horizontal forces to the right, and downward vertical forces are assumed to be positive. The axial forces are related to themselves and the applied forces by means of the geometry of the structure. The relationships may be obtained using Newton's laws, which say that at equilibrium, the net force at each node is zero. For the pin-jointed truss, resolution of forces in a horizontal direction at node *B*, for example, gives (Figures 8.3 and 8.4)

$$-Q_1 \sin\theta_4 + Q_3 \cos\theta_3 + Q_5 + Q_6 \cos\theta_2 + Q_7 \cos\theta_1 + \alpha W = 0,$$

or

$$-Q_1/\sqrt{65} + 7Q_3/\sqrt{113} + Q_5 + Q_6/\sqrt{65} + 5Q_7/\sqrt{89} + \alpha W = 0.$$

Figure 8.3   Pin-jointed truss: forces and angles.

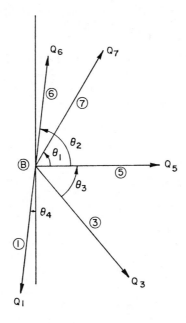

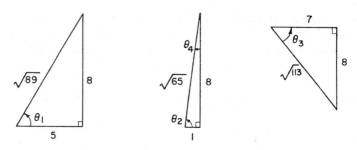

Figure 8.4 Pin-jointed truss: trigonometric ratios.

In the vertical direction,

$$Q_1 \cos\theta_4 + Q_3 \sin\theta_3 - Q_6 \sin\theta_2 - Q_7 \sin\theta_1 = 0,$$

or

$$8Q_1/\sqrt{65} + 8Q_3/\sqrt{113} - 8Q_6/\sqrt{65} - 8Q_7/\sqrt{89} = 0.$$

Establishing these relationships as well at nodes C, D, and E leads to the equilibrium conditions for the truss, which may be summarized in matrix form as

$$C'Q + F_0 + \alpha F = 0, \tag{8.12}$$

where F_0 is an 8-vector of dead loads (in our example, $F_0 = 0$) and $F = (W, 0, W, 2W, 0, 2W, 0, 0)'$ is a vector of live loads (i.e., those which are to be increased proportionally). The matrix C (untransposed) in (8.12) is called the compatibility matrix. The complete compatibility matrix for the pin-jointed truss example is given in Table 8.2.

TABLE 8.2   Compatibility Matrix for Truss Example

i/j	2	3	4	5	6	7	8
$C = (c_{ij})$							
1	$-1/\sqrt{65}$	$8/\sqrt{65}$	0	0	0	0	0	0
2	0	0	0	0	0	$-7/\sqrt{113}$	$8/\sqrt{113}$
3	$7/\sqrt{113}$	$8/\sqrt{113}$	0	0	0	0	0
4	0	0	0	0	0	$1/\sqrt{65}$	$8/\sqrt{65}$
5	1	0	0	0	0	-1	0
6	$1/\sqrt{65}$	$-8/\sqrt{65}$	$-1/\sqrt{65}$	$8/\sqrt{65}$	0	0	0
7	$5/\sqrt{89}$	$-8/\sqrt{89}$	0	0	$-5/\sqrt{89}$	$8/\sqrt{89}$	0	0
8	0	0	$5/\sqrt{89}$	$8/\sqrt{89}$	0	0	$-5/\sqrt{89}$	$-8/\sqrt{89}$
9	0	0	0	0	$1/\sqrt{65}$	$8/\sqrt{65}$	$-1/\sqrt{65}$	$-8/\sqrt{65}$
10	0	0	1	0	-1	0	0

The elastic range of force Q_i is bounded by known yield limits K_i^1 and $-K_i^2$, which depend on the material properties of the truss elements. For proper behavior, Q_i must satisfy

$$-K_i^2 \leq Q_i \leq K_i^1.$$

For the planar truss, these yield conditions are simple lower and upper bounds on the axial force Q_i. For more general structures, the yield conditions will be general inequalities involving a generalized stress vector Q_i. We write the bounds above in the more suggestive form

$$N_i'Q_i \leq K_i,$$

where for our truss,

$$N_i = [1, -1] \quad \text{and} \quad K_i = \begin{bmatrix} K_i^1 \\ K_i^2 \end{bmatrix}.$$

Letting $N = [N_i]$ and $K = [K_i]$, the yield conditions are

$$N'Q \leq K. \tag{8.13}$$

Rigid-plastic collapse load analysis seeks the largest value of α for which (8.12) and (8.13) are satisfied. This is precisely the linear programming problem

$$\max\{\alpha \mid C'Q + \alpha F = -F_0, \quad N'Q \leq K\}. \tag{8.14}$$

The dual of (8.14) is

$$\min\{K'\lambda - F_0'\dot{u} \mid F'\dot{u} = 1, \quad N\lambda + C\dot{u} = 0, \quad \lambda \geq 0\}. \tag{8.15}$$

The optimal values of λ and $\dot{u}$ are, respectively, plastic multiplier rates and displacement rates that characterize the rigid-plastic collapse mechanism of the structure.

For the example truss, nodes B through E move in response to the live loads. Let u denote the 8-vector of element nodal displacements and q denote the 10-vector of element axial deformations. The relationship between q and u may be ascertained from the geometry of the structure. From Figure 8.5, at node B for member 1

$$q_1 = -u_1 \sin\theta_4 + u_2 \cos\theta_4$$

$$= -u_1/\sqrt{65} + 8u_2/\sqrt{65}.$$

Similarly, at node E for member 2

$$q_2 = u_7 \sin\theta_3 - u_8 \cos\theta_3$$

$$= -7u_7/\sqrt{113} + 8u_8/\sqrt{113},$$

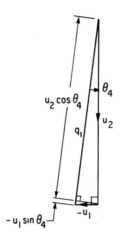

Figure 8.5 Pin-jointed truss: displacement analysis.

and, at node B again, for member 3

$$q_3 = 7u_1/\sqrt{113} + 8u_2/\sqrt{113},$$

and so on. More generally, the reader may verify that

$$q = Cu.$$

In (8.15), the first constraint condition states that plastic mechanism collapse must be associated with a positive workrate of the externally applied live loads. The second set of constraints state that the plastic flow rates for the elements of the truss must be kinematically consistent with the displacement rates of the nodes of the truss. The final set of sign restrictions ensure that plastic flow is associated with dissipation of mechanical energy. Furthermore, the complementary slackness portion of the optimality conditions for (8.14) states that

$$\dot{\lambda}'(N'Q - K) = 0.$$

This means that the rate of plastic flow can be strictly positive only if the stress in the associated element has reached its elastic limit.

Rigid-plastic analysis may be performed using either the primal problem (8.14) or the dual problem (8.15). The primal and dual problems are, in fact, mathematical statements of the Static and Kinematic theorems of plastic analysis, respectively.

With, for example, $W = 1$ and each $K_i^1 = K_i^2 = 1$, the data for the truss analysis are complete. Solving (8.14) gives the maximum value of α as $s = 0.349$. Figure 8.6 shows the nodal displacement rates and the plastic deformation rates for the elements that are active in the collapse mechanism of the truss.

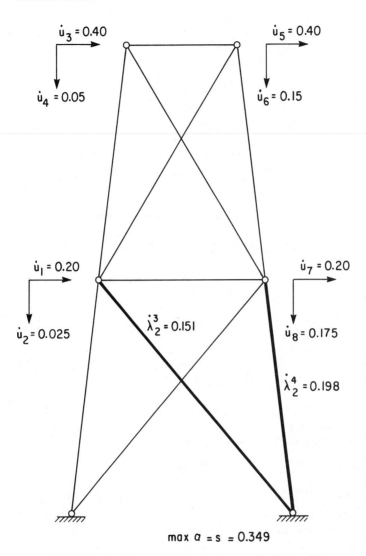

Figure 8.6   Pin-jointed truss: displacement rates at incipient collapse.

8.5 ECONOMIC MODELS

Economic planning must be based on predictions of future economic behavior. In turn, prediction must be based on an economic model. In the 1930s, Wassily Leontief formulated a linear model of an economy composed of n industries. We first present the basic Leontief model and then an extension of it given in Benenson and Glassey [1975A]. The latter is a model of the U.S. economy and was motivated by the OPEC embargo of the mid-1970s. The essential idea of the model was to predict the energy substitution and consumption corresponding to a decrease in crude oil imports. This was formulated as a parametric linear programming problem.

The Leontief model supposes that each industry produces a single good (output) by just one process. That good is produced in sufficient quantity to meet the input requirements of the other industries and to meet the exogenous or net demand of the customer. Let

$$y_{ij} = \text{amount of good } i \text{ required by industry } j,$$
$$(= \text{ input from industry } i \text{ to industry } j),$$
$$b_i = \text{ exogenous demand for good } i,$$
$$x_i = \text{ total amount produced by industry } i.$$

The amount produced by industry i to meet the total demand is therefore

$$x_i = \sum_{j=1}^{n} y_{ij} + b_i. \tag{8.16}$$

The y_{ij} can be thought of as production functions. In general, the y_{ij}'s are nonlinear and difficult, if not impossible, to obtain explicitly. Leontief made the simplifying assumption that the amount of good i input to industry j is proportional to the output of industry j; that is,

$$y_{ij} = a_{ij}x_j, \quad \text{for all } i, j.$$

In matrix notation, (8.16) becomes

$$x = Ax + b. \tag{8.17}$$

The matrix A is sometimes called the direct requirements matrix or input-output matrix, and the vector x of total amounts produced is called the gross output vector. The economic conservation law expressed by (8.17) can be interpreted in the following way. Multiplying the direct requirements matrix A by the gross output vector x yields the vector of total requirements of goods and services input to the producing sectors of the economy. The conservation law simply states that production (gross output) equals consumption by the producing sectors plus consumption by the final demand sectors.

Next we consider prices for the goods produced in the economy and show that they are, in some sense, "dual" to the quantities of goods produced. Let p_j denote the unit price of good j. The total cost of all goods required to manufacture one unit of good j is

$$\sum_{i=1}^{n} a_{ij}p_i.$$

The difference between p_j and this quantity is called the *value added* v_j:

$$v_j = p_j - \sum_{i=1}^{n} a_{ij} p_i,$$

or

$$p = A'p + v. \qquad (8.18)$$

The value added is composed of labor costs and factors such as profit.

Inspection of (8.17) and (8.18) reveals a certain symmetry. Replacing x with p, A with A', and b with v in (8.17) results in (8.18). The quantities x and p are uniquely determined by (8.17) and (8.18). Solving yields

$$x = (I - A)^{-1}b, \quad p = (I - A')^{-1}v.$$

The model specified by (8.17) and (8.18) is rigid. For specified demand b and value added v, the amounts produced and their associated prices are completely specified. A more flexible model is described by Benenson and Glassey to analyze the effect of fuel and energy use in the United States. Their model is designed to answer the general question: What are the impacts of an energy shortage? More specifically, for each energy sector (coal, crude petroleum, refined petroleum, crude and refined petroleum imports, electric utilities, and gas utilities) and for any degree of hypothesized shortage, the quantities analyzed are:

1. pattern of energy used throughout the economy;
2. pattern of fuel substitution within the electric utilities and iron and steel sectors;
3. direct and indirect impacts on industrial production by sector;
4. impact on total employment and on specific occupations within each producing sector; and
5. impact on Gross National Product (GNP).

The basic model is the linear programming problem

$$
\begin{aligned}
\text{maximize:} \quad & \sum_i v_i x_i \qquad (= \text{GNP}) \\
\text{subject to:} \quad & (I - A)x + (I - B)M - y = 0, \\
& Lx \leq \bar{L}, \\
& x \leq \bar{x}, \\
& M \leq \bar{M}, \\
& - y \leq -\bar{y}, \\
& x, M \geq 0,
\end{aligned}
\right\} \qquad (8.19)
$$

where:

x = gross output vector (production),

A = direct requirements matrix,

L = labor requirements matrix,

y = final demand vector,

M = vector of import activity variables,

B = service requirements matrix for import activities,

$\overline{M}$ = vector of upper bounds on import activities,

$\overline{x}$ = vector of upper bounds on gross output, representing productive capacity of each sector,

$\overline{L}$ = upper bound on total employment representing the labor supply,

$\underline{y}$ = vector of lower bounds for final demand,

v = vector of value-added coefficients.

The demand vector y is variable to allow changes in consumption patterns. For a shortage of crude oil, for example, this would allow a shift in consumption to commodities that require smaller amounts of crude oil in their consumption. The lower bound $\underline{y}$ on final demand prevents final demand from varying unrealistically from its observed value of the data base year.

Imports M are mentioned explicitly for three reasons. Domestic production technology is more accurately represented by the input-output coefficients. The constraints on domestic capacity, employment, and availability of commodity input limits domestic output, not imports. Finally, the effect of import quotas or embargos (e.g., petroleum) can be analyzed.

The service requirements matrix B plays a rôle for imports M which is analogous to that of the direct requirements matrix A for output x. Since imports have few domestic input activities other than transportation, warehousing, and insurance, B has nonzero elements in only a small number of rows.

The constraints of (8.19) will be satisfied by a wide variety of final demands and production and import activities. The model assumes that the economy acts in such a way as to maximize the Gross National Product

$$\sum_i v_i x_i.$$

It is shown that GNP can be closely approximated by total final demand

$$\sum_i y_i$$

and the latter is used as the objective function. The model contains additional refinements which we describe briefly as follows.

Fuel substitution in the electric utilities sector, generating technologies using coal, gas, oil, combinations thereof, and hydroelectric and nuclear power, are explicitly represented. As one type of energy shortage is posited, the model simulates the ability of the electric utility sector to substitute abundant for scarce energy sources, limited by plant capacities for switching from one fuel type to another.

Fuel substitution in the iron and steel sectors, iron and steel production, one of the largest energy consumers in the United States, is formulated using five possible technologies. The first produces iron and steel output in agreement with the data base year of minimum total cost. The remaining four processes produce the same amount of output and minimize the use of one of four types of energy input (coal, oil, natural gas, and electricity). If, for example, oil becomes a scarce commodity in the national model, the process that minimizes the use of oil will be chosen in the submodel.

The model also makes a distinction between peak power and total energy demand and disaggregates the output of the crude petroleum and natural gas sector into two products: crude petroleum, and natural gas liquids and crude natural gas.

In the fuel substitution in the electric utilities submodel, nonfuel inputs are assumed with equal intensity by each generation technology and so are allocated to a transmission and distribution activity which produces "sold electricity" from "generated electricity."

The analysis begins by projecting the most recent data available to several future base years (1975, 1980, 1985 were used in the study). For each base year a base case is constructed. Using projected capacities and incorporating the various submodel constraints the linear programming problem (8.19) is solved to obtain the base case solution. Parametric linear programming is then used to simulate shortages in each of the energy sectors and in petroleum imports. For example, if x_{16} denotes the electric utilities gross domestic output and $\bar{x}_{16}$ its projected capacity for the base year under consideration, the capacity constraint $x_{16} \leq \bar{x}_{16}$ is replaced with the parametric constraint

$$x_{16} \leq (100 - t)\bar{x}_{16}/100$$

and all other constraints remain unchanged. As t varies from 0 to 100, the optimal solution to the parametric linear program traces the effect on the remaining sectors of the economy of reducing electric utilities gross output from 0 to 100% of its estimated capacity.

For each base year the Benenson-Glassey study parametrically reduced capacities on:

1. electric utilities, including summer and winter peaks, the gas-coal, gas-oil, and coal-oil burning capacities, the gas, coal, and oil capacities, and hydroelectric and nuclear capacity;

2. nuclear generating capacity;
3. crude petroleum and natural gas;
4. refined petroleum;
5. imports of crude petroleum and refined petroleum products;
6. gas utilities; and
7. coal.

From the solutions of the linear programming problems, they were able to discern a general pattern in the behavior of the model components which we repeat here. A scarce energy resource is first reduced to its final demand lower bound and then imported to its upper bound; next fuel substitution in the electric utilities and iron and steel sectors takes place. The substitution continues until either the scarce fuel use is reduced to zero or the substitute fuel is burned to its capacity. In most cases a substitute fuel is used. When its capacity is reached, noticeable effects begin to occur in the other sectors of the economy. The changes in the sector's output depend on the intensity of use of the scarce fuel in the sector's production process and the sector's contribution to the GNP.

There are several types of limitations to this type of model, some inherent and some removable. The inherent limitations include the assumption that the economy is perfectly competitive and in equilibrium. The removable limitations include the assumptions of single technology production, homogeneous commodities, and uniform prices across sectors. The first assumption is that each sector uses only one technology to produce a single output so that no substitution of inputs is possible. This limitation can be overcome by disaggregation of activities as for the iron and steel and electric utilities sectors. The second assumption requires, for example, that steel is steel and not one of the several hundred grades, shapes, or sizes actually marketed. Again, disaggregation of commodities can be used to overcome this limitation. The assumption of uniform prices across sectors implies that all buyers pay the same price for a given commodity. However, in practice prices may fluctuate due to regionality, quantity discounts, or special contracts. Goods differentiated by such side conditions may be considered as distinct economic goods.

An inherent and serious limitation of the model is that the demand schedule for each product is L-shaped and independent of other product prices. To explain why this assumption is implicit in the linear model, we return to the interpretation of dual variables as shadow prices as discussed in Section 3.4.

The analysis does not depend on the structure of (8.19) so we rewrite it in the more compact form

$$\max\{c'x \mid Ax \le b\}. \tag{8.20}$$

Regard (8.20) as the primal problem. Let x and u be optimal solutions for

the primal and dual, respectively. From the Strong Duality Theorem
(Theorem 3.8)

$$\sum_{i=1}^{n} c_i x_i = \sum_{i=1}^{m} b_i u_i. \tag{8.21}$$

Since $c'x$ is measured in dollars, so too must each term $b_i u_i$. Therefore, the
units of u_i must be dollars per unit of resource i. Thus u_i acts as a price.

Suppose that constraint i is active. Then it is limiting the maximum
profit of the business. If Δb_i additional units of resource i become available
at a unit price of p_i they would be purchased only if they increased total profit. The cost to the business of the additional units is $\Delta b_i p_i$ and the additional profit they bring is $\Delta b_i u_i$. Therefore, they are purchased only if the net
profit for the business is nonnegative (i.e., $u_i \geq p_i$). Conversely, the industry
that produces the scarce resource can sell its product only if it offers it for
sale at an attractive price (i.e., $p_i \leq u_i$). Because of the price interpretation
of the multipliers u_i, in an economic context they are frequently called *shadow prices* or *imputed costs*.

If $u_i > p_i$, the manufacturers will purchase additional units of resource
i to the limit at which they can use it. Eventually, some other resource limits
the production process (i.e., a new constraint becomes active and changes the
u_i's). This forces the resource seller to reevaluate his prices. To sell his
resource, the new price $\hat{p}_i$ must be lower than the new value of $\hat{u}_i$. If he can
reduce his price below u_i he will do so. Eventually, the selling and buying
comes to a halt. At this equilibrium point, the price p_i of each scarce
resource is equal to u_i. If p_i were less than u_i, additional units of the ith
resource would be purchased. If p_i were greater than u_i, the manufacturer
would reduce the amount of ith resource purchased. Thus, $p_i = u_i$
corresponds to the balanced state of economic equilibrium.

Return now to the specific model (8.19). Let c now denote the vector
of prices of commodities produced by the economy. It can be shown under
appropriate assumptions that GNP can be expressed as $c'y$, where y denotes
the final demand vector. Thus the objective function can be replaced with
$c'y$. Let u and w denote the multiplier vectors associated with the first and
fifth sets of constraints of (8.19). Part of the dual feasibility portion of the
optimality conditions for this problem states that

$$c_i = p_i - w_i. \tag{8.22}$$

The associated complementary slackness condition states that

$$w_i(y_i - \bar{y}_i) = 0.$$

We may interpret (8.22) as a demand schedule of the ith commodity. p_i is
the unit equilibrium price of the ith commodity. When demand is above its
lower limit, the equilibrium price remains constant at the nominal value c_i
($y_i > \bar{y}_i$ implies that $w_i = 0$). The demand schedule is thus L-shaped as
shown in Figure 8.7.

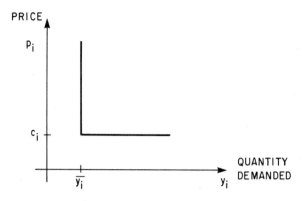

Figure 8.7 Linear economic model: demand schedule.

In economic terms, the schedule is made up of a completely inelastic (vertical) segment and a completely elastic (horizontal) segment. It implies that changes in price greater than c_i will have no effect on quantity y_i consumed (inelastic segment) and that any quantity greater than $\bar{y}_i$ may be purchased at the nominal price c_i. This crude assumption is inherent in the (linear) model as previously stated and cannot be improved upon without significantly increasing the number of variables. A similar demand schedule can be constructed for labor. The implication of it is that if labor demand is not precisely equal to labor supply, then the price of labor is zero.

The L-shaped demand schedule is unrealistic because we expect that larger quantities can be purchased at a lower unit price and that smaller quantities bring a higher unit price. That is, we expect demand curves to be downward sloping.

In addition, (8.22) implies that product prices are independent of one another. This unrealistic consequence of the model and the implicit assumption of L-shaped demand curves can be removed by replacing the linear objective function of (8.19) with a quadratic objective function (see Benenson and Glassey, [1975B]).

EXERCISES

8.1. Given observations (t_i, v_i), $i = 1, \ldots, k$, we wish to find the coefficients α and β of the line $f(t) = \alpha + \beta t$ for which $f(t)$ best "fits" the observations. One way of doing this is by making the error

$$\max_{i=1,\ldots,k} |f(t_i) - v_i|$$

as small as possible.

(a) Formulate this problem as a linear programming problem.
(b) Suppose that β, t, and t_i, $i = 1, \ldots, k$ are now n-vectors. Formulate the resulting problem as an LP.

8.2. Because of traffic density changes, a bus company has variable demand for drivers. The following data are given.

Time	Drivers Required
0100–0500	15
0500–0900	30
0900–1300	26
1300–1700	32
1700–2100	30
2100–0100	19

Every driver works on an 8-hour shift. A shift can start at 0100, 0500, 0900, 1300, 1700, or 2100. Each driver must work the full 8 hours. Find the schedule with the minimum number of drivers.

APPENDIX A

DEGENERACY

All algorithms for the solution of a linear programming problem presented thus far require a nondegeneracy assumption in order to guarantee finite termination. The purpose of this appendix is to present a modification of the algorithms for which finite termination is guaranteed with no assumptions.

A.1 DEGENERACY

In this section we give a detailed description of the implementation of Bland's rules [Bland, 1977] which ensures finite termination in the presence of active constraints having linearly dependent gradients. Since every algorithm in Chapters 2 and 4 can be viewed as a special case of Algorithm 3 as applied to the model problem

$$
\left.
\begin{aligned}
\text{minimize:} \quad & c'x \\
\text{subject to:} \quad & a_i'x \leq b_i, \quad i = 1, \ldots, m, \\
& a_i'x = b_i, \quad i = m + 1, \ldots, m + r,
\end{aligned}
\right\} \quad \text{(A.1)}
$$

it is sufficient to discuss the appropriate modifications of Algorithm 3.

The possibility of cycling occurs only if, at some iteration j, $\sigma_j = 0$, and s_j is determined by Step 1.2 of Algorithm 3. In this case s_j is rejected. A new search direction is computed by Bland's rules (i.e., $s_j = c_{kj}$), where k is such that

$$
\alpha_{kj} = \min \{ \alpha_{ij} \mid \text{all } i \text{ with } c'c_{ij} > 0 \text{ and } 1 \leq \alpha_{ij} \leq m \}.
$$

Thus the modified algorithm can be stated as follows

ALGORITHM A

Model Problem:

$$\min\{c'x \mid a_i'x \le b_i, i = 1, \ldots, m; a_i'x = b_i, i = m + 1, \ldots, m + r\}$$

Initialization:

Start with any feasible point x_0, $J_0 = \{\alpha_{10}, \ldots, \alpha_{n0}\}$, and D_0^{-1}, where $D_0' = \left[d_1, \ldots, d_n \right]$ is nonsingular, $d_i = a_{\alpha_{i0}}$ for all i with $1 \le \alpha_{i0} \le m + r$, and each of $m + 1, \ldots, m + r$ is in J_0. Compute $c'x_0$ and set $j = 0$.

Step 1:   Computation of Search Direction s_j.

Same as Step 1 of Algorithm 3.

Step 1.1:

Set $\gamma_j = 1$ and continue as in Step 1.1 of Algorithm 3.

Step 1.2:

Set $\gamma_j = 2$ and continue as in Step 1.2 of Algorithm 3.

Step 1.3:

Set $\gamma_j = 3$ and compute

$$v_i = c'c_{ij}, \quad \text{for all } i \text{ with } 1 \le \alpha_{ij} \le m.$$

If $v_i \le 0$ for all i with $1 \le \alpha_{ij} \le m$, stop with optimal solution x_j. Otherwise, determine k such that

$$\alpha_{kj} = \min\{\alpha_{ij} \mid \text{all } i \text{ with } v_i > 0 \text{ and } 1 \le \alpha_{ij} \le m\}.$$

Set $s_j = c_{kj}$ and go to Step 2.

Step 2:   Computation of Step Size σ_j.

Compute σ_j as in Step 2 of Algorithm 3.
If $\sigma_j = 0$ and $\gamma_j = 2$, then go to Step 1.3. Otherwise, go to Step 3.

Step 3:   Update.

Same as Step 3 of Algorithm 3.

The following theorem establishes the main properties of Algorithm A.

Theorem A.1.

Let Algorithm A be applied to the model problem (A.1). Then Algorithm A terminates after a finite number of steps with either an optimal solution or the information that the problem is unbounded from below.

Proof:

By construction, each x_j obtained by Algorithm A is feasible. If termination occurs with "problem unbounded from below," then

$c's_j > 0$, $a_i's_j \geq 0$, $i = 1, \ldots, m$; $a_i's_j = 0$, $i = m + 1, \ldots, m + r$.

Thus $x_j - \sigma s_j$ is feasible for every $\sigma > 0$ and $c'(x_j - \sigma s_j) = c'x_j - \sigma c's_j$ is a strictly decreasing function of σ. The objective function is indeed unbounded from below on the set of feasible solutions. If termination occurs with "optimal solution x_j," we have $c'c_{ij} = 0$ for all i with $\alpha_{ij} = 0$, $c'c_{ij} \leq 0$ for all i with $1 \leq \alpha_{ij} \leq m$, and

$$c = \sum (c'c_{ij})a_{\alpha_{ij}},$$

where the summation is over all i with $1 \leq \alpha_{ij} \leq m + r$. Let x be any feasible solution. Then $x = x_j + (x - x_j)$ and

$$a_i'x = a_i'x_j + a_i'(x - x_j) = b_i;$$

i.e.,

$$a_i'(x - x_j) = 0, \quad i = m + 1, \ldots, m + r.$$

Furthermore, for $1 \leq \alpha_{ij} \leq m$,

$$a_{\alpha_{ij}}'x = a_{\alpha_{ij}}'x_j + a_{\alpha_{ij}}'(x - x_j) \leq b_i;$$

i.e.,

$$a_{\alpha_{ij}}'(x - x_j) \leq 0.$$

Therefore,

$$c'x = c'x_j + c'(x - x_j) = c'x_j + \sum (c'c_{ij})a_{\alpha_{ij}}'(x - x_j) \geq c'x_j.$$

Thus $c'x_j \leq c'x$ and x_j is indeed an optimal solution.

To prove finite termination of the algorithm, choose any $j_1 < j_2$. Let J_{j_1} and J_{j_2} be the index set used at iteration j_1 and j_2, respectively. Furthermore, denote by S_1 and S_2 the set of all positive elements contained in J_{j_1} and J_{j_2}, respectively. We will show by contradiction that $S_1 \neq S_2$.

Suppose that $S_1 = S_2$. Because the number of positive elements of the index sets J_j cannot decrease and increases whenever s_j is determined by Step 1.1, it follows that, for every j with $j_1 \leq j < j_2$, s_j is determined either by Step 1.2 or Step 1.3. In particular, if we denote the columns of D_{j_1}' and $D_{j_1}^{-1}$ by $d_{11}, \ldots, d_{n1}$ and $c_{11}, \ldots, c_{n1}$, respectively, and the elements of J_{j_1} by $\alpha_{11}, \ldots, \alpha_{n1}$, we have

$$d_{i1} \equiv a_{\alpha_{i1}}, \quad \text{for all } i \text{ with } \alpha_{i1} > 0,$$

$$c = \sum_{i=1}^{n} v_i d_{i1}, \quad \text{with } v_i = c'c_{i1}, \tag{A.2}$$

and $v_i = 0$ for all i with $\alpha_{i1} = 0$. Let

$$T = \{x \mid a'_{\alpha_{i1}}x = b_i \text{ for all } i \text{ with } \alpha_{i1} \geq 1\}.$$

Then $x_{j_1} \in T$. For every $x \in T$ we have $a'_{\alpha_{i1}}(x - x_j) = 0$ for all i with $\alpha_{i1} \geq 1$. Thus it follows from (A.2) that $c'x = c'x_{j_1}$ for every $x \in T$. Because $S_1 = S_2$ we have $x_{j_2} \in T$. Therefore,

$$c'x_{j_1} = c'x_{j_2}. \tag{A.3}$$

Since $c'x_{j+1} \leq c'x_j$ for every j and $c'x_{j+1} < c'x_j$ whenever s_j is determined by Step 1.2 we deduce from (A.3) that s_j is computed from Step 1.3 and $x_{j+1} = x_j$ for all j with $j_1 \leq j < j_2$.

Define $I \subset \{1, \ldots, m\}$ such that $i \in I$ if and only if $i \notin J_j$ and $i \in J_{j+1}$ for at least one j with $j_1 \leq j < j_2$; that is, $i \in I$ if and only if a_i is not in D'_j but is a column of D'_{j+1} for at least one j with $j_1 \leq j < j_2$. Let

$$\rho = \max\{i \mid i \in I\}.$$

Since $S_1 = S_2$ it follows that there are integers ν and η such that $j_1 \leq \nu$, $\eta < j_2$,

$$\rho \notin J_\nu, \quad \rho \in J_{\nu+1}, \tag{A.4}$$

and

$$\rho \in J_\eta, \quad \rho \notin J_{\eta+1}. \tag{A.5}$$

Let

$$v = (v_1, \ldots, v_n)' = D_\eta s_\nu \quad \text{and} \quad u = (u_1, \ldots, u_n)' = (D_\eta^{-1})'c.$$

Then

$$v'u = s'_\nu(D'_\eta)(D_\eta^{-1})'c = s'_\nu c = c's_\nu > 0,$$

and there is an i_0 such that $v_{i_0}u_{i_0} > 0$. Let $\tau = \alpha_{i_0\eta}$ and $D'_\eta = [d_{1\eta}, \ldots, d_{n\eta}]$. Then $v_{i_0} = d'_{i_0\eta}s_\nu$. If $\tau = 0$, then $d_{i_0\eta}$ is a column of D'_j for $j = j_1, j_1 + 1, \ldots, j_2$ and $v_{i_0} = 0$. Thus $\tau \geq 1$ and $d_{i_0\eta} = a_\tau$. Furthermore, $u_{i_0} = c'_{i_0\eta}c = c'c_{i_0\eta}$. Therefore,

$$(a'_\tau s_\nu)(c'c_{i_0\eta}) > 0. \tag{A.6}$$

Assume first that $\tau > \rho$. Then the definition of ρ implies that a_τ is a column of D'_j for every j with $j_1 \leq j \leq j_2$. Thus $a'_\tau s_\nu = 0$ and $(a'_\tau s_\nu)(c'c_{i_0\eta}) = 0$. Now suppose that $\tau = \rho$. Then it follows from (A.4) that $a'_\tau s_\nu < 0$ and from (A.5) that $c'c_{i_0\eta} = c's_\nu > 0$. Thus $(a'_\tau s_\nu)(c'c_{i_0\eta}) < 0$. Finally, suppose that $\tau < \rho$. Since by (A.4) the vector a_ρ is chosen to be incorporated into the matrix $D'_{\nu+1}$ it follows from the definition of ρ and the selection rule in Step 2 of Algorithm A that $a'_\tau s_\nu \geq 0$. Furthermore, the method for choosing k in

Step 1.3 of Algorithm A and the fact that a_ρ is deleted from D'_η imply that $c'c_{i_0\eta} \leq 0$. This argument shows that the inequality (A.6) cannot be true. This contradiction derives from the assumption that $S_1 = S_2$.

Therefore, we have proven that no two index sets contain the same positive elements. Because every J_j contains the numbers $m + 1, \ldots, m + r$ the potential number of index sets and thus the potential number of iterations is bounded by the number of different subsets of $\{1, \ldots, m\}$ with k elements where $k = 0, 1, \ldots, n - r$. This completes the proof. ■

APPENDIX B

COMPUTER PROGRAMS

In this appendix, computer programs are presented which implement the algorithms set out in this book. Also given is the output from applying these programs to example problems from the main text. The programs are self-contained and short. For example, the program implementing Algorithm 1 is less than 200 statements. Each program is written as a set of subroutines corresponding to the steps of the algorithm.

The purpose of the computer programs is to increase the reader's understanding of the various algorithms. They can be used to solve the example problems and variations of them, the exercise problems, and the application problems. Readers whose interest in linear programming stems from engineering or business applications may wish to use the programs as subroutines in a larger program, one which performs an analysis of which linear programming is just one part. This capability is especially useful because many commercially available LP software packages are written as stand-alone systems which are difficult, if not impossible, to interface with other programs.

There are many computing languages available with a variety of attractive features. We will use FORTRAN and adhere to the conventions of FORTRAN 77 (ANSI Standard) [American National Standards Institute, 1978]. Our reasons for choosing FORTRAN are that it has been in common use for over two decades and that FORTRAN 77 compilers are widely available both for mainframe and microcomputers.

The programs have been written in a simple and direct manner and closely reflect the formal statement of the algorithm. We have specifically avoided programming techniques which depend on internal conventions of

FORTRAN (such as storing matrices by column, for example). Occasionally, the commitment to simplicity was made with some cost to computational efficiency. For example, in Sections B.1.4 to B.1.6, the Phase 1 and Phase 2 data are stored separately. The doubling of the storage requirement could be avoided by clever programming with just one copy of the data.

Table B.1 shows the name, figure reference, and purpose of each subroutine presented in this appendix.

TABLE B.1   Name and Purpose of FORTRAN Subroutines

Subroutine Name	Figure	Purpose
ALG1	B.4	Algorithm 1
ALG2	B.12	Algorithm 2
ALG3	B.17	Algorithm 3
STEP11	B.5	Step 1 of Algorithm 1
STEP12	B.13	Step 1 of Algorithm 2
STEP13	B.18	Step 1 of Algorithm 3
STEP2	B.6	Step 2 of Algorithms 1, 2, and 3
STEP3	B.7	Step 3 of Algorithms 1, 2, and 3
PHI	B.8	Procedure Φ
VERIFY	B.2	checks input data for inconsistencies
SOLN	B.3	prints optimal solution from ALG1, ALG2, ALG3, P1P2A, P1P2B, P1P2C, and ALG5
PDSOLN	B.32	prints optimal primal and dual solutions from ALG1, ALG2, ALG3, P1P2A, P1P2B, P1P2C, and ALG5
P1P2A	B.22	Phase 1−Phase 2 procedure for problem with no equality constraints; $D_0^{-1} = I, J_0 = \{0, \ldots, 0\}$ as initial data for Phase 2
P1P2B	B.26	Phase 1−Phase 2 procedure for problem with no equality constraints; initial data for Phase 2 extracted from final data of Phase 1
P1P2C	B.29	Phase 1−Phase 2 procedure for problem with equality constraints; initial data for Phase 2 extracted from final data of Phase 1
SIMPLX	B.36	revised simplex method
RSTEP1	B.37	Step 1 of the revised simplex method
RSTEP2	B.38	Step 2 of the revised simplex method
RSTEP3	B.39	Step 3 of the revised simplex method
PHIPRM	B.40	$[\Phi((B^{-1})', A_k, l)]'$; i.e., update of B^{-1}
P1P2S	B.45	Phase 1−Phase 2 revised simplex method
RSOLN	B.41	prints optimal primal and dual solutions from SIMPLX and DSMPLX

TABLE B.1   *(continued)*

Subroutine Name	Figure	Purpose
ALG4	B.49	Algorithm 4
PSTEP1	B.50	Step 1 of Algorithm 4
PSTEP2	B.51	Step 2 of Algorithm 4
PSTEP3	B.52	Step 3 of Algorithm 4
PSTEP4	B.53	Step 4 of Algorithm 4
PSTEP5	B.54	Step 5 of Algorithm 4
SUMM	B.55	prints solution summary from ALG4
ALG5	B.59	Algorithm 5
DSTEP1	B.60	Step 1 of Algorithm 5
DSTEP2	B.61	Step 2 of Algorithm 5
DSTEP3	B.62	Step 3 of Algorithm 5
DSMPLX	B.66	dual simplex method
DRSTP1	B.67	Step 1 of the dual simplex method
DRSTP2	B.68	Step 2 of the dual simplex method
DRSTP3	B.69	Step 3 of the dual simplex method
USMPLX	B.73	upper-bounded simplex method
USTEP1	B.74	Step 1 of the upper-bounded simplex method
USTEP2	B.75	Step 2 of the upper-bounded simplex method
USTEP3	B.76	Step 3 of the upper-bounded simplex method
USOLN	B.77	prints optimal primal and dual solutions from USMPLX
GUB	B.81	generalized upper-bounded simplex method
GSTEP1	B.83	Step 1 of the generalized upper-bounded simplex method
GSTEP2	B.84	Step 2 of the generalized upper-bounded simplex method
GSTEP3	B.85	Step 3 of the generalized upper-bounded simplex method
GSOLN	B.86	prints optimal primal and dual solutions from GUB
ELT	B.82	given l, ELT finds ν such that $l \in S_\nu$

B.1   THE ALGORITHMS OF CHAPTER 2

In this section, we give computer programs for the algorithms presented in Chapter 2: Algorithms 1, 2, 3 (ALG1, ALG2, ALG3, respectively), and three variations of the Phase 1–Phase 2 procedure (P1P2A, P1P2B, P1P2C). The first Phase 1–Phase 2 procedure is that of Section 2.5. It is formulated for inequality constraints only and has two variations. The first variation uses the initial data $D_0^{-1} = I$ and $J_0 = \{0, \ldots, 0\}$ as initial data for the Phase 2 problem. The second variation extracts J_0 and D_0^{-1} from the final data of the initial $(n + 1)$-dimensional Phase 1 problem. The third variation allows explicit linear equality constraints and extracts the n-dimensional initial data J_0 and D_0^{-1} for the Phase 2 problem from the final data of the $(n + 1)$-dimensional Phase 1 problem of Section 2.6.

Each computer program is written using several subroutines. Each subroutine corresponds to a step in an algorithm or some well-defined function, and may be used by several algorithms. The name and purpose of each subroutine is given in Table B.2. The most important FORTRAN identifiers and their meanings are summarized in Table B.3.

TABLE B.2   Name and Purpose of FORTRAN Subroutines for the Algorithms of Chapter 2

Subroutine Name	Figure	Purpose
	B.1	main program for Algorithm 1
	B.11	main program for Algorithm 2
	B.16	main program for Algorithm 3
	B.21	main program for Phase 1–Phase 2 Procedure A
	B.25	main program for Phase 1–Phase 2 Procedure B
	B.28	main program for Phase 1–Phase 2 Procedure C
ALG1	B.4	Algorithm 1
ALG2	B.12	Algorithm 2
ALG3	B.17	Algorithm 3
P1P2A	B.22	Phase 1–Phase 2 procedure for problem with no equality constraints; $D_0^{-1} = I, J_0 = \{0, \ldots, 0\}$ as initial data for Phase 2
P1P2B	B.26	Phase 1–Phase 2 procedure for problem with no equality constraints; initial data for Phase 2 extracted from final data of Phase 1
P1P2C	B.29	Phase 1–Phase 2 procedure for problem with equality constraints; initial data for Phase 2 extracted from final data of Phase 1
STEP11	B.5	Step 1 of Algorithm 1
STEP12	B.13	Step 1 of Algorithm 2
STEP13	B.18	Step 1 of Algorithm 3
STEP2	B.6	Step 2 of Algorithms 1, 2, and 3
STEP3	B.7	Step 3 of Algorithms 1, 2, and 3
PHI	B.8	Procedure Φ
VERIFY	B.2	checks input data for inconsistencies
SOLN	B.3	prints optimal solution from ALG1, ALG2, ALG3, P1P2A, P1P2B, P1P2C, and ALG5
PDSOLN	B.32	prints optimal primal and dual solutions from ALG1, ALG2, ALG3, P1P2A, P1P2B, P1P2C, and ALG5

**TABLE B.3   Meaning of FORTRAN Identifiers
for the Algorithms of Chapter 2**

FORTRAN Identifier	Notation in Main Text
N	n
M	m
R	r
A	A
A(I,1),...,A(I,N)	a_i'
B	b
C	c
X	x_j
K	k
ELL	l
S	s_j
ITER	j
JJ	J_j
DINV	D_j^{-1}
SIGJ	σ_j
ALPHA	α_j
RHO	ρ
OBJ	$c'x_j$
MTOT	$m + r$
JOUT	$= \begin{cases} 1, & \text{optimal solution} \\ 2, & \text{problem is unbounded from below} \\ 3, & \text{problem has no feasible solution} \end{cases}$

B.1.1   Algorithm 1

The main program for Algorithm 1 (Figure B.1) reads the data for A, b, c, x_0, J_0, D_0^{-1}, m, and n into A, B, C, X, JJ, DINV, M, and N, respectively. The READ statements are format-free. It is assumed that the data has been placed on a file associated with logical unit 37 (IIN = 37). All output is written to a file associated with logical unit 38 (IOUT = 38). IIN and IOUT have been placed in the COMMON block UNITS so that they are accessible to other subroutines.

A call is made to VERIFY which looks for inconsistencies in the problem data. If none are found, PASS is returned with value .TRUE. and the program proceeds. Otherwise, VERIFY writes a list of the difficulties to unit 38, and the main program halts. Strictly speaking, VERIFY is not necessary. However, it is possible to make mistakes in preparing x_0, J_0, and D_0^{-1}. The following programs will do *something* with inconsistent data, but the solutions produced may lead to erroneous conclusions. We have found it best to use VERIFY.

```
      IMPLICIT DOUBLE PRECISION (A-H,O-Z)
      DIMENSION A(5,2),B(5),C(2),X(2),DINV(2,2),
     1   S(2),V(2),WORK(2),JJ(2)
      INTEGER R
      LOGICAL PASS
      COMMON /UNITS/ IIN,IOUT
      IIN = 37
      IOUT = 38
      READ(IIN,*) M,N
      MTOT = M
      R = 0
      READ(IIN,*) (C(I),I=1,N)
      READ(IIN,*) (B(I),I=1,M)
      DO 100 I=1,M
  100 READ(IIN,*) (A(I,J),J=1,N)
      READ(IIN,*) (X(I),I=1,N)
      READ(IIN,*) (JJ(I),I=1,N)
      DO 200 I=1,N
  200 READ(IIN,*) (DINV(I,J),J=1,N)
      CALL VERIFY(X,DINV,JJ,A,B,N,M,R,MTOT,PASS)
      IF(.NOT.PASS) STOP
      CALL ALG1(A,B,C,X,DINV,S,V,WORK,JJ,OBJ,N,M,
     1   MTOT,JOUT)
      CALL SOLN(JOUT,X,N,OBJ)
      STOP
      END
```

Figure B.1   Main program for Algorithm 1.

SUBROUTINE VERIFY (Figure B.2) decides if x_0 is feasible and if D_0^{-1} really is the inverse of D_0 as implicitly defined by J_0. Any difficulties are reported on unit 38. The DO 100 and DO 200 loops compute $a_i' x_0$. If $a_i' x_0 \leq b_i + 10^{-5}$, for $1 \leq i \leq m$ or $|a_i' x_0 - b_i| \leq 10^{-5}$ for $m + 1 \leq i \leq m + r$, then x_0 is considered feasible with respect to constraint i. If the condition is not satisfied, an appropriate message is written to unit 38 and PASS is set to .FALSE..

Let $D_0' = [d_1, \ldots, d_n]$, $D_0^{-1} = [c_{10}, \ldots, c_{n0}]$, and $J_0 = \{\alpha_{10}, \ldots, \alpha_{n0}\}$. If $1 \leq \alpha_{i0} \leq m + r$, then $d_i = a_{\alpha_{i0}}$ and, by definition of the inverse matrix,

$$a_{\alpha_{i0}}' c_{j0} = \begin{cases} 1, & \text{for } j = i, \\ 0, & \text{for } j = 1, \ldots, n, \ j \neq i. \end{cases}$$

The DO 300, 400, and 500 loops compute the left-hand side of the above and compare it with the right-hand side. If agreement within 10^{-5} is obtained, the test is passed. Otherwise, an appropriate message is issued.

```
      SUBROUTINE VERIFY(X,DINV,JJ,A,B,N,M,R,MTOT,
     1   PASS)
      IMPLICIT DOUBLE PRECISION(A-H,O-Z)
      DIMENSION X(N),DINV(N,N),JJ(N),A(MTOT,N),
     1   B(MTOT)
      INTEGER R
      LOGICAL PASS
      COMMON /UNITS/ IIN,IOUT
      TOL = 1.D-5
      PASS = .TRUE.
C   CHECK FEASIBILITY FIRST
      DO 200 I=1,MTOT
      SUM = 0.D0
      DO 100 J=1,N
  100 SUM = SUM + A(I,J)*X(J)
      IF((I.LE.M).AND.(SUM.LE.(B(I)+TOL))) GO TO 200
      IF((I.GT.M).AND.(DABS(SUM-B(I)).LE.TOL))
     1                                         GO TO 200
      PASS = .FALSE.
      WRITE(IOUT,8000) I,SUM,B(I)
  200 CONTINUE
C
C   NOW CHECK DINV AND JJ
C
      DO 500 I=1,N
      INDEX = JJ(I)
      IF(INDEX.EQ.0) GO TO 500
C   FORM INNER PRODUCT OF GRADIENT OF CONSTRAINT INDEX
C   WITH EVERY COLUMN OF DINV
      DO 400 J=1,N
      SUM = 0.D0
      DO 300 IRUN =1,N
  300 SUM = SUM + A(INDEX,IRUN)*DINV(IRUN,J)
      IF((I.EQ.J).AND.(DABS(SUM-1.D0).LE.TOL))
     1          GO TO 400
      IF((I.NE.J).AND.(DABS(SUM).LE.TOL)) GO TO 400
      PASS = .FALSE.
      WRITE(IOUT,8010) I,J,INDEX,SUM
  400 CONTINUE
  500 CONTINUE
      RETURN
C
 8000 FORMAT(6X,'CONSTRAINT',I3,' IS VIOLATED:',
     1   ' LHS =',G15.4,'RHS =',G15.4)
 8010 FORMAT(6X,'ERROR IN ROW',I3,' COLUMN',I3,
     1   ' (CONSTRAINT',I3,')':,/,8X,'INNER ',
     2   'PRODUCT =',G15.4)
      END
```

Figure B.2 Subroutine VERIFY.

Note that if all the tests of **VERIFY** are passed, then **VERIFY** is "silent;" that is, it produces no output. Also note that $(R =) r = 0$ for Algorithm 1 and that $\alpha_{i0} \geq 1$, $i = 1, \ldots, n$. **VERIFY** is written so that it can also check the initial data for Algorithms 2 and 3.

A call is made to **ALG1** which is responsible for solving the linear programming problem using Algorithm 1. If an optimal solution is determined, its value is returned in **X**, **OBJ** contains the optimal objective function value, and the output condition code 1 is returned in **JOUT**. A returned value of 2 for **JOUT** means that the problem is unbounded from below. Following the call to **ALG1** is a call to **SOLN**. SUBROUTINE **SOLN** (Figure B.3) prints the optimal solution according to the value of **JOUT**.

```
      SUBROUTINE SOLN(JOUT,X,N,OBJ)
      IMPLICIT DOUBLE PRECISION (A-H,O-Z)
      DIMENSION X(N)
      COMMON /UNITS/ IIN,IOUT
      IF(JOUT.GE.2) GO TO 200
      WRITE(IOUT,8000) OBJ
      WRITE(IOUT,8010)
      DO 100 I=1,N
  100 WRITE(IOUT,8020) I,X(I)
      RETURN
  200 CONTINUE
      IF(JOUT.EQ.2) WRITE(IOUT,8030)
      IF(JOUT.EQ.3) WRITE(IOUT,8040)
      RETURN
C
 8000 FORMAT(//,6X,'OPTIMAL OBJECTIVE FUNCTION ',
     1   'VALUE IS',F12.5)
 8010 FORMAT(//,18X,'OPTIMAL SOLUTION')
 8020 FORMAT(18X,'X(',I2,') =',F14.7)
 8030 FORMAT(//,6X,'PROBLEM IS UNBOUNDED FROM ',
     1   'BELOW')
 8040 FORMAT(//,6X,'PROBLEM HAS NO FEASIBLE ',
     1   'SOLUTION')
      END
```

Figure B.3 Subroutine SOLN.

SUBROUTINE **ALG1** (Figure B.4) organizes the calls to subroutines **STEP11**, **STEP2**, and **STEP3**, which implement Steps 1, 2, and 3, respectively, of Algorithm 1. The name **STEP11** is used rather than **STEP1** because we will have analogous subroutines for Algorithms 2 and 3. **STEP12** and **STEP13** will be used for Step 1 of Algorithms 2 and 3, respectively.

```
      SUBROUTINE ALG1(A,B,C,X,DINV,S,V,WORK,JJ,OBJ,
     1  N,M,MTOT,JOUT)
      IMPLICIT DOUBLE PRECISION (A-H,O-Z)
      DIMENSION A(MTOT,N),B(MTOT),C(N),X(N),
     1  DINV(N,N),S(N),V(N),WORK(N),JJ(N)
      INTEGER ELL
      COMMON /UNITS/ IIN,IOUT
C
C
C  INITIALIZE
      WRITE(IOUT,8000) (I,I=1,N)
      ITER = 0
      SUM = 0.D0
      DO 100 I=1,N
  100 SUM = SUM + C(I)*X(I)
      OBJ = SUM
C
C
  200 CONTINUE
      CALL STEP11(C,S,V,DINV,N,K,JOUT)
      IF(JOUT.EQ.1) WRITE(IOUT,8010) ITER,OBJ,K,0,
     1  (JJ(I),I=1,N)
      IF(JOUT.EQ.1) RETURN
C
C
      CALL STEP2(A,B,X,S,JJ,SIGJ,ELL,N,M,MTOT,JOUT)
      WRITE(IOUT,8010) ITER,OBJ,K,ELL,(JJ(I),I=1,N)
      IF(JOUT.EQ.2) RETURN
C
C
      CALL STEP3(X,C,S,DINV,A,WORK,SIGJ,OBJ,JJ,ELL,
     1  K,N,M,MTOT,ITER)
      GO TO 200
C
C
 8000 FORMAT(/,20X,'ALGORITHM 1',//,40X,'ACTIVE SET'
     1  ,/,7X,'ITER',6X,'OBJECTIVE',4X,'K',4X,'L',
     2  2X,(T39,10I3))
 8010 FORMAT(6X,I5,F15.5,2I5,2X,(T39,10I3))
      END
```

Figure B.4 Subroutine ALG1.

MTOT $(m + r)$ is used to specify the total number of rows of A (A). In this case, MTOT is identical to M (m). However, for Algorithm 3 it will be different.   ALG1 initializes by setting the iteration counter ITER (j) to zero, using the DO 100 loop to compute $c'x_0$ and place it in OBJ. The call to STEP11 computes the search direction S (s_j) and column index K (k). If the optimality criterion is satisfied, STEP11 returns with JOUT = 1. In this case, the final objective function value, column index K, and active set JJ (J_j) are printed and control returns to the main program. Otherwise STEP2 is called.

STEP2 computes the maximum feasible step size SIGJ (σ_j) and associated index ELL (l). The results for the current iteration, j, $c'x_j$, k, l, and J_j, are then printed. If $a_i's_j \geq 0$ for $i = 1, \ldots, m$, the problem is unbounded from below, STEP2 returns with JOUT = 2, and control returns to the main program.

Control now passes to STEP3, which updates X, JJ, DINV, and ITER $(x_j, J_j, D_j^{-1}, \text{and } j, \text{respectively})$. This completes one iteration, and the next begins when control transfers back to statement 200.

SUBROUTINE STEP11 (Figure B.5) implements Step 1 of Algorithm 1 as follows. The DO 200 loop computes $(V(I) =) v_i = c'c_{ij}, i = 1, \ldots, n$. The DO 300 loop computes the largest of the V(I) and its associated index K. The analytic optimality test, $v_k \leq 0$, is weakened to the numerical test $v_k \leq 10^{-6}$, to compensate for arithmetic roundoff errors. We do not claim that TOLCON = 1.D-6 is the universal convergence tolerance for all linear programming problems. A linear program whose variables represent intergalactic distances would require a different convergence tolerance than one concerning atoms in a molecule. Nonetheless, if $v_k < 10^{-6}$, JOUT is set to 1 and control returns to ALG1. If $v_k \geq 10^{-6}$, the DO 500 loop copies the kth column of D_j^{-1} into s_j.

Figure B.5   Subroutine STEP11.

```
      SUBROUTINE STEP11(C,S,V,DINV,N,K,JOUT)
      IMPLICIT DOUBLE PRECISION (A-H,O-Z)
      DIMENSION C(N),S(N),V(N),DINV(N,N)
      TOLCON = 1.D-6
      JOUT = 0
      DO 200 I=1,N
      SUM = 0.D0
      DO 100 J=1,N
  100 SUM = SUM + C(J)*DINV(J,I)
```

```
200 V(I) = SUM
    VMAX = - 1.D30
    DO 300 I=1,N
    IF(V(I).LE.VMAX) GO TO 300
    VMAX = V(I)
    K = I
300 CONTINUE
    IF(VMAX.GE.TOLCON) GO TO 400
    JOUT = 1
    RETURN
400 CONTINUE
    DO 500 I=1,N
500 S(I) = DINV(I,K)
    RETURN
    END
```

Figure B.5 *(continued)*

Note that the DO 300 loop updates VMAX (v_k) only if V(I) is strictly greater than VMAX. Thus k is indeed the smallest index such that

$$v_k = \max\{v_i \mid i = 1, \ldots, n\}.$$

SUBROUTINE STEP2 (Figure B.6) implements Step 2 of Algorithm 1. The DO 400 loop considers each constraint I (i) in turn. The DO 100 loop checks to see if constraint I is in the active set. If it is, constraint I is not considered further. Otherwise, the DO 200 loop computes BOT ($= a_i's_j$). If BOT $\geq -10^{-6}$, constraint I is no longer a candidate for the maximum feasible step size. Here we have strengthened the analytic test $a_i's_j < 0$ to $a_i's_j < -10^{-6}$ to account for numerical roundoff error. The remarks made in STEP11 concerning tolerances are also relevant here.

If BOT $< -10^{-6}$, then RATIO = TOP/BOT (TOP $= a_i'x_j - b_i$) is compared with the current estimate of the maximum feasible step size. If RATIO is strictly smaller, SIGJ is updated with RATIO, and the associated index is stored in ELL.

If ELL remains at its initial value of zero, then $a_i's_j \geq -10^{-6}$ for $i = 1, \ldots, m$, indicating that the problem is unbounded from below. This information is communicated to ALG1 by setting JOUT = 2.

Note that SIGJ is updated only if RATIO is strictly less than the current minimum. Thus ELL (l) is indeed the smallest index such that

$$\sigma_j = \frac{a_l'x_j - b_l}{a_l's_j}.$$

```
      SUBROUTINE STEP2(A,B,X,S,JJ,SIGJ,ELL,N,M,MTOT,
    1   JOUT)
      IMPLICIT DOUBLE PRECISION (A-H,O-Z)
      DIMENSION A(MTOT,N),B(MTOT),X(N),S(N),JJ(N)
      INTEGER ELL
      JOUT = 0
      EPS = 1.D-6
      SIGJ = 1.D35
      ELL = 0
      DO 400 I=1,M
      DO 100 INDEX=1,N
      IF(JJ(INDEX).EQ.I) GO TO 400
  100 CONTINUE
      BOT = 0.D0
      DO 200 INDEX=1,N
  200 BOT = BOT + A(I,INDEX)*S(INDEX)
      IF(BOT.GE.-EPS) GO TO 400
      TOP = - B(I)
      DO 300 INDEX=1,N
  300 TOP = TOP + A(I,INDEX)*X(INDEX)
      RATIO = TOP/BOT
      IF(RATIO.GE.SIGJ) GO TO 400
      SIGJ = RATIO
      ELL = I
  400 CONTINUE
      IF(ELL.EQ.0) JOUT = 2
      RETURN
      END
```

Figure B.6 Subroutine STEP2.

SUBROUTINE STEP3 (Figure B.7) implements Step 3 of Algorithm 1. The DO 100 loop computes $x_{j+1} = x_j - \sigma_j s_j$. The DO 200 loop computes $c' x_{j+1}$ and stores it in OBJ. The DO 300 loop copies a_l into WORK, and then a call is made to PHI which replaces D_j^{-1} with $\Phi(D_j^{-1}, a_l, k)$. The active set is updated by changing α_{kj} to l and leaving all other elements unchanged. Finally, the iteration counter is increased by one.

Figure B.7 Subroutine STEP3.

```
      SUBROUTINE STEP3(X,C,S,DINV,A,WORK,SIGJ,OBJ,
    1   JJ,ELL,K,N,M,MTOT,ITER)
      IMPLICIT DOUBLE PRECISION (A-H,O-Z)
      DIMENSION X(N),C(N),S(N),WORK(N),DINV(N,N),
    1   A(MTOT,N),JJ(N)
```

```
         INTEGER ELL
         DO 100 I=1,N
   100   X(I) = X(I) - SIGJ*S(I)
         SUM = 0.D0
         DO 200 I=1,N
   200   SUM = SUM + C(I)*X(I)
         OBJ = SUM
         DO 300 I=1,N
   300   WORK(I) = A(ELL,I)
         CALL PHI(DINV,WORK,K,N)
         JJ(K) = ELL
         ITER = ITER + 1
         RETURN
         END
```

Figure B.7   *(continued)*

SUBROUTINE PHI (Figure B.8) implements Procedure Φ. The DO
100 loop computes $d'c_k$. If $d'c_k = 0$, Procedure Φ is undefined because the
new matrix whose inverse is sought is singular. If $|d'c_k|$ is very small, we
suspect that the new matrix is close to being singular. If $|d'c_k| \leq 10^{-6}$ in
SUBROUTINE PHI, an appropriate error message is issued, and execution is
halted. In theory this can never happen, and in practice, this *should* never
happen. However, it *may* happen for a variety of reasons: the gradient of the
new active constraint may be "almost" or numerically linearly dependent on
the gradients of the other active constraints, the problem data may be in-
correctly communicated, or there may be an error in the calling sequences for
the various higher level subroutines. It is more graceful to trap the problem
here and issue an appropriate message than to suffer an overflow error at the
system level.

Figure B.8   Subroutine PHI.

```
         SUBROUTINE PHI(DINV,D,K,N)
         IMPLICIT DOUBLE PRECISION (A-H,O-Z)
         DIMENSION DINV(N,N),D(N)
         COMMON /UNITS/ IIN,IOUT
         TOL = 1.D-6
         SUM = 0.D0
         DO 100 I=1,N
   100   SUM = SUM + D(I)*DINV(I,K)
         IF(DABS(SUM).GE.TOL) GO TO 200
         WRITE(IOUT,8000) SUM
         STOP
```

```
 200  CONTINUE
      SUM = 1.D0/SUM
      DO 300 I=1,N
 300  DINV(I,K) = DINV(I,K)*SUM
      DO 600 J=1,N
      IF(J.EQ.K) GO TO 600
      TEMP = 0.D0
      DO 400 I=1,N
 400  TEMP = TEMP + D(I)*DINV(I,J)
      DO 500 I=1,N
 500  DINV(I,J) = DINV(I,J) - TEMP*DINV(I,K)
 600  CONTINUE
      RETURN
8000  FORMAT(6X,'**** ERROR **** NEW MATRIX WOULD ',
    1  'BE SINGULAR, INNER PRODUCT =',G15.6)
      END
```

Figure B.8 *(continued)*

We illustrate ALG1 by using it to solve the problem of Example 2.2. The complete program consists of the main program of Figure B.1 and subroutines VERIFY, ALG1, SOLN, STEP11, STEP2, STEP3, and PHI. The data file for Example 2.2 is shown in Figure B.9. The output file (from unit 38) is shown in Figure B.10.

Figure B.9 Example 2.2 data file for ALG1.

```
5 2
-5. 2.
2. 14. 36. 0. 0.
-2. 1.
1. 2.
4. 3.
-1. 0.
0. -1.
2. 6.
1 2
-0.4 0.2
0.2 0.4
```

ALGORITHM 1

				ACTIVE	SET
ITER	OBJECTIVE	K	L	1	2
0	2.00000	1	3	1	2
1	-22.00000	2	5	3	2
2	-45.00000	1	0	3	5

OPTIMAL OBJECTIVE FUNCTION VALUE IS   -45.00000

OPTIMAL SOLUTION
X(1) =   9.0000000
X(2) =   0.0000000

Figure B.10   Example 2.2 output from ALG1.

B.1.2   Algorithm 2

We obtain a computer program for Algorithm 2 by modifying the one for Algorithm 1. Algorithm 2 may be initialized with $D_0^{-1} = I$ and $J_0 = \{0, \ldots, 0\}$. The main program for Algorithm 2 (Figure B.11) performs this initialization with the DO 200 and DO 300 loops. The remaining READ statements are the same as those for Algorithm 1 (Figure B.1). The call to ALG1 is now a call to ALG2. Note that because the initial data for DINV and JJ are constructed rather than read in, the essential role of VERI-FY is to check the feasibility of x_0.

Figure B.11   Main program for Algorithm 2.

```
IMPLICIT DOUBLE PRECISION (A-H,O-Z)
DIMENSION A(5,2),B(5),C(2),X(2),DINV(2,2),
1   S(2),V(2),WORK(2),JJ(2)
INTEGER R
LOGICAL PASS
COMMON /UNITS/ IIN,IOUT
IIN = 37
IOUT = 38
READ(IIN,*) M,N
MTOT = M
R = 0
READ(IIN,*) (C(I),I=1,N)
READ(IIN,*) (B(I),I=1,M)
DO 100 I=1,M
```

```
100  READ(IIN,*)  (A(I,J),J=1,N)
     READ(IIN,*)  (X(I),I=1,N)
     DO 200 I=1,N
     DO 200 J=1,N
200  DINV(I,J) = 0.D0
     DO 300 I=1,N
     DINV(I,I) = 1.D0
300  JJ(I) = 0
     CALL VERIFY(X,DINV,JJ,A,B,N,M,R,MTOT,PASS)
     IF(.NOT.PASS) STOP
     CALL ALG2(A,B,C,X,DINV,S,V,WORK,JJ,OBJ,N,M,
    1  MTOT,JOUT)
     CALL SOLN(JOUT,X,N,OBJ)
     STOP
     END
```

Figure B.11   *(continued)*

Other than the initial data requirement, Algorithm 2 differs from Algorithm 1 only in Step 1. Thus SUBROUTINE ALG2 (Figure B.12) differs from SUBROUTINE ALG1 only in that the name has been changed from ALG1 to ALG2 and that the call to STEP11 has been changed to STEP12. Of course, the "ALGORITHM 1" in FORMAT statement 8000 has been changed to "ALGORITHM 2."

Figure B.12   Subroutine ALG2.

```
     SUBROUTINE ALG2(A,B,C,X,DINV,S,V,WORK,JJ,
    1  OBJ,N,M,MTOT,JOUT)
     IMPLICIT DOUBLE PRECISION (A-H,O-Z)
     DIMENSION A(MTOT,N),B(MTOT),C(N),X(N),
    1 DINV(N,N),S(N),V(N),WORK(N),JJ(N)
     INTEGER ELL
     COMMON /UNITS/ IIN,IOUT
C
C
C   INITIALIZE
     WRITE(IOUT,8000) (I,I=1,N)
     ITER = 0
     SUM = 0.D0
     DO 100 I=1,N
100  SUM = SUM + C(I)*X(I)
     OBJ = SUM
```

```
  200 CONTINUE
C
C

      CALL STEP12(C,S,V,DINV,N,K,JOUT,JJ)
      IF(JOUT.EQ.1) WRITE(IOUT,8010) ITER,OBJ,K,0,
     1 (JJ(I),I=1,N)
      IF(JOUT.EQ.1) RETURN
C
C

      CALL STEP2(A,B,X,S,JJ,SIGJ,ELL,N,M,MTOT,JOUT)
      WRITE(IOUT,8010) ITER,OBJ,K,ELL,(JJ(I),I=1,N)
      IF(JOUT.EQ.2) RETURN
C
C

      CALL STEP3(X,C,S,DINV,A,WORK,SIGJ,OBJ,JJ,ELL,
     1 K,N,M,MTOT,ITER)
      GO TO 200
C
C
 8000 FORMAT(/,20X,'ALGORITHM 2',//,40X,'ACTIVE SET'
     1 ,/,7X,'ITER',6X,'OBJECTIVE',4X,'K',4X,'L',
     2 2X,(T39,10I3))
 8010 FORMAT(6X,I5,F15.5,2I5,2X,(T39,10I3))
      END
```

Figure B.12   *(continued)*

SUBROUTINE STEP12 (Figure B.13) implements Step 1 of Algorithm 2. The DO 100 loop looks for an α_{ij} which has value zero. If one is found, control transfers to statement 200 which begins Step 1.1. If none is found, control transfers to statement 900 which begins Step 1.2.

Figure B.13   Subroutine STEP12.

```
      SUBROUTINE STEP12(C,S,V,DINV,N,K,JOUT,JJ)
      IMPLICIT DOUBLE PRECISION (A-H,O-Z)
      DIMENSION C(N),S(N),V(N),DINV(N,N),JJ(N)
C

      TOLCON = 1.D-6
      JOUT = 0
      DO 100 I=1,N
      IF(JJ(I).EQ.0) GO TO 200
```

```
    100 CONTINUE
        GO TO 900
C
C
C  STEP 1.1
    200 CONTINUE
        DO 400 I=1,N
        IF(JJ(I).NE.0) GO TO 400
        SUM = 0.D0
        DO 300 J=1,N
    300 SUM = SUM + C(J)*DINV(J,I)
        V(I) = SUM
    400 CONTINUE
        VMAXA = -1.D0
        K = 0
        DO 500 I=1,N
        IF(JJ(I).NE.0) GO TO 500
        IF(DABS(V(I)).LE.VMAXA) GO TO 500
        VMAXA = DABS(V(I))
        K = I
    500 CONTINUE
        IF(VMAXA.LE.TOLCON) GO TO 900
        IF(V(K).LT.0.D0) GO TO 700
        DO 600 I=1,N
    600 S(I) = DINV(I,K)
        RETURN
    700 CONTINUE
        DO 800 I=1,N
    800 S(I) = - DINV(I,K)
        RETURN
C
C
C  STEP 1.2
    900 CONTINUE
        DO 1100 I=1,N
        IF(JJ(I).EQ.0) GO TO 1100
        SUM = 0.D0
        DO 1000 J=1,N
   1000 SUM = SUM + C(J)*DINV(J,I)
        V(I) = SUM
```

Figure B.13   *(continued)*

```
1100  CONTINUE
      VMAX = - 1.D30
      DO 1200 I=1,N
      IF(JJ(I).EQ.0) GO TO 1200
      IF(V(I).LE.VMAX) GO TO 1200
      VMAX = V(I)
      K = I
1200  CONTINUE
      IF(VMAX.GE.TOLCON) GO TO 1300
      JOUT = 1
      RETURN
1300  CONTINUE
      DO 1400 I=1,N
1400  S(I) = DINV(I,K)
      RETURN
      END
```

Figure B.13 *(continued)*

Step 1.1 begins by computing $v_i = c'c_{ij}$ for all i with $\alpha_{ij} = 0$. The DO 500 loop computes the smallest index k such that

$$|v_k| = \max\{ |v_k| \mid \text{all } i \text{ with } \alpha_{ij} = 0 \}.$$

If (VMAXA =) $|v_k| < 10^{-6}$, control transfers to Step 1.2. Otherwise, the DO 600 loop sets $s_j = c_{kj}$ if $v_k > 0$, and the DO 800 loop sets $s_j = -c_{kj}$ if $v_k < 0$.

Step 1.2 begins with statement 900. The DO 1100 loop computes $v_i = c'c_{ij}$ for those i with $\alpha_{ij} \geq 1$. The DO 1200 loop finds the largest of these v_i and its associated index k. If $v_k \geq 10^{-6}$, the DO 1400 loop sets $s_j = c_{kj}$ and control returns to the calling routine ALG2. If $v_k < 10^{-6}$, the optimality criterion is satisfied: JOUT is set to 1 and control returns to ALG2. Note that the computations for Step 1.2 are identical to those for Step 1 of Algorithm 1 (STEP11) except that all computations involving v_i are restricted to those v_i for which $\alpha_{ij} \geq 1$.

We illustrate ALG2 by using it to solve the problem of Example 2.5. The complete program consists of the main program of Figure B.11 and subroutines VERIFY, ALG2, SOLN, STEP12, STEP2, STEP3, and PHI. The data file for Example 2.5 is shown in Figure B.14. The output file (from unit 38) is shown in Figure B.15.

```
5  2
-6.  -8.
25.  29.  45.  0.  0.
-3.  5.
1.  3.
3.  2.
-1.  0.
0.  -1.
2.  3.
```

Figure B.14   Example 2.5 data file for ALG2.

ALGORITHM 2

				ACTIVE SET	
ITER	OBJECTIVE	K	L	1	2
0	-36.00000	2	1	0	0
1	-61.60000	1	2	0	1
2	-94.00000	2	3	2	1
3	-114.00000	2	0	2	3

OPTIMAL OBJECTIVE FUNCTION VALUE IS   -114.00000

OPTIMAL SOLUTION
X(1) =   11.0000000
X(2) =   6.0000000

Figure B.15   Example 2.5 output from ALG2.

B.1.3   Algorithm 3

We obtain a computer program for Algorithm 3 by modifying those for Algorithms 1 and 2. In addition to the usual problem data, Algorithm 3 requires R (r), the number of equality constraints. The main program for Algorithm 3 (Figure B.16) reads R as well as the remaining data. The total number of constraints, MTOT, is computed as M+R. Values for D_0^{-1} and J_0 are read in, and the call is made to ALG3. Note that the argument list for ALG3 has been expanded to include R.

```
      IMPLICIT DOUBLE PRECISION (A-H,O-Z)
      DIMENSION A(3,2),B(3),C(2),X(2),DINV(2,2),
     1 S(2),V(2),WORK(2),JJ(2)
      INTEGER R
      LOGICAL PASS
      COMMON /UNITS/ IIN,IOUT
      IIN = 37
      IOUT = 38
      READ(IIN,*) M,N,R
      MTOT = M + R
      READ(IIN,*) (C(I),I=1,N)
      READ(IIN,*) (B(I),I=1,MTOT)
      DO 100 I=1,MTOT
  100 READ(IIN,*) (A(I,J),J=1,N)
      READ(IIN,*) (X(I),I=1,N)
      READ(IIN,*) (JJ(I),I=1,N)
      DO 200 I=1,N
  200 READ(IIN,*) (DINV(I,J),J=1,N)
      CALL VERIFY(X,DINV,JJ,A,B,N,M,R,MTOT,PASS)
      IF(.NOT.PASS) STOP
      CALL ALG3(A,B,C,X,DINV,S,V,WORK,JJ,OBJ,N,M,
     1 R,MTOT,JOUT)
      CALL SOLN(JOUT,X,N,OBJ)
      STOP
      END
```

Figure B.16   Main program for Algorithm 3.

SUBROUTINE ALG3 (Figure B.17) is obtained from SUBROUTINE ALG2 by changing the name from ALG2 to ALG3, expanding the argument list to include R (which is specified as an INTEGER variable), changing the call to STEP12 to a call to STEP13, expanding the argument list of STEP13 to include M, and changing "ALGORITHM 2" in FORMAT statement 8000 to "ALGORITHM 3."

```
      SUBROUTINE ALG3(A,B,C,X,DINV,S,V,WORK,JJ,
     1  OBJ,N,M,R,MTOT,JOUT)
      IMPLICIT DOUBLE PRECISION (A-H,O-Z)
      DIMENSION A(MTOT,N),B(MTOT),C(N),X(N),
     1 DINV(N,N),S(N),V(N),WORK(N),JJ(N)
      INTEGER ELL,R
      COMMON /UNITS/ IIN,IOUT
C
C
C  INITIALIZE
      WRITE(IOUT,8000) (I,I=1,N)
      ITER = 0
      SUM = 0.D0
      DO 100 I=1,N
  100 SUM = SUM + C(I)*X(I)
      OBJ = SUM
  200 CONTINUE
C
C
      CALL STEP13(C,S,V,DINV,N,M,K,JOUT,JJ)
      IF(JOUT.EQ.1) WRITE(IOUT,8010) ITER,OBJ,K,0,
     1   (JJ(I),I=1,N)
      IF(JOUT.EQ.1) RETURN
C
C
      CALL STEP2(A,B,X,S,JJ,SIGJ,ELL,N,M,MTOT,JOUT)
      WRITE(IOUT,8010) ITER,OBJ,K,ELL,(JJ(I),I=1,N)
      IF(JOUT.EQ.2) RETURN
C
C
      CALL STEP3(X,C,S,DINV,A,WORK,SIGJ,OBJ,JJ,ELL,
     1   K,N,M,MTOT,ITER)
      GO TO 200
C
C
 8000 FORMAT(/,20X,'ALGORITHM 3',//,40X,'ACTIVE SET'
     1   ,/,7X,'ITER',6X,'OBJECTIVE',4X,'K',4X,'L',
     2   2X,(T39,10I3))
 8010 FORMAT(6X,I5,F15.5,2I5,2X,(T39,10I3))
      END
```

Figure B.17   Subroutine ALG3.

SUBROUTINE STEP13 (Figure B.18) is derived from STEP12 and contains only three changes. The IF statements at the top of the DO 1100 and DO 1200 loops have been modified so that any v_i with $\alpha_{ij} > m$ is not processed further. Finally, M has been added to the argument list.

Figure B.18   Subroutine STEP13.

```
      SUBROUTINE STEP13(C,S,V,DINV,N,M,K,JOUT,JJ)
      IMPLICIT DOUBLE PRECISION (A-H,O-Z)
      DIMENSION C(N),S(N),V(N),DINV(N,N),JJ(N)
C
      TOLCON = 1.D-6
      JOUT = 0
      DO 100 I=1,N
      IF(JJ(I).EQ.0) GO TO 200
  100 CONTINUE
      GO TO 900
C
C
C   STEP 1.1
  200 CONTINUE
      DO 400 I=1,N
      IF(JJ(I).NE.0) GO TO 400
      SUM = 0.D0
      DO 300 J=1,N
  300 SUM = SUM + C(J)*DINV(J,I)
      V(I) = SUM
  400 CONTINUE
      VMAXA = -1.D0
      K = 0
      DO 500 I=1,N
      IF(JJ(I).NE.0) GO TO 500
      IF(DABS(V(I)).LE.VMAXA) GO TO 500
      VMAXA = DABS(V(I))
      K = I
  500 CONTINUE
      IF(VMAXA.LE.TOLCON) GO TO 900
      IF(V(K).LT.0.D0) GO TO 700
      DO 600 I=1,N
```

```
 600  S(I) = DINV(I,K)
      RETURN
 700  CONTINUE
      DO 800 I=1,N
 800  S(I) = - DINV(I,K)
      RETURN
C
C
C  STEP 1.2
 900  CONTINUE
      DO 1100 I=1,N
      IF((JJ(I).EQ.0).OR.(JJ(I).GT.M)) GO TO 1100
      SUM = 0.D0
      DO 1000 J=1,N
1000  SUM = SUM + C(J)*DINV(J,I)
      V(I) = SUM
1100  CONTINUE
      VMAX = - 1.D30
      DO 1200 I=1,N
      IF((JJ(I).EQ.0).OR.(JJ(I).GT.M)) GO TO 1200
      IF(V(I).LE.VMAX) GO TO 1200
      VMAX = V(I)
      K = I
1200  CONTINUE
      IF(VMAX.GE.TOLCON) GO TO 1300
      JOUT = 1
      RETURN
1300  CONTINUE
      DO 1400 I=1,N
1400  S(I) = DINV(I,K)
      RETURN
      END
```

Figure B.18 *(continued)*

ALG3 is illustrated by solving the problem of Example 2.10. The complete program consists of the main program of Figure B.16 and subroutines VERIFY, ALG3, SOLN, STEP13, STEP2, STEP3, and PHI. The data file for Example 2.10 is shown in Figure B.19. The output file (from unit 38) is shown in Figure B.20.

```
2  2  1
-1.  0.
5.  0.  3.
2.  1.
-1.  0.
1.  1.
1.  2.
3  0
1.  -1.
0.  1.
```

Figure B.19 Example 2.10 data file for ALG3.

ALGORITHM 3

				ACTIVE	SET
ITER	OBJECTIVE	K	L	1	2
0	-1.00000	2	1	3	0
1	-2.00000	2	0	3	1

OPTIMAL OBJECTIVE FUNCTION VALUE IS   -2.00000

```
OPTIMAL  SOLUTION
X( 1)  =      2.0000000
X( 2)  =      1.0000000
```

Figure B.20 Example 2.10 output from ALG3.

B.1.4   Phase 1-Phase 2 Procedure, Version A

The Phase 1–Phase 2 procedure of Section 2.5 gives the complete capability to solve the LP

$$\min \{ c'x \mid Ax \leq b \}.$$

There are two ways that the Phase 1–Phase 2 procedure can be performed. The first uses the feasible point x_0 obtained from the Phase 1 problem, together with $D_0^{-1} = I$ and $J_0 = \{0, \ldots, 0\}$, as the initial data for the Phase 2 problem. This way (Version A) is used in the computer program of this section. The second way is to extract a D_0^{-1} and J_0 from the final data for the Phase 1 problem. These data reflect those constraints which are active at x_0. This second way (Version B) is used in Section B.1.5.

The main program (Figure B.21) reads the problem data and is similar to the previous main programs. In addition to the one- and two-dimensional arrays normally required, we require a second set for the Phase 1 problem. We have named these quantities by appending a "1" to the name of their Phase 2 counterpart. Thus A represents the original problem constraint matrix, and A1 represents the constraint matrix for the Phase 1 problem. The number of variables and constraints for the Phase 1 problem are each one greater than for the Phase 2 problem. This is reflected in the DIMENSION statements and in the statements which compute N1 and M1.

```
      IMPLICIT DOUBLE PRECISION (A-H,O-Z)
      DIMENSION A(4,2),B(4),C(2),X(2),DINV(2,2),
     1  S(2),V(2),WORK(2),JJ(2)
      DIMENSION A1(5,3),B1(5),C1(3),X1(3),
     1  DINV1(3,3),S1(3),V1(3),WORK1(3),JJ1(3)
      COMMON /UNITS/ IIN,IOUT
      IIN = 37
      IOUT = 38
      READ(IIN,*) M,N
      READ(IIN,*) (C(I),I=1,N)
      READ(IIN,*) (B(I),I=1,M)
      DO 100 I=1,M
  100 READ(IIN,*) (A(I,J),J=1,N)
      N1 = N + 1
      MTOT = M
      M1 = M + 1
      MTOT1 = M1
      CALL P1P2A(A,B,C,X,DINV,S,V,WORK,JJ,OBJ,N,
     1  M,MTOT,A1,B1,C1,X1,DINV1,S1,V1,WORK1,JJ1,
     2  OBJ1,N1,M1,MTOT1,JOUT)
      CALL SOLN(JOUT,X,N,OBJ)
      STOP
      END
```

Figure B.21   Main program for Phase 1–Phase 2, Version A.

After the data is read in, a call is made to P1P2A which controls the Phase 1–Phase 2 computations. Note that because of the Phase 1 quantities, the argument list for P1P2A is almost twice the size of that for ALG2.

SUBROUTINE P1P2A (Figure B.22) first constructs the data for the Phase 1 problem. It is solved using ALG2 and, if appropriate, ALG2 is again used to solve the Phase 2 problem.

```
      SUBROUTINE P1P2A(A,B,C,X,DINV,S,V,WORK,JJ,
     1   OBJ,N,M,MTOT,A1,B1,C1,X1,DINV1,S1,V1,
     2   WORK1,JJ1,OBJ1,N1,M1,MTOT1,JOUT)
      IMPLICIT DOUBLE PRECISION (A-H,O-Z)
      DIMENSION A(MTOT,N),B(MTOT),C(N),X(N),
     1   DINV(N,N),S(N),V(N),WORK(N),JJ(N)
      DIMENSION A1(MTOT1,N1),B1(MTOT1),C1(N1),
     1   X1(N1),DINV1(N1,N1),S1(N1),V1(N1),
     2   WORK1(N1),JJ1(N1)
      COMMON /UNITS/ IIN,IOUT
C
C
      TOLFES = 1.D-6
      WRITE(IOUT,8020)
      WRITE(IOUT,8000)
C
C   CONSTRUCT PHASE 1 DATA
C   INITIAL POINT CONSTRAINT MATRIX
      DO 200 I=1,M
      DO 100 J=1,N
  100 A1(I,J) = A(I,J)
  200 A1(I,N1) = - 1.D0
      DO 300 I=1,N
  300 A1(M1,I) = 0.D0
      A1(M1,N1) = -1.D0
C
C   INITIAL POINT RIGHT HAND SIDE
      DO 400 I=1,M
  400 B1(I) = B(I)
      B1(M1) = 0.D0
C
C   INITIAL POINT OBJECTIVE FUNCTION
      DO 500 I=1,N
  500 C1(I) = 0.D0
      C1(N1) = 1.D0
C
C   INITIAL FEASIBLE POINT
      DO 600 I=1,N
  600 X1(I) = 0.D0
      ALPHA = - 1.D30
      DO 700 I=1,M
      IF(ALPHA.GT.(-B(I))) GO TO 700
      ALPHA = - B(I)
```

Figure B.22 Subroutine P1P2A.

```
  700 CONTINUE
      ALPHA = DMAX1(0.D0,ALPHA)
      X1(N1) = ALPHA
C
C  D INVERSE AND INDEX SET
      DO 800 I=1,N1
      DO 800 J=1,N1
  800 DINV1(I,I) = 0.D0
      DO 900 I=1,N1
      DINV1(I,I) = 1.D0
  900 JJ1(I) = 0
C
C  INITIAL POINT DATA IS NOW COMPLETE:
C  USE ALGORITHM 2 TO SOLVE THE INITIAL POINT
C  PROBLEM
C
      CALL ALG2(A1,B1,C1,X1,DINV1,S1,V1,WORK1,
     1 JJ1,OBJ,N1,M1,MTOT1,JOUT)
      IF(OBJ.LE.TOLFES) GO TO 1000
C  NO FEASIBLE SOLUTION
      JOUT = 3
      RETURN
 1000 CONTINUE
      WRITE(IOUT,8010)
      DO 1100 I=1,N
 1100 X(I) = X1(I)
      DO 1200 I=1,N
      DO 1200 J=1,N
 1200 DINV(I,J) = 0.D0
      DO 1300 I=1,N
      DINV(I,I) = 1.D0
 1300 JJ(I) = 0
      CALL ALG2(A,B,C,X,DINV,S,V,WORK,JJ,OBJ,N,M,
     1 MTOT,JOUT)
      RETURN
C
 8000 FORMAT(//,20X,'PHASE 1 LP')
 8010 FORMAT(//,20X,'PHASE 2 LP')
 8020 FORMAT(10X,'PHASE 1 - PHASE 2 PROCEDURE',
     1 ' VERSION A')
      END
```

Figure B.22   *(continued)*

The DO 100, 200, and 300 loops construct the Phase 1 constraint matrix. The DO 400 and DO 500 loops construct the right-hand side and objective function coefficients, respectively, for the Phase 1 problem. The DO 600 loop sets the first N (n) components of the initial feasible solution to zero. The DO 700 loop computes α_0 and then replaces it with zero if it is negative. Next, the last component of the initial feasible solution is set equal to α_0. Finally, the DO 800 and DO 900 loops set $D_0^{-1} = I$ and $J_0 = \{0, \ldots, 0\}$ for the Phase 1 problem.

The Phase 1 problem is then solved using Algorithm 2 by calling ALG2. The optimal Phase 1 objective function value is returned in OBJ. If OBJ $> 10^{-6}$, the original problem has no feasible solution. This is communicated to P1P2A by setting JOUT $= 3$ before returning.

If OBJ $\leq 10^{-6}$, a feasible solution for the original problem has been obtained. This feasible solution is copied into X using the DO 1100 loop. The DO 1200 and DO 1300 loops set $D_0^{-1} = I$ and $J_0 = \{0, \ldots, 0\}$. SUBROUTINE ALG2 is then called to solve the Phase 2 problem using Algorithm 2.

We illustrate P1P2A by applying it to

$$\text{minimize:} \quad -x_2$$

subject to:

$$x_1 + x_2 \leq 10, \quad (1)$$
$$-x_1 - x_2 \leq -8, \quad (2)$$
$$x_1 - x_2 \leq 6, \quad (3)$$
$$-x_1 + x_2 \leq -4. \quad (4)$$

Note that these constraints were used in Example 2.9 to illustrate the Phase 1 problem.

The complete program consists of the main program of Figure B.21 and subroutines P1P2A, ALG2, SOLN, STEP12, STEP2, STEP3, and PHI. The data file for the example is shown in Figure B.23. The output file is shown in Figure B.24. The reader may wish to compare the Phase 1 portion of the output with Example 2.9.

Figure B.23   Data file for P1P2A.

```
4  2
0.  -1.
10.  -8.  6.  -4.
1.  1.
-1.  -1.
1.  -1.
-1.  1.
```

PHASE 1 - PHASE 2 PROCEDURE VERSION A

PHASE 1 LP

ALGORITHM 2

				ACTIVE	SET	
ITER	OBJECTIVE	K	L	1	2	3
0	8.00000	3	2	0	0	0
1	8.00000	1	3	0	0	2
2	1.00000	2	5	3	0	2
3	0.00000	1	0	3	5	2

PHASE 2 LP

ALGORITHM 2

				ACTIVE	SET
ITER	OBJECTIVE	K	L	1	2
0	-1.00000	2	1	0	0
1	-3.00000	1	4	0	1
2	-3.00000	1	0	4	1

OPTIMAL OBJECTIVE FUNCTION VALUE IS   -3.00000

OPTIMAL SOLUTION
X(1) =   7.0000000
X(2) =   3.0000000

Figure B.24   Output from P1P2A.

B.1.5   Phase 1-Phase 2 Procedure, Version B

In this section, we modify the Phase 1–Phase 2 procedure of Section B.1.4 so that the D_0^{-1} and J_0 used for the initial data of the Phase 2 problem are extracted from the final data of the Phase 1 problem. These data reflect those constraints which are active at x_0. Most of the computer program is unchanged from that of Section B.1.4. The main program (Figure B.25) differs only in that the call to **P1P2A** has been changed to a call to **P1P2B**.

```
      IMPLICIT DOUBLE PRECISION (A-H,O-Z)
      DIMENSION A(4,2),B(4),C(2),X(2),DINV(2,2),
     1  S(2),V(2),WORK(2),JJ(2)
      DIMENSION A1(5,3),B1(5),C1(3),X1(3),
     1  DINV1(3,3),S1(3),V1(3),WORK1(3),JJ1(3)
      COMMON /UNITS/ IIN,IOUT
      IIN = 37
      IOUT = 38
      READ(IIN,*) M,N
      READ(IIN,*) (C(I),I=1,N)
      READ(IIN,*) (B(I),I=1,M)
      DO 100 I=1,M
  100 READ(IIN,*) (A(I,J),J=1,N)
      N1 = N + 1
      MTOT = M
      M1 = M + 1
      MTOT1 = M1
      CALL P1P2B(A,B,C,X,DINV,S,V,WORK,JJ,OBJ,N,
     1  M,MTOT,A1,B1,C1,X1,DINV1,S1,V1,WORK1,JJ1,
     2  OBJ1,N1,M1,MTOT1,JOUT)
      CALL SOLN(JOUT,X,N,OBJ)
      STOP
      END
```

Figure B.25   Main program for Phase 1–Phase 2, Version B.

SUBROUTINE P1P2B (Figure B.26) is obtained by modifying P1P2A as follows. The coding to set up and solve the Phase 1 problem is unchanged. The changes are introduced following the DO 1100 loop. The DO 1200 loop checks if $m + 1$ is in $\hat{J}_0$. If it is, control transfers to statement 1500. If not, Exercise 2.26 is used as follows to perform one update of $\hat{D}_0^{-1}$ so that $m + 1$ is in the modified $\hat{J}_0$. The DO 1300 loop computes the index k of Exercise 2.26. The next few statements replace column k of $\hat{D}_0'$ with $(0', -1)'$, compute the new inverse matrix, and set the kth element of the new index set to $m + 1$.

After statement 1500, we are sure that $m + 1$ is an element of $\hat{J}_0$. The DO 1700 and DO 1600 loops extract D_0^{-1} and J_0 from $\hat{D}_0^{-1}$ and $\hat{J}_0$, respectively, as follows. The first n elements of each column of $\hat{D}_0^{-1}$ are copied into the next available column of D_0^{-1}. When column j of $\hat{D}_0^{-1}$ is encountered with $\hat{\alpha}_{j0} = m + 1$, column j is not copied. J_0 is constructed from $\hat{J}_0$ in a similar manner.

```
      SUBROUTINE P1P2B(A,B,C,X,DINV,S,V,WORK,JJ,
     1  OBJ,N,M,MTOT,A1,B1,C1,X1,DINV1,S1,V1,
     2  WORK1,JJ1,OBJ1,N1,M1,MTOT1,JOUT)
      IMPLICIT DOUBLE PRECISION (A-H,O-Z)
      DIMENSION A(MTOT,N),B(MTOT),C(N),X(N),
     1  DINV(N,N),S(N),V(N),WORK(N),JJ(N)
      DIMENSION A1(MTOT1,N1),B1(MTOT1),C1(N1),
     1  X1(N1),DINV1(N1,N1),S1(N1),V1(N1),
     2  WORK1(N1),JJ1(N1)
      COMMON /UNITS/ IIN,IOUT
C
C

      TOLFES = 1.D-6
      WRITE(IOUT,8020)
      WRITE(IOUT,8000)
C
C  CONSTRUCT PHASE 1 DATA
C  INITIAL POINT CONSTRAINT MATRIX
      DO 200 I=1,M
      DO 100 J=1,N
  100 A1(I,J) = A(I,J)
  200 A1(I,N1) = - 1.D0
      DO 300 I=1,N
  300 A1(M1,I) = 0.D0
      A1(M1,N1) = -1.D0
C
C  INITIAL POINT RIGHT HAND SIDE
      DO 400 I=1,M
  400 B1(I) = B(I)
      B1(M1) = 0.D0
C
C  INITIAL POINT OBJECTIVE FUNCTION
      DO 500 I=1,N
  500 C1(I) = 0.D0
      C1(N1) = 1.D0
C
C  INITIAL FEASIBLE POINT
      DO 600 I=1,N
  600 X1(I) = 0.D0
      ALPHA = - 1.D30
```

Figure B.26 Subroutine P1P2B.

```
      DO 700 I=1,M
      IF(ALPHA.GT.(-B(I))) GO TO 700
      ALPHA = - B(I)
  700 CONTINUE
      ALPHA = DMAX1(0.D0,ALPHA)
      X1(N1) = ALPHA
C
C   D INVERSE AND INDEX SET
      DO 800 I=1,N1
      DO 800 J=1,N1
  800 DINV1(I,I) = 0.D0
      DO 900 I=1,N1
      DINV1(I,I) = 1.D0
  900 JJ1(I) = 0
C
C   INITIAL POINT DATA IS NOW COMPLETE:
C   USE ALGORITHM 2 TO SOLVE THE INITIAL POINT
C   PROBLEM
C
      CALL ALG2(A1,B1,C1,X1,DINV1,S1,V1,WORK1,
     1   JJ1,OBJ,N1,M1,MTOT1,JOUT)
      IF(OBJ.LE.TOLFES) GO TO 1000
C   NO FEASIBLE SOLUTION
      JOUT = 3
      RETURN
 1000 CONTINUE
      WRITE(IOUT,8010)
      DO 1100 I=1,N
 1100 X(I) = X1(I)
C   IS M + 1 IN THE ACTIVE SET?
      DO 1200 I=1,N1
      IF(JJ1(I).EQ.M1) GO TO 1500
 1200 CONTINUE
C   M + 1 IS NOT IN THE ACTIVE SET;
C   APPLY EXERCISE 2.26
      CCK0 = -1.D0
      K = 0
      DO 1300 I=1,N1
      IF(DABS(DINV1(N1,I)).LT.CCK0) GO TO 1300
      CCK0 = DABS(DINV1(N1,I))
      K = I
```

Figure B.26 *(continued)*

```
     1300 CONTINUE
          DO 1400 I=1,N
     1400 WORK1(I) = 0.D0
          WORK1(N1) = -1.D0
          CALL PHI(DINV1,WORK1,K,N1)
          JJ1(K) = M1
C  M + 1 IS NOW IN THE ACTIVE SET
     1500 CONTINUE
C
C   EXTRACT DINV FROM DINV1 AND JJ FROM JJ1
C
          NXTCOL = 1
          DO 1700 J=1,N1
          IF(JJ1(J).EQ.M1) GO TO 1700
          DO 1600 I=1,N
     1600 DINV(I,NXTCOL) = DINV1(I,J)
          JJ(NXTCOL) = JJ1(J)
          NXTCOL = NXTCOL + 1
     1700 CONTINUE
C
C   INITIAL DATA IS NOW COMPLETE
C
          CALL ALG2(A,B,C,X,DINV,S,V,WORK,JJ,OBJ,N,M,
         1    MTOT,JOUT)
          RETURN
C
     8000 FORMAT(//,20X,'PHASE 1 LP')
     8010 FORMAT(//,20X,'PHASE 2 LP')
     8020 FORMAT(10X,'PHASE 1 - PHASE 2 PROCEDURE',
         1    ' VERSION B')
          END
```

Figure B.26 *(continued)*

We illustrate P1P2B by applying it to the problem of Section B.1.4. The complete program consists of the main program of Figure B.25 and subroutines P1P2B, ALG2, SOLN, STEP12, STEP2, STEP3, and PHI. The data file remains as in Figure B.23. The output file is shown in Figure B.27.

PHASE 1 - PHASE 2 PROCEDURE VERSION B

PHASE 1 LP

ALGORITHM 2

| | | | | ACTIVE SET | | |
ITER	OBJECTIVE	K	L	1	2	3
0	8.00000	3	2	0	0	0
1	8.00000	1	3	0	0	2
2	1.00000	2	5	3	0	2
3	0.00000	1	0	3	5	2

PHASE 2 LP

ALGORITHM 2

| | | | | ACTIVE SET | |
ITER	OBJECTIVE	K	L	1	2
0	-1.00000	1	4	3	2
1	-2.00000	2	1	4	2
2	-3.00000	1	0	4	1

OPTIMAL OBJECTIVE FUNCTION VALUE IS   -3.00000

OPTIMAL SOLUTION
X(1) =   7.0000000
X(2) =   3.0000000

Figure B.27   Output from P1P2B.

B.1.6   Phase 1-Phase 2 Procedure, Version C

The Phase 1−Phase 2 procedures of Sections B.1.4 and B.1.5 are for problems which have inequality constraints only. In this section, we give a computer program for a model problem which has both inequality and equality constraints. It implements the Phase 1−Phase 2 procedure of Section 2.6

and solves the Phase 1 problem (2.26) rather than the Phase 1 problem (2.20) used by P1P2A and P1P2B.

The main program (Figure B.28) reads in the problem data including R (r), the number of problem equality constraints. N1 and M1 denote the number of variables and constraints, respectively, for the Phase 1 problem (2.26). A call is made to P1P2C which solves the Phase 1 problem using Algorithm 2 and the Phase 2 problem using Algorithm 3.

```
      IMPLICIT DOUBLE PRECISION (A-H,O-Z)
      DIMENSION A(5,3),B(5),C(3),X(3),DINV(3,3),
     1   S(3),V(3),WORK(3),JJ(3)
      DIMENSION A1(6,4),B1(6),C1(4),X1(4),
     1   DINV1(4,4),S1(4),V1(4),WORK1(4),JJ1(4)
      INTEGER R
      COMMON /UNITS/ IIN,IOUT
      IIN = 37
      IOUT = 38
      READ(IIN,*) M,N,R
      MTOT = M + R
      READ(IIN,*) (C(I),I=1,N)
      READ(IIN,*) (B(I),I=1,MTOT)
      DO 100 I=1,MTOT
  100 READ(IIN,*) (A(I,J),J=1,N)
      N1 = N + 1
      M1 = M + R + 1
      MTOT1 = M1
      CALL P1P2C(A,B,C,X,DINV,S,V,WORK,JJ,OBJ,N,
     1   M,R,MTOT,A1,B1,C1,X1,DINV1,S1,V1,WORK1,JJ1,
     2   OBJ1,N1,M1,MTOT1,JOUT)
      CALL SOLN(JOUT,X,N,OBJ)
      STOP
      END
```

Figure B.28   Main program for Phase 1–Phase 2, Version C.

SUBROUTINE P1P2C (Figure B.29) is similar to P1P2A and P1P2B. The argument list for P1P2C has been expanded to include R. P1P2C begins by replacing any constraint $a_i'x = b_i$ with $m + 1 \le i \le m + r$ and $b_i < 0$ by $(-a_i)'x = (-b_i)$. This is done by the DO 30 and DO 40 loops. The construction of the Phase 1 constraint matrix is changed somewhat because α does not appear in constraints $m + 1, \ldots, m + r$ of the Phase 1 problem (2.26). The corresponding values of A1 are set to zero by the DO

250 loop. The DO 540 and DO 530 loops compute the Phase 1 objective coefficients for (2.26) and store them in C1. The construction of the initial feasible point, D_0^{-1}, and J_0 are the same as for P1P2A and P1P2B. Note that the total number of inequality constraints for the Phase 1 problem is $m + r + 1$. This value was computed in the main program and assigned to M1.

Figure B.29   Subroutine P1P2C.

```
      SUBROUTINE P1P2C(A,B,C,X,DINV,S,V,WORK,JJ,
     1   OBJ,N,M,R,MTOT,A1,B1,C1,X1,DINV1,S1,V1,
     2   WORK1,JJ1,OBJ1,N1,M1,MTOT1,JOUT)
      IMPLICIT DOUBLE PRECISION (A-H,O-Z)
      DIMENSION A(MTOT,N),B(MTOT),C(N),X(N),
     1   DINV(N,N),S(N),V(N),WORK(N),JJ(N)
      DIMENSION A1(MTOT1,N1),B1(MTOT1),C1(N1),
     1   X1(N1),DINV1(N1,N1),S1(N1),V1(N1),
     2   WORK1(N1),JJ1(N1)
      INTEGER R,RHO
      DOUBLE PRECISION LAMBDA
      COMMON /UNITS/ IIN,IOUT
C
      TOLFES = 1.D-6
      EPS = 1.D-5
      WRITE(IOUT,8020)
      WRITE(IOUT,8000)
C
C   CHECK FOR B(I).GE.0 FOR EQUALITY CONSTRAINTS
C
      IF(R.EQ.0) GO TO 50
      DO 40 I=1,R
      IF(B(M+I).GE.0.D0) GO TO 40
      DO 30 J=1,N
   30 A(M+I,J) = - A(M+I,J)
      B(M+I) = - B(M+I)
   40 CONTINUE
   50 CONTINUE
C
C
C   CONSTRUCT PHASE 1 DATA
C   INITIAL POINT CONSTRAINT MATRIX
      DO 200 I=1,MTOT
      DO 100 J=1,N
```

```
      100 A1(I,J) = A(I,J)
      200 CONTINUE
          DO 240 I=1,M
      240 A1(I,N1) = - 1.D0
          IF(R.EQ.0) GO TO 260
          DO 250 I=1,R
      250 A1(M+I,N1) = 0.D0
      260 CONTINUE
          DO 300 I=1,N
      300 A1(M1,I) = 0.D0
          A1(M1,N1) = -1.D0
C
C  INITIAL POINT RIGHT HAND SIDE
          DO 400 I=1,MTOT
      400 B1(I) = B(I)
          B1(M1) = 0.D0
C
C  INITIAL POINT OBJECTIVE FUNCTION
          DO 500 I=1,N
      500 C1(I) = 0.D0
          C1(N1) = 1.D0
          IF(R.EQ.0) GO TO 550
          DO 540 I=1,R
          DO 530 J=1,N
      530 C1(J) = C1(J) - A(M+I,J)
      540 CONTINUE
      550 CONTINUE
C
C  INITIAL FEASIBLE POINT
          DO 600 I=1,N
      600 X1(I) = 0.D0
          ALPHA = - 1.D30
          DO 700 I=1,M
          IF(ALPHA.GT.(-B(I))) GO TO 700
          ALPHA = - B(I)
      700 CONTINUE
          ALPHA = DMAX1(0.D0,ALPHA)
          X1(N1) = ALPHA
C
C  D INVERSE AND INDEX SET
          DO 800 I=1,N1
          DO 800 J=1,N1
```

Figure B.29   *(continued)*

```
    800 DINV1(I,J) = 0.D0
        DO 900 I=1,N1
        DINV1(I,I) = 1.D0
    900 JJ1(I) = 0
C
C   INITIAL POINT DATA IS NOW COMPLETE:
C   USE ALGORITHM 2 TO SOLVE THE INITIAL POINT
C   PROBLEM
C
        CALL ALG2(A1,B1,C1,X1,DINV1,S1,V1,WORK1,
       1   JJ1,OBJ,N1,M1,MTOT1,JOUT)
        SUMB = 0.D0
        IF(R.EQ.0) GO TO 920
        DO 910 I=1,R
    910 SUMB = SUMB + B(M+I)
    920 CONTINUE
        IF(DABS(OBJ+SUMB).LE.TOLFES) GO TO 1000
C   NO FEASIBLE SOLUTION
        JOUT = 3
        RETURN
   1000 CONTINUE
        WRITE(IOUT,8010)
        DO 1100 I=1,N
   1100 X(I) = X1(I)
C   IS M + 1 IN THE ACTIVE SET?
        DO 1200 I=1,N1
        IF(JJ1(I).EQ.M1) GO TO 1500
   1200 CONTINUE
C   M + 1 IS NOT IN THE ACTIVE SET;
C   APPLY EXERCISE 2.26
        CCK0 = -1.D0
        K = 0
        DO 1300 I=1,N1
        IF(DABS(DINV1(N1,I)).LT.CCK0) GO TO 1300
        CCK0 = DABS(DINV1(N1,I))
        K = I
   1300 CONTINUE
        DO 1400 I=1,N
   1400 WORK1(I) = 0.D0
        WORK1(N1) = -1.D0
        CALL PHI(DINV1,WORK1,K,N1)
        JJ1(K) = M1
```

Figure B.29   *(continued)*

```
C   M + 1 IS NOW IN THE ACTIVE SET
 1500 CONTINUE
      IF(R.EQ.0) GO TO 2200
      IBOT = M + 1
      ITOP = M + R
      DO 2100 RHO=IBOT,ITOP
C   IS RHO IN THE ACTIVE SET?
      DO 1600 I=1,N1
      IF(JJ1(I).EQ.RHO) GO TO 2100
 1600 CONTINUE
C   EQUALITY CONSTRAINT RHO IS NOT IN
C   TRY TO PUT IT IN
      LAMBDA = -1.D0
      K = 0
      DO 1800 I=1,N1
      IF(JJ1(I).GT.M) GO TO 1800
      SUM = 0.D0
      DO 1700 J=1,N
 1700 SUM = SUM + A(RHO,J)*DINV1(J,I)
      IF(DABS(SUM).LE.LAMBDA) GO TO 1800
      LAMBDA = DABS(SUM)
      K = I
 1800 CONTINUE
      IF(LAMBDA.GE.EPS) GO TO 1900
C   HERE, EQUALITY CONSTRAINT RHO IS REDUNDANT
C   NOTE THIS ON THE OUTPUT UNIT AND CONTINUE
      WRITE(IOUT,8030) RHO
      GO TO 2100
 1900 CONTINUE
      DO 2000 I=1,N
 2000 WORK1(I) = A(RHO,I)
      WORK1(N1) = 0.D0
      CALL PHI(DINV1,WORK1,K,N1)
      JJ1(K) = RHO
 2100 CONTINUE
C
C   ALL POSSIBLE EQUALITY CONSTRAINTS ARE NOW IN
C   THE ACTIVE SET
C
 2200 CONTINUE
C
```

Figure B.29   *(continued)*

```
C   EXTRACT DINV FROM DINV1 AND JJ FROM JJ1
C
      NXTCOL = 1
      DO 2400 J=1,N1
      IF(JJ1(J).EQ.M1) GO TO 2400
      DO 2300 I=1,N
 2300 DINV(I,NXTCOL) = DINV1(I,J)
      JJ(NXTCOL) = JJ1(J)
      NXTCOL = NXTCOL + 1
 2400 CONTINUE
C
C   INITIAL DATA NOW COMPLETE
C
      CALL ALG3(A,B,C,X,DINV,S,V,WORK,JJ,OBJ,N,M,
     1 R,MTOT,JOUT)
      RETURN
C
 8000 FORMAT(//,20X,'PHASE 1 LP')
 8010 FORMAT(//,20X,'PHASE 2 LP')
 8020 FORMAT(10X,'PHASE 1 - PHASE 2 PROCEDURE',
     1 ' VERSION C')
 8030 FORMAT(6X,'EQUALITY CONSTRAINT',I3,
     1 ' IS REDUNDANT')
      END
```

Figure B.29 *(continued)*

ALG2 is now used to solve the Phase 1 problem. Upon return from ALG2, OBJ contains the optimal Phase 1 objective function value $d'x^* + \alpha^*$. The feasibility test requires the computation of

$$\sum_{i=m+1}^{m+r} b_i .$$

The DO 910 loop computes and stores this in SUMB. If OBJ+SUMB $> 10^{-6}$, then the Phase 2 problem has no feasible solution. In this case, JOUT is set to 3 and control returns to the main program. If OBJ+SUMB $\leq 10^{-6}$, the first n components contain a feasible point for the Phase 2 problem. These are extracted using the DO 1100 loop. As in P1P2B, the DO 1200 through DO 1500 loops ensure that $m + r + 1 \in \hat{J}_0$. Next, it is necessary to see that all nonredundant equality constraints are in the active set. The DO 2100 loop examines each equality constraint in turn. The DO 1600 loop checks to see if each equality constraint RHO

(ρ) is an element of $\hat{J}_0$. If it is, constraint ρ is not considered further in the DO 2100 loop. If $\rho \notin \hat{J}_0$, we proceed as in the discussion at the end of Section 2.6. The DO 1800 loop computes the largest (in absolute value) of those λ_i with $0 \leq \alpha_{i0} \leq m$. Its value is stored in LAMBDA, and K is the associated index.

If (LAMBDA =) $|\lambda_k| < 10^{-5}$, then constraint ρ is redundant and an appropriate message is printed. Since constraint ρ is superfluous, we wish to remove it from further consideration. Removing a row from A in FORTRAN is awkward because A has fixed dimension in the main program. However, a little thought shows that explicit removal is not necessary. Leaving the data for constraint ρ exactly where it is will have exactly the right effect since ρ is not in the active set. Examination of the statement of Algorithm 3 shows that it does not require a_ρ or b_ρ for any equality constraint ρ. This convention must be remembered in any computations following the call to ALG3. If any equality constraint ρ is such that $\rho \notin J_j$, then that equality constraint is redundant (for example, see Section B.1.7).

If $|\lambda_k| \geq 10^{-5}$, the DO 2000 loop copies $(a_\rho', 0)'$ into WORK1, the call to PHI replaces $\hat{D}_0^{-1}$ with

$$\Phi\left(\hat{D}_0^{-1}, \begin{bmatrix} a_\rho \\ 0 \end{bmatrix}, k \right),$$

and the next statement replaces the kth element of $\hat{J}_0$ with ρ.

At the end of the DO 2100 loop, all possible equality constraints are in the active set and all superfluous ones have been effectively deleted. The DO 2400 loop extracts J_0 and D_0^{-1} from $\hat{J}_0$ and $\hat{D}_0^{-1}$, respectively, exactly as in P1P2B. The initial data are then complete, and P1P2C continues by calling ALG3 to solve the Phase 2 problem.

We illustrate P1P2C by applying it to the problem of Example 2.11. The complete program consists of the main program of Figure B.28 and subroutines P1P2C, ALG2, ALG3, SOLN, STEP12, STEP13, STEP2, STEP3, and PHI. The data and output files are shown in Figures B.30 and B.31, respectively.

Figure B.30   Example 2.11 data file for P1P2C.

```
3  3  2
0.  -7.  3.
-3.  1.  0.  27.  12.
-1.  -2.  -1.
-1.  0.  0.
0.  0.  -1.
1.  3.  2.
2.  -1.  3.
```

PHASE 1 - PHASE 2 PROCEDURE VERSION C

PHASE 1 LP

ALGORITHM 2

| | | | | ACTIVE SET | | | |
ITER	OBJECTIVE	K	L	1	2	3	4
0	3.00000	3	5	0	0	0	0
1	-17.00000	2	4	0	0	5	0
2	-36.00000	4	6	0	4	5	0
3	-39.00000	2	0	0	4	5	6

PHASE 2 LP

ALGORITHM 3

| | | | | ACTIVE SET | | |
ITER	OBJECTIVE	K	L	1	2	3
0	-19.09091	1	3	0	4	5
1	-42.00000	1	0	3	4	5

OPTIMAL OBJECTIVE FUNCTION VALUE IS   -42.00000

OPTIMAL SOLUTION
X(1) =   9.0000000
X(2) =   6.0000000
X(3) =   0.

Figure B.31   Example 2.11 output from P1P2C.

B.1.7 Computation of Optimal Solution for the Dual Problem

It was shown in Section 3.2 that an optimal solution for the dual problem could be constructed from the final data of Algorithms 1, 2, and 3. Indeed, an explicit formula is given by (3.32). In this section, we modify SUBROUTINE SOLN to include this computation and print the results. The resulting subroutine is called PDSOLN (Figure B.32), and may be used after either ALG1, ALG2, ALG3, P1P2A, P1P2B, or P1P2C is called.

```
      SUBROUTINE PDSOLN(JOUT,X,N,M,R,OBJ,C,DINV,JJ)
      IMPLICIT DOUBLE PRECISION (A-H,O-Z)
      DIMENSION X(N),C(N),DINV(N,N),JJ(N)
      INTEGER R
      COMMON /UNITS/ IIN,IOUT
C
      MTOT = M + R
      ZERO = 0.D0
      IF(JOUT.GE.2) GO TO 600
      WRITE(IOUT,8000) OBJ
C
C  WRITE OPTIMAL PRIMAL SOLUTION
      WRITE(IOUT,8010)
      DO 100 I=1,N
  100 WRITE(IOUT,8020) I,X(I)
C
C  WRITE OPTIMAL DUAL SOLUTION
      WRITE(IOUT,8030)
      DO 500 I=1,MTOT
      DO 200 J=1,N
      IF(JJ(J).NE.I) GO TO 200
      K = J
      GO TO 300
  200 CONTINUE
C  HERE, CONSTRAINT I MUST BE INACTIVE, OR,
C  REDUNDANT
      IF(I.LE.M) WRITE(IOUT,8040) I,ZERO
      IF(I.GT.M) WRITE(IOUT,8050) I
      GO TO 500
  300 CONTINUE
      SUM = 0.D0
      DO 400 J=1,N
  400 SUM = SUM + C(J)*DINV(J,K)
      SUM = - SUM
      WRITE(IOUT,8040) I,SUM
  500 CONTINUE
      RETURN
  600 CONTINUE
      IF(JOUT.EQ.2) WRITE(IOUT,8060)
      IF(JOUT.EQ.3) WRITE(IOUT,8070)
      RETURN
C
```

Figure B.32 Subroutine PDSOLN.

```
8000 FORMAT(//,6X,'OPTIMAL OBJECTIVE FUNCTION ',
    1    'VALUE IS',F12.5)
8010 FORMAT(//,17X,'OPTIMAL PRIMAL SOLUTION')
8020 FORMAT(18X,'X(',I2,') =',F14.7)
8030 FORMAT(//,18X,'OPTIMAL DUAL SOLUTION')
8040 FORMAT(18X,'U(',I2,') =',F14.7)
8050 FORMAT(10X,'CONSTRAINT',I2,5X,
    1    'IS REDUNDANT')
8060 FORMAT(//,6X,'PROBLEM IS UNBOUNDED FROM ',
    1    'BELOW')
8070 FORMAT(//,6X,'PROBLEM HAS NO FEASIBLE ',
    1    'SOLUTION')
     END
```

Figure B.32   *(continued)*

The argument R for **PDSOLN** should have value 0 for Algorithms 1 and 2, and value r for Algorithm 3. After writing the optimal primal solution, the DO 500 loop computes the optimal dual variable associated with each constraint I (i). The DO 200 loop checks if $i \in J_j$. If $i \notin J_j$, there are two possibilities. If $i \leq m$, then inequality constraint i is inactive, $u_i = 0$ [from (3.32)], and this is printed. If $i > m$, equality constraint i is redundant (see Section B.1.6), and this information is printed. If $i \in J_j$, $u_i = -c'c_{kj}$ where k is such that $i = \alpha_{kj}$. This computation is performed by the DO 400 loop.

PDSOLN is illustrated by replacing the call to **SOLN** in Figure B.1. The resulting main program is shown in Figure B.33. Using the data for Example 2.2 (Figure B.9) produces the output shown in Figure B.34.

Figure B.33   Main program for PDSOLN.

```
IMPLICIT DOUBLE PRECISION (A-H,O-Z)
DIMENSION A(5,2),B(5),C(2),X(2),DINV(2,2),
1   S(2),V(2),WORK(2),JJ(2)
INTEGER R
LOGICAL PASS
COMMON /UNITS/ IIN,IOUT
IIN = 37
IOUT = 38
READ(IIN,*) M,N
MTOT = M
R = 0
READ(IIN,*) (C(I),I=1,N)
READ(IIN,*) (B(I),I=1,M)
DO 100 I=1,M
```

```
100 READ(IIN,*) (A(I,J),J=1,N)
    READ(IIN,*) (X(I),I=1,N)
    READ(IIN,*) (JJ(I),I=1,N)
    DO 200 I=1,N
200 READ(IIN,*) (DINV(I,J),J=1,N)
    CALL VERIFY(X,DINV,JJ,A,B,N,M,R,MTOT,PASS)
    IF(.NOT.PASS) STOP
    CALL ALG1(A,B,C,X,DINV,S,V,WORK,JJ,OBJ,N,M,
   1  MTOT,JOUT)
    CALL PDSOLN(JOUT,X,N,M,R,OBJ,C,DINV,JJ)
    STOP
    END
```

Figure B.33 *(continued)*

ALGORITHM 1

| | | | | ACTIVE SET | |
ITER	OBJECTIVE	K	L	1	2
0	2.00000	1	3	1	2
1	-22.00000	2	5	3	2
2	-45.00000	1	0	3	5

OPTIMAL OBJECTIVE FUNCTION VALUE IS   -45.00000

OPTIMAL PRIMAL SOLUTION
X(1) =   9.0000000
X(2) =   0.0000000

OPTIMAL DUAL SOLUTION
U(1) =   0.0000000
U(2) =   0.0000000
U(3) =   1.2500000
U(4) =   0.0000000
U(5) =   5.7500000

Figure B.34 Example 2.2 output from PDSOLN.

B.2 THE SIMPLEX METHOD

In this section, we give computer programs for the revised simplex method
and the Phase 1–Phase 2 revised simplex method as formulated in Sections

4.1 and 4.2, respectively. The name and purpose of the various subroutines are shown in Table B.4. The most important FORTRAN identifiers and their meanings are summarized in Table B.5.

**TABLE B.4   Name and Purpose of FORTRAN Subroutines
for the Revised Simplex Method**

Subroutine Name	Figure	Purpose
	B.35	main program for the revised simplex method
	B.44	main program for the Phase 1−Phase 2 revised simplex method
SIMPLX	B.36	revised simplex method
RSTEP1	B.37	Step 1 of the revised simplex method
RSTEP2	B.38	Step 2 of the revised simplex method
RSTEP3	B.39	Step 3 of the revised simplex method
PHIPRM	B.40	$[\Phi((B^{-1})', A_k, l)]'$; i.e., update of B^{-1}
P1P2S	B.45	Phase 1−Phase 2 revised simplex method
RSOLN	B.41	prints optimal primal and dual solutions from SIMPLX and DSMPLX

**TABLE B.5   Meaning of FORTRAN Identifiers
for the Revised Simplex Method**

FORTRAN Identifier	Notation in Main Text
N	n
M	m
A	A
A(1,K),...,A(M,K)	A_k'
B	b
K	k
ELL	l
C	c
XB	x_B
BINV	B^{-1}
SB	s_B
SIGB	σ_B
IB	I_B
OBJ	$c_B' x_B$
JOUT	$= \begin{cases} 1, & \text{optimal solution} \\ 2, & \text{problem is unbounded from below} \\ 3, & \text{problem has no feasible solution} \end{cases}$

B.2.1 The Revised Simplex Method

The main program for the revised simplex method (Figure B.35) reads the data for m, n, c, b, A, x_B, and I_B into M, N, C, B, A, XB, and IB, respectively. As with the previous programs, the READ statements are format-free. The input data is assumed to be on a file associated with logical unit 37. All output is written to a file associated with logical unit 38.

```
      IMPLICIT DOUBLE PRECISION(A-H,O-Z)
      DIMENSION A(2,5),B(2),C(5),XB(2),BINV(2,2),
     1  SB(2),WORK(2),IB(2),U(2)
      COMMON /UNITS/ IIN,IOUT
      IIN = 37
      IOUT = 38
      READ(IIN,*) M,N
      READ(IIN,*) (C(I),I=1,N)
      READ(IIN,*) (B(I),I=1,M)
      DO 100 I=1,M
  100 READ(IIN,*) (A(I,J),J=1,N)
      READ(IIN,*) (XB(I),I=1,M)
      READ(IIN,*) (IB(I),I=1,M)
      DO 200 I=1,M
  200 READ(IIN,*) (BINV(I,J),J=1,M)
      CALL SIMPLX(A,B,C,XB,BINV,SB,U,WORK,IB,OBJ,
     1  N,M,JOUT)
      CALL RSOLN(JOUT,XB,U,C,A,IB,M,N,OBJ)
      STOP
      END
```

Figure B.35 Main program for the revised simplex method.

A call is made to SIMPLX which is responsible for solving the problem using the revised simplex method. If an optimal solution is obtained, the optimal basic variables are returned in XB, OBJ contains the optimal objective function value, and the output condition code 1 is returned in JOUT. A returned value of 2 for JOUT means that the problem is unbounded from below. The optimal primal and dual solutions are then printed by calling RSOLN.

It would be useful to have a subroutine analogous to VERIFY (Figure B.2) which checks the consistency of the input data for SIMPLX. The reader may wish to write such a subroutine and call it immediately before the call to SIMPLX.

SUBROUTINE SIMPLX (Figure B.36) organizes the calls to RSTEP1, RSTEP2, and RSTEP3 which implement Steps 1, 2, and 3, respectively, of the revised simplex method.   SIMPLX initializes by setting the iteration counter ITER to zero, and using the DO 100 loop to compute $c_B' x_B$ and place it in OBJ. The call to RSTEP1 computes the search direction SB (s_B) and the index K (k) of the new basic variable. If the optimality condition is satisfied, RSTEP1 returns with JOUT = 1. In this case, the final objective function value, column index k, and index set IB (I_B) are printed and control returns to the main program. Otherwise, RSTEP2 is called.

Figure B.36   Subroutine SIMPLX.

```
      SUBROUTINE SIMPLX(A,B,C,XB,BINV,SB,U,WORK,
     1  IB,OBJ,N,M,JOUT)
      IMPLICIT DOUBLE PRECISION(A-H,O-Z)
      DIMENSION A(M,N),B(M),C(N),XB(M),BINV(M,M),
     1  SB(M),U(M),WORK(M),IB(M)
      INTEGER ELL
      COMMON /UNITS/ IIN,IOUT
C
C
C  INITIALIZE
      WRITE(IOUT,8000) (I,I=1,M)
      ITER = 0
      SUM = 0.D0
      DO 100 I=1,M
  100 SUM = SUM + C(IB(I))*XB(I)
      OBJ = SUM
C
C
  200 CONTINUE
      CALL RSTEP1(A,C,SB,U,BINV,IB,N,M,K,JOUT)
      IF(JOUT.EQ.1) WRITE(IOUT,8010) ITER,OBJ,K,0,
     1  (IB(I),I=1,M)
      IF(JOUT.EQ.1) RETURN
C
C
      CALL RSTEP2(XB,SB,SIGB,ELL,M,JOUT)
      WRITE(IOUT,8010) ITER,OBJ,K,ELL,(IB(I),I=1,M)
      IF(JOUT.EQ.2) RETURN
C
C
```

```
      CALL  RSTEP3(XB,C,B,BINV,A,WORK,OBJ,IB,ELL,K,N,
    1    M,ITER)
      GO TO 200
C
 8000 FORMAT(/,17X,'REVISED SIMPLEX METHOD',//,40X,
    1    'BASIC SET',/,7X,'ITER',6X,'OBJECTIVE',4X,
    2    'K',4X,'L',2X,(T39,10I3))
 8010 FORMAT(6X,I5,F15.5,2I5,2X,(T39,10I3))
      END
```

Figure B.36   *(continued)*

RSTEP2 computes the maximum feasible step size **SIGB** (σ_B) and associated index **ELL** (l). The results for the current iteration, j, $c_B'x_B$, k, l, and I_B, are then printed. If $s_B \le 0$, the problem is unbounded from below, RSTEP2 returns with **JOUT** = 2, and control returns to the main program.

Control now passes to **RSTEP3** which updates **XB**, **IB**, **BINV**, and **ITER** (x_B, I_B, B^{-1}, and j, respectively). This completes one iteration and the next begins when control transfers back to statement **200**.

SUBROUTINE **RSTEP1** (Figure B.37) implements Step 1 of the revised simplex method as follows. The **DO** 100 and **DO** 200 loops calculate **U** ($u = -(B^{-1})'c_B$). The **DO** 500 loop considers each variable **I** (i) in turn. The **DO** 300 loop asks if $i \in I_B$; if so, variable i is not considered further. Otherwise, the **DO** 400 loop computes v_i and stores it in **SUM**. Next, v_i is compared to the current estimate of **VKMIN** (v_k), and v_k is replaced with v_i if appropriate. At the end of the **DO** 500 loop, **VKMIN** is v_k where

$$v_k = \min\{c_i + A_i'u \mid i \notin I_B\},$$

and **K** (k) is the associated index. The optimality test $v_k \ge 0$ has been weakened to $v_k > -10^{-6}$ to account for numerical roundoff errors as discussed in Section B.1. If this test is satisfied, **JOUT** is set to 1 and control returns to the calling program. Otherwise, the **DO** 700 loop and **DO** 800 loops set $s_B = B^{-1}A_k$ and return.

Figure B.37   Subroutine RSTEP1.

```
      SUBROUTINE  RSTEP1(A,C,SB,U,BINV,IB,N,M,K,JOUT)
      IMPLICIT DOUBLE PRECISION(A-H,O-Z)
      DIMENSION A(M,N),C(N),SB(M),U(M),BINV(M,M),
    1    IB(M)
C
```

```
        TOLCON = 1.D-6
        JOUT = 0
C
C    COMPUTE U
        DO 200 J=1,M
        SUM = 0.D0
        DO 100 I=1,M
    100 SUM = SUM + BINV(I,J)*C(IB(I))
    200 U(J) = - SUM
        K = 0
        VKMIN = 1.D30
        DO 500 I=1,N
C    CHECK IF I IS IN IB
        DO 300 J=1,M
        IF(I.EQ.IB(J)) GO TO 500
    300 CONTINUE
C    I IS NOT IN IB
        SUM = C(I)
        DO 400 J=1,M
    400 SUM = SUM + A(J,I)*U(J)
        IF(SUM.GE.VKMIN) GO TO 500
        VKMIN = SUM
        K = I
    500 CONTINUE
        IF(VKMIN.LE.-TOLCON) GO TO 600
        JOUT = 1
        RETURN
    600 CONTINUE
C
C    FORM SB
        DO 800 I=1,M
        SUM = 0.D0
        DO 700 J=1,M
    700 SUM = SUM + BINV(I,J)*A(J,K)
    800 SB(I) = SUM
        RETURN
        END
```

Figure B.37 *(continued)*

SUBROUTINE RSTEP2 (Figure B.38) implements Step 2 of the revised simplex method. The DO 100 loop computes σ_B and l in a straightforward manner. Note that the requirement $(SB(I) =)\ (s_B)_i > 0$ has been strengthened to $(s_B)_i \geq 10^{-6}$ to account for roundoff errors.

```
         SUBROUTINE RSTEP2(XB,SB,SIGB,ELL,M,JOUT)
         IMPLICIT DOUBLE PRECISION(A-H,O-Z)
         DIMENSION XB(M),SB(M)
         INTEGER ELL
         EPS = 1.D-6
         ELL = 0
         SIGB = 1.D30
         DO 100 I=1,M
         IF(SB(I).LT.EPS) GO TO 100
         RATIO = XB(I)/SB(I)
         IF(RATIO.GE.SIGB) GO TO 100
         SIGB = RATIO
         ELL = I
 100     CONTINUE
         IF(ELL.EQ.0) JOUT = 2
         RETURN
         END
```

Figure B.38   Subroutine RSTEP2.

SUBROUTINE RSTEP3 (Figure B.39) implements Step 3 of the revised simplex method. It begins by copying A_k into **WORK** and then calling PHIPRM which updates B^{-1}. The **DO** 300 loop computes the new $x_B = B^{-1}b$. I_B is then updated by setting $\beta_l = k$. The **DO** 400 loop computes the new objective function value $c'_B x_B$ and stores it in **OBJ**. Finally, the iteration counter is increased by one.

Figure B.39   Subroutine RSTEP3.

```
         SUBROUTINE RSTEP3(XB,C,B,BINV,A,WORK,OBJ,IB,
     1   ELL,K,N,M,ITER)
         1MPLICIT DOUBLE PRECISION(A-H,O-Z)
         DIMENSION XB(M),C(N),B(M),BINV(M,M),A(M,N),
     1   WORK(M),IB(M)
         INTEGER ELL
         DO 100 I=1,M
 100     WORK(I) = A(I,K)
         CALL PHIPRM(BINV,WORK,ELL,M)
         DO 300 I=1,M
         SUM = 0.D0
         DO 200 J=1,M
 200     SUM = SUM + BINV(I,J)*B(J)
 300     XB(I) = SUM
         IB(ELL) = K
         SUM = 0.D0
         DO 400 I=1,M
```

```
400 SUM = SUM + C(IB(I))*XB(I)
    OBJ = SUM
    ITER = ITER + 1
    RETURN
    END
```

Figure B.39   *(continued)*

SUBROUTINE   PHIPRM   (Figure   B.40)   replaces B^{-1} with
$[\Phi((B^{-1})', A_k, l)]'$. One way to perform this computation is to copy $(B^{-1})'$
into a second matrix, call PHI (Figure B.8) with the second matrix as an argument, then copy the transpose of the result into B^{-1}. This way is rather
awkward and requires an additional matrix. Thus it seems preferable to have
a subroutine which performs the computations more directly.   SUBROUTINE
PHIPRM does this. It is similar to PHI but modifies rows rather than
columns.

Figure B.40   Subroutine PHIPRM.

```
    SUBROUTINE PHIPRM(BINV,D,ELL,M)
    IMPLICIT DOUBLE PRECISION(A-H,O-Z)
    DIMENSION BINV(M,M),D(M)
    INTEGER ELL
    COMMON /UNITS/ IIN,IOUT
    TOL = 1.D-6
    SUM = 0.D0
    DO 100 I=1,M
100 SUM = SUM + BINV(ELL,I)*D(I)
    IF(DABS(SUM).GE.TOL) GO TO 200
    WRITE(IOUT,8000) SUM
    STOP
200 CONTINUE
    SUM = 1.D0/SUM
    DO 300 I=1,M
300 BINV(ELL,I) = SUM*BINV(ELL,I)
    DO 600 J=1,M
    IF(J.EQ.ELL) GO TO 600
    TEMP =   0.D0
    DO 400 I=1,M
400 TEMP = TEMP + BINV(J,I)*D(I)
    DO 500 I=1,M
500 BINV(J,I) = BINV(J,I) - TEMP*BINV(ELL,I)
600 CONTINUE
    RETURN
8000 FORMAT(6X,'**** ERROR **** NEW MATRIX WOULD ',
   1   'BE SINGULAR, INNER PRODUCT =',G15.6)
    END
```

SUBROUTINE RSOLN (Figure B.41) prints the optimal primal and dual solutions as determined by SIMPLX. It first uses the output condition JOUT to see if an optimal solution has been found. If no solution has been found, control transfers to statement 800 where an appropriate message is printed. If an optimal solution has been found, the DO 300 loop constructs the optimal solution for the primal. Each component x_i is considered in turn. The DO 100 loop decides if $i \in I_B$. If so, i and the corresponding component of x_B are printed. If not, x_i is a nonbasic variable and the value zero is printed.

Figure B.41 Subroutine RSOLN.

```
      SUBROUTINE RSOLN(JOUT,XB,U,C,A,IB,M,N,OBJ)
      IMPLICIT DOUBLE PRECISION(A-H,O-Z)
      DIMENSION XB(M),U(M),C(N),A(M,N),IB(M)
      COMMON /UNITS/ IIN,IOUT
C
      ZERO = 0.D0
      IF(JOUT.GE.2) GO TO 800
      WRITE(IOUT,8000) OBJ
      WRITE(IOUT,8010)
      DO 300 I=1,N
C IS X(I) BASIC?
      DO 100 J=1,M
      INDEX = J
      IF(IB(J).EQ.I) GO TO 200
  100 CONTINUE
      WRITE(IOUT,8020) I,ZERO
      GO TO 300
  200 CONTINUE
      WRITE(IOUT,8020) I,XB(INDEX)
  300 CONTINUE
C
C
      WRITE(IOUT,8030)
      WRITE(IOUT,8040) (I,U(I),I=1,M)
      WRITE(IOUT,8050)
      DO 700 I=1,N
C   IS X(I) BASIC?
      DO 400 J=1,M
      IF(IB(J).EQ.I) GO TO 600
  400 CONTINUE
C   X(I) IS NON-BASIC
      SUM = C(I)
      DO 500 J=1,M
```

```
      500 SUM = SUM + A(J,I)*U(J)
          WRITE(IOUT,8060) I,SUM
          GO TO 700
    C  X(I) IS BASIC
      600 WRITE(IOUT,8060) I,ZERO
      700 CONTINUE
          RETURN
      800 CONTINUE
          IF(JOUT.EQ.2) WRITE(IOUT,8070)
          IF(JOUT.EQ.3) WRITE(IOUT,8080)
          RETURN
    C
     8000 FORMAT(//,6X,'OPTIMAL OBJECTIVE FUNCTION ',
        1  'VALUE IS',F12.5)
     8010 FORMAT(//,17X,'OPTIMAL PRIMAL SOLUTION')
     8020 FORMAT(18X,'X(',I2,') =',F14.7)
     8030 FORMAT(//,18X,'OPTIMAL DUAL SOLUTION',
        1  /,18X,'EQUALITY CONSTRAINTS')
     8040 FORMAT(18X,'U(',I2,') =',F14.7)
     8050 FORMAT(/,15X,'NONNEGATIVITY CONSTRAINTS')
     8060 FORMAT(18X,'V(',I2,') =',F14.7)
     8070 FORMAT(//,6X,'PROBLEM IS UNBOUNDED FROM ',
        1  'BELOW')
     8080 FORMAT(//,6X,'PROBLEM HAS NO FEASIBLE ',
        1  'SOLUTION')
          END
```

Figure B.41   *(continued)*

Next, the optimal solution for the dual problem is computed and print-ed. The optimal dual variables U (u) associated with the problem's equality constraints are printed directly. The optimal dual variables associated with the nonnegativity constraints, v are next computed as follows. Each nonnega-tivity constraint $x_i \geq 0$ is considered in turn. The DO 400 loop determines whether or not the corresponding primal variable is basic or not. If it is, a value of zero is printed for v_i. Otherwise, the DO 500 loop computes $v_i = c_i + A_i'u$.

We illustrate SIMPLX by using it to solve the problem of Example 4.1. The complete program consists of the main program of Figure B.35 and sub-routines SIMPLX, RSTEP1, RSTEP2, RSTEP3, PHIPRM, and RSOLN. The data file for Example 4.1 is shown in Figure B.42. The output file (from unit 38) is shown in Figure B.43.

```
2 5
2.  14.  36.  0.  0.
5.  2.
-2.  1.  4.  -1.  0.
-1.  -2.  -3.  0.  1.
5.  12.
2 5
1.  0.
2.  1.
```

Figure B.42   Example 4.1 data file for SIMPLX.

Figure B.43   Example 4.1 output from SIMPLX.

REVISED SIMPLEX METHOD

				BASIC	SET
ITER	OBJECTIVE	K	L	1	2
0	70.00000	3	1	2	5
1	45.00000	2	0	3	5

OPTIMAL OBJECTIVE FUNCTION VALUE IS   45.00000

OPTIMAL PRIMAL SOLUTION
$X(1) = 0.0000000$
$X(2) = 0.0000000$
$X(3) = 1.2500000$
$X(4) = 0.0000000$
$X(5) = 5.7500000$

OPTIMAL DUAL SOLUTION
EQUALITY CONSTRAINTS
$U(1) = -9.0000000$
$U(2) = 0.0000000$

NONNEGATIVITY CONSTRAINTS
$V(1) = 20.0000000$
$V(2) = 5.0000000$
$V(3) = 0.0000000$
$V(4) = 9.0000000$
$V(5) = 0.0000000$

B.2.2   The Phase 1-Phase 2 Revised Simplex Method

The main program for the Phase 1−Phase 2 revised simplex method is shown in Figure B.44. It reads the problem data, calls P1P2S, and then calls RSOLN to print the optimal primal and dual solutions.

```
      IMPLICIT DOUBLE PRECISION(A-H,O-Z)
      DIMENSION A(2,4),B(2),C(4),XB(2),BINV(2,2),
     1   SB(2),WORK(2),IB(2),U(2)
      DIMENSION A1(2,6),C1(6)
      COMMON /UNITS/ IIN,IOUT
      IIN = 37
      IOUT = 38
      READ(IIN,*) M,N
      N1 = N + M
      READ(IIN,*) (C(I),I=1,N)
      READ(IIN,*) (B(I),I=1,M)
      DO 100 I=1,M
  100 READ(IIN,*) (A(I,J),J=1,N)
      CALL P1P2S(A,B,C,XB,BINV,SB,U,WORK,IB,OBJ,
     1   N,M,JOUT,A1,C1,N1)
      CALL RSOLN(JOUT,XB,U,C,A,IB,M,N,OBJ)
      STOP
      END
```

Figure B.44   Main program for the Phase 1−Phase 2 revised simplex method.

SUBROUTINE P1P2S (Figure B.45) organizes the Phase 1−Phase 2 revised simplex method. The DO 100 and DO 200 loops check the sign of each b_i. If b_i is negative, the ith equality constraint is multiplied by -1. The Phase 1 constraint matrix is denoted by A1. The DO 300 and DO 400 loops copy the first N columns of A into A1. The DO 500, 600, and 700 loops put the (m,m) identity matrix into columns $n+1,\ldots,n+m$ of A1. These correspond to the m artificial variables for the Phase 1 problem. The DO 800 and DO 900 loops form the objective function coefficients for the Phase 1 problem. The DO 1000, 1100, and 1200 loops set $B^{-1} = I$, $I_B = \{n+1,\ldots,n+m\}$, and $x_B = b$. The Phase 1 data is then complete, and the problem is solved with a call to SIMPLX. If the Phase 1 objective function is greater than 10^{-6}, the Phase 2 problem has no feasible solution. This is communicated to the main program by assigning JOUT the value 3. Otherwise, the final data from the Phase 1 problem is examined before proceeding to the Phase 2 problem.

```
      SUBROUTINE P1P2S(A,B,C,XB,BINV,SB,U,WORK,
     1  IB,OBJ,N,M,JOUT,A1,C1,N1)
      IMPLICIT DOUBLE PRECISION(A-H,O-Z)
      DIMENSION A(M,N),B(M),C(N),XB(M),BINV(M,M),
     1  SB(M),U(M),WORK(M),IB(M),A1(M,N1),C1(N1)
      INTEGER ELL
      COMMON /UNITS/ IIN,IOUT
C
      TOLFES = 1.D-6
      ZERO = 0.D0
      EPS = 1.D-5
      WRITE(IOUT,8000)
      WRITE(IOUT,8010)
C
C   CONSTRUCT PHASE 1 DATA
C   FIRST CHECK FOR B(I) .LT. 0
      DO 200 I=1,M
      IF(B(I).GE.ZERO) GO TO 200
      B(I) = - B(I)
      DO 100 J=1,N
  100 A(I,J) = - A(I,J)
  200 CONTINUE
C
C   COPY A INTO A1
      DO 400 I=1,M
      DO 300 J=1,N
  300 A1(I,J) = A(I,J)
  400 CONTINUE
C
C   PUT IDENTITY MATRIX INTO LAST M COLUMNS
      DO 600 I=1,M
      DO 500 J=1,M
  500 A1(I,N+J) = ZERO
  600 CONTINUE
      DO 700 I=1,M
  700 A1(I,N+I) = 1.D0
C
C   PHASE 1 OBJECTIVE FUNCTION
      DO 800 I=1,N
  800 C1(I) = ZERO
      DO 900 I=1,M
```

Figure B.45 Subroutine P1P2S.

```
  900 C1(N+I) = 1.D0
C
C  SET BINV TO IDENTITY
      DO 1100 I=1,M
      DO 1000 J=1,M
 1000 BINV(I,J) = ZERO
 1100 CONTINUE
      DO 1200 I=1,M
      BINV(I,I) = 1.D0
      IB(I) = N + I
 1200 XB(I) = B(I)
C
C  PHASE 1 DATA IS NOW COMPLETE
C
      CALL SIMPLX(A1,B,C1,XB,BINV,SB,U,WORK,IB,
     1  OBJ,N1,M,JOUT)
      IF(OBJ.LE.TOLFES) GO TO 1300
      JOUT = 3
      RETURN
 1300 CONTINUE
C
C  ARE ANY ARTIFICIALS STILL BASIC?
      DO 1800 ELL=1,M
      IF(IB(ELL).LE.N) GO TO 1800
C  HERE, AN ARTIFICIAL IS BASIC IN ROW ELL
      BIG = -1.D0
      K = 0
      DO 1500 I=1,N
      SUM = 0.D0
      DO 1400 J=1,M
 1400 SUM = SUM + BINV(ELL,J)*A(J,I)
      IF(DABS(SUM).LE.BIG) GO TO 1500
      BIG = DABS(SUM)
      K = I
 1500 CONTINUE
      IF(BIG.GE.EPS) GO TO 1600
C  HERE, CONSTRAINT ELL IS REDUNDANT
C  WRITE THE GOOD NEWS AND QUIT
      WRITE(IOUT,8030) ELL
      STOP
 1600 CONTINUE
      DO 1700 I=1,M
```

Figure B.45   *(continued)*

```
 1700  WORK(I)  =  A(I,K)
       CALL  PHIPRM(BINV,WORK,ELL,N)
       IB(ELL)  =  K
 1800  CONTINUE
C
C   PHASE  2
       WRITE(IOUT,8020)
       CALL  SIMPLX(A,B,C,XB,BINV,SB,U,WORK,IB,OBJ,
     1  N,M,JOUT)
       RETURN
C
 8000  FORMAT(12X,'PHASE  1  -  PHASE  2  SIMPLEX  ',
     1  'METHOD')
 8010  FORMAT(//,23X,'PHASE  1  LP')
 8020  FORMAT(//,23X,'PHASE  2  LP')
 8030  FORMAT(6X,'CONSTRAINT',I3,'IS  REDUNDANT')
       END
```

Figure B.45 *(continued)*

The DO 1800 loop examines each element of I_B. If no artificial variables are in the basis, the DO 1800 loop does nothing and the subroutine proceeds to call SIMPLX to solve the Phase 2 problem. Otherwise, suppose that $\beta_l > n$. Then an artificial variable is basic in row l. The DO 1400 and DO 1500 loops compute k such that

$$|(B^{-1}A_k)_l| = \max\{|(B^{-1}A_k)_i| \mid i = 1, \ldots, n\}$$

[see discussion following (4.25)]. The left-hand side of the above is stored in BIG. If $BIG > 10^{-5}$, then variable k is made basic in row l. B^{-1} and I_B are updated accordingly. If $BIG < 10^{-5}$, then the lth equality constraint is redundant. The program prints this information and then stops. In order to solve the problem at this point, the lth equality constraint must be deleted. Including this capability with P1P2S would make the program somewhat lengthy so we have chosen not to do it.

Once the DO 1800 loop is satisfied, all artificial variables are out of the basis and P1P2S proceeds by using the final data for the Phase 1 problem as initial data for the Phase 2 problem. The Phase 2 problem is solved with a call to SIMPLX.

P1P2S is illustrated by solving the problem of Example 4.3. The complete program consists of the main program of Figure B.44, and subroutines P1P2S, SIMPLX, RSTEP1, RSTEP2, RSTEP3, PHIPRM, and RSOLN. The data file for Example 4.3 is shown in Figure B.46. The output file is shown in Figure B.47.

```
2  4
1.  1.  0.  0.
5.  1.
1.  1.  1.  0.
1.  -1.  0.  -1.
```

Figure B.46 Example 4.3 data file for P1P2S.

Figure B.47 Example 4.3 output from P1P2S.

PHASE 1 - PHASE 2 SIMPLEX METHOD

PHASE 1 LP

REVISED SIMPLEX METHOD

				BASIC SET	
ITER	OBJECTIVE	K	L	1	2
0	6.00000	1	2	5	6
1	4.00000	2	1	5	1
2	0.00000	3	0	2	1

PHASE 2 LP

REVISED SIMPLEX METHOD

				BASIC SET	
ITER	OBJECTIVE	K	L	1	2
0	5.00000	3	1	2	1
1	1.00000	4	0	3	1

OPTIMAL OBJECTIVE FUNCTION VALUE IS   1.00000

```
        OPTIMAL PRIMAL SOLUTION
        X( 1) =    1.0000000
        X( 2) =    0.0000000
        X( 3) =    4.0000000
        X( 4) =    0.0000000
```

OPTIMAL DUAL SOLUTION
EQUALITY CONSTRAINTS
U(1) =   0.0000000
U(2) =   -1.0000000

NONNEGATIVITY CONSTRAINTS
V(1) =   0.0000000
V(2) =   2.0000000
V(3) =   0.0000000
V(4) =   1.0000000

Figure B.47   *(continued)*

B.3   ALGORITHM 4

Here we give a computer program which implements Algorithm 4 for the solution of the parametric linear programming problem. The name and purpose of the various subroutines are shown in Table B.6. Table B.7 shows the FORTRAN identifiers used for the various quantities required by Algorithm 4. The main program (Figure B.48) reads the problem data, calls ALG4, and then stops. Note that ALG4 requires two temporary vectors, BTEMP and WORK, of dimension m and n, respectively.

**TABLE B.6   Name and Purpose of FORTRAN Subroutines
for Algorithm 4**

Subroutine Name	Figure	Purpose
	B.48	main program for Algorithm 4
ALG4	B.49	Algorithm 4
PSTEP1	B.50	Step 1 of Algorithm 4
PSTEP2	B.51	Step 2 of Algorithm 4
PSTEP3	B.52	Step 3 of Algorithm 4
PSTEP4	B.53	Step 4 of Algorithm 4
PSTEP5	B.54	Step 5 of Algorithm 4
PHI	B.8	Procedure Φ
STEP2	B.6	Step 2 of Algorithms 1, 2, and 3 Used by PSTEP3
SUMM	B.55	prints solution summary from ALG4

TABLE B.7   Meaning of FORTRAN Identifiers for Algorithm 4

FORTRAN Identifier	Notation in Main Text
N	n
M	m
A	A
B	b
C	c
Q	q
P	p
XJ	$x_j(t_j)$
XJ1	$x_j(t_{j+1})$
K	k
ELL	l
ITER	j
JJ	J_j
DINV	D_j^{-1}
H1	h_{1j}
H2	h_{2j}
G1	g_{1j}
G2	g_{2j}
SIGJ	σ_j
U	u_j
TJ1TIL	$\tilde{t}_{j+1}$
DTJTIL	$\delta\tilde{t}_j$
TJ1STR	t_{j+1}^{*}
DTJSTR	δt_j^{*}
TBOT	$\underline{t}$
TTOP	$\bar{t}$
KOUT	$= \begin{cases} 1, & \text{if } t_{j+1} = \tilde{t}_{j+1} \\ 2, & \text{if } t_{j+1} = t_{j+1}^{*} \end{cases}$
JOUT	$= \begin{cases} 1, & \text{if } t_{j+1} = \bar{t} \\ 2, & \text{problem is unbounded from below for } t > t_{j+1} \\ 3, & \text{problem has no feasible solution for } t > t_{j+1} \end{cases}$

```
      IMPLICIT DOUBLE PRECISION(A-H,O-Z)
      DIMENSION A(6,2),B(6),P(6),C(2),Q(2),XJ(2),
    1    XJ1(2),H1(2),H2(2),U(2),G1(2),G2(2),S(2),
    2    BTEMP(6),JJ(2),DINV(2,2),WORK(2)
      COMMON /UNITS/ IIN,IOUT
      IIN = 37
      IOUT = 38
      READ(IIN,*) M,N
      MTOT = M
      READ(IIN,*) TBOT,TTOP
      READ(IIN,*) (C(I),I=1,N)
      READ(IIN,*) (Q(I),I=1,N)
      READ(IIN,*) (B(I),I=1,M)
      READ(IIN,*) (P(I),I=1,M)
      DO 100 I=1,M
  100 READ(IIN,*) (A(I,J),J=1,N)
      READ(IIN,*) (XJ(I),I=1,N)
      DO 200 I=1,N
  200 READ(IIN,*) (DINV(I,J),J=1,N)
      READ(IIN,*) (JJ(I),I=1,N)
      CALL ALG4(A,B,C,P,Q,XJ,XJ1,H1,H2,U,G1,G2,S,
    1    BTEMP,TBOT,TTOP,N,M,MTOT,JOUT,JJ,DINV,
    2    WORK)
      STOP
      END
```

Figure B.48   Main program for Algorithm 4.

ALG4 assumes that the parametric LP has already been solved for $t = \underline{t}$ and that the optimal solution x_0 and its associated data D_0^{-1} and J_0 are input in XJ, DINV, and JJ, respectively. This assumption could be avoided by using the Phase 1—Phase 2 subroutine P1P2B (Section B.1.5) before calling ALG4.

SUBROUTINE ALG4 (Figure B.49) organizes the calls to the steps of Algorithm 4. Subroutines PSTEP1,...,PSTEP5 implement Steps 1,...,5, respectively, of Algorithm 4. ALG4 begins by setting the iteration counter to zero and setting $t_0 = \underline{t}$. A call is then made to PSTEP1 which computes h_{1j} and h_{2j}. After calling PSTEP2, two tests are made. As the algorithm progresses, the output termination indicator JOUT has value 0. If in Step 2 $t_{j+1} = \bar{t}$, then JOUT is set to 1, and in ALG4 control transfers directly to statement 300 which prints the final solution summary and terminates. If JOUT has value 0, then a second test is performed. In PSTEP2, KOUT is assigned the value 1 if $t_{j+1} = \tilde{t}_{j+1}$, and the value 2 if $t_{j+1} = t_{j+1}^*$. Control transfers to **PSTEP3** or **PSTEP4** accordingly. Following the execution of **PSTEP3** or **PSTEP4**, a call is made to **SUMM** which prints a summary of the optimal primal and dual solutions as well as changes in the active set for the current critical interval. In **PSTEP3** or **PSTEP4**, it may be determined that the problem is either unbounded from below or has no feasible solution for $t > t_{j+1}$. This is communicated to **ALG4** by setting **JOUT** to 2 or 3, respectively. A positive value for **JOUT** indicates that the parametric analysis is complete and control transfers to the top of the loop to evaluate the next critical interval.

Figure B.49 Subroutine ALG4.

```
      SUBROUTINE ALG4(A,B,C,P,Q,XJ,XJ1,H1,H2,U,G1,G2,
     1  S,BTEMP,TBOT,TTOP,N,M,MTOT,JOUT,JJ,DINV,
     2  WORK)
      IMPLICIT DOUBLE PRECISION(A-H,O-Z)
      DIMENSION A(MTOT,N),B(MTOT),C(N),Q(N),P(MTOT),
     1  XJ(N),XJ1(N),H1(N),H2(N),G1(N),G2(N),S(N),
     2  BTEMP(MTOT),JJ(N),U(N),DINV(N,N),WORK(N)
      INTEGER ELL
      COMMON /UNITS/ IIN,IOUT
C
      WRITE(IOUT,8000)
      ITER = 0
      TJ = TBOT
C
  100 CONTINUE
      CALL PSTEP1(B,P,DINV,H1,H2,N,M,MTOT,JJ)
C
      CALL PSTEP2(A,B,C,Q,P,XJ,XJ1,H1,H2,U,G1,G2,
     1  DINV,JJ,TJ,TJ1,TTOP,ELL,K,N,M,MTOT,KOUT,
     2  JOUT)
C
      IF(JOUT.EQ.1) GO TO 300
      IF(KOUT.EQ.2) GO TO 200
C   HERE, KOUT .EQ. 1
C
C
      CALL PSTEP3(XJ1,XJ,A,B,P,DINV,BTEMP,S,TJ1,
     1  N,M,MTOT,ELL,K,JOUT,JJ)
      GO TO 300
C
C
```

```
      200 CALL PSTEP4(A,XJ,XJ1,G1,G2,DINV,TJ1,ELL,
    1    K,N,MTOT,JOUT)
C
C
      300 CALL SUMM(H1,H2,G1,G2,JJ,K,ELL,KOUT,TJ,TJ1,
    1    N,M,ITER,JOUT)
          IF(JOUT.NE.0) RETURN
          CALL PSTEP5(DINV,A,JJ,WORK,N,MTOT,TJ,TJ1,
    1    K,ELL,ITER)
          GO TO 100
C
     8000 FORMAT(18X,'ALGORITHM 4')
          END
```

Figure B.49 *(continued)*

SUBROUTINE PSTEP1 (Figure B.50) implements Step 1 of Algorithm 4. In the statement of Algorithm 4, equivalent expressions for h_{1j} (H1) and h_{2j} (H2) are

$$h_{1j} = \sum_{i=1}^{n} b_{\alpha_{ij}} c_{ij}$$

and

$$h_{2j} = \sum_{i=1}^{n} p_{\alpha_{ij}} c_{ij},$$

where $D_j^{-1} = [c_{1j}, \ldots, c_{nj}]$. PSTEP1 computes H1 and H2 in this form rather than forming b_j and p_j explicitly.

Figure B.50 Subroutine PSTEP1.

```
      SUBROUTINE PSTEP1(B,P,DINV,H1,H2,N,M,MTOT,JJ)
      IMPLICIT DOUBLE PRECISION(A-H,O-Z)
      DIMENSION B(MTOT),P(MTOT),H1(N),H2(N),JJ(N),
    1    DINV(N,N)
C   COMPUTATION OF H1 AND H2
          DO 100 I=1,N
          H1(I) = 0.D0
      100 H2(I) = 0.D0
          DO 300 J=1,N
          DO 200 I=1,N
          H1(I) = H1(I) + B(JJ(J))*DINV(I,J)
      200 H2(I) = H2(I) + P(JJ(J))*DINV(I,J)
      300 CONTINUE
          RETURN
          END
```

SUBROUTINE PSTEP2 (Figure B.51) implements Step 2 of Algorithm 4. The DO 100 and DO 200 loops compute g_{1j} (G1) and g_{2j} (G2). If $g_{2j} > 10^{-6}$, then $\tilde{t}_{j+1}$ (TJ1TIL) is set to 10^{30} and control transfers to Step 2.2. Otherwise, the DO 500 loop computes u_j (U), and the DO 600 loop computes k (K) and $\delta \tilde{t}_j$ (DTJTIL).

Figure B.51 Subroutine PSTEP2.

```
      SUBROUTINE PSTEP2(A,B,C,Q,P,XJ,XJ1,H1,H2,U,
     1   G1,G2,DINV,JJ,TJ,TJ1,TTOP,ELL,K,N,M,MTOT,
     2   KOUT,JOUT)
      IMPLICIT DOUBLE PRECISION(A-H,O-Z)
      DIMENSION A(MTOT,N),B(MTOT),C(N),Q(N),P(MTOT),
     1   XJ(N),XJ1(N),H1(N),H2(N),U(N),G1(N),G2(N),
     2   DINV(N,N),JJ(N)
      INTEGER ELL
C
      TOL = 1.D-6
C   COMPUTATION OF THE CRITICAL VALUE TJ1
C
C   STEP 2.1
C
C   COMPUTE G1 AND G2
      DO 200 J=1,N
      SUM1 = 0.D0
      SUM2 = 0.D0
      DO 100 I=1,N
      SUM1 = SUM1 - DINV(I,J)*C(I)
  100 SUM2 = SUM2 - DINV(I,J)*Q(I)
      G1(J) = SUM1
  200 G2(J) = SUM2
C   SEE IF G2 IS .GE. 0
      DO 300 I=1,N
      IF(G2(I).LE.-TOL) GO TO 400
  300 CONTINUE
      TJ1TIL = 1.D30
      DTJTIL = 1.D30
      K = 0
      GO TO 700
  400 CONTINUE
C   COMPUTE U
      DO 500 J=1,N
  500 U(J) = G1(J) + TJ*G2(J)
C   FIND SMALLEST RATIO
      DTJTIL = 1.D30
      DO 600 J=1,N
      IF(G2(J).GE.-TOL) GO TO 600
```

```
       RATIO = - U(J)/G2(J)
       IF(RATIO.GE.DTJTIL) GO TO 600
       DTJTIL = RATIO
       K = J
  600  CONTINUE
       TJ1TIL = TJ + DTJTIL
C
C   STEP 2.2
C
  700  CONTINUE
       DTJSTR = 1.D30
       ELL = 0
       DO 1100 I=1,M
       DO 800 INDEX=1,N
       IF(JJ(INDEX).EQ.I) GO TO 1100
  800  CONTINUE
       BOT = - P(I)
       DO 900 INDEX=1,N
  900  BOT = BOT + A(I,INDEX)*H2(INDEX)
       IF(BOT.LE.TOL) GO TO 1100
       TOP = B(I) + TJ*P(I)
       DO 1000 INDEX=1,N
 1000  TOP = TOP - A(I,INDEX)*XJ(INDEX)
       RATIO = TOP/BOT
       IF(RATIO.GE.DTJSTR) GO TO 1100
       DTJSTR = RATIO
       ELL = I
 1100  CONTINUE
       IF(ELL.EQ.0) TJ1STR = 1.D30
       IF(ELL.GT.0) TJ1STR = TJ + DTJSTR
C
C   STEP 2.3
C
       TJ1 = DMIN1(TJ1TIL,TJ1STR,TTOP)
       IF(TJ1.EQ.TJ1TIL) KOUT = 1
       IF(TJ1.EQ.TJ1STR) KOUT = 2
       JOUT = 0
       IF(TJ1.EQ.TTOP) JOUT = 1
       IF(TJ1.EQ.TTOP) RETURN
       DO 1200 J=1,N
 1200  XJ1(J) = H1(J) + TJ1*H2(J)
       RETURN
       END
```

Figure B.51   *(continued)*

Statement 700 begins Step 2.2. The DO 1100 loop computes l (ELL) and δt_j^* (DTJSTR). Prior to the loop, ELL is initialized to zero. If it is unchanged after the loop, then $a_i' h_{2j} - p_i \leq 10^{-6}$ for $i = 1, \ldots, m$. In this case, t_{j+1}^* (TJ1STR) is set to 10^{30}.

Step 2.3 computes t_{j+1}. KOUT is set to 1 or 2 according to whether $t_{j+1} = \tilde{t}_{j+1}$ or t_{j+1}^*, respectively. JOUT is assigned the value 1 if $t_{j+1} = \bar{t}$, and value 0 otherwise. Finally, if $t_{j+1} < \bar{t}$, $x_j(t_{j+1})$ is computed and stored in XJ1 using the DO 1200 loop.

SUBROUTINE PSTEP3 (Figure B.52) implements Step 3 of Algorithm 4. Note that this step is identical to Step 2 of Algorithms 1, 2, and 3, with x_j replaced by $x_j(t_{j+1})$, s_j replaced by c_{kj}, and b_i replaced by $b_i + t_{j+1} p_i$, $i = 1, \ldots, m$. PSTEP3 makes this replacement and proceeds by calling STEP2 (Figure B.5). The DO 100 loop copies the modified right-hand side into BTEMP and c_{kj} into S. If $a_i' c_{kj} \geq -10^{-6}$, STEP2 returns with JOUT assigned the value 2. In the present context, this means that the problem is unbounded from below for $t > t_{j+1}$. This information is communicated to ALG4 through JOUT. Otherwise, PSTEP3 concludes by using the DO 300 loop to compute $x_{j+1}(t_{j+1})$ and storing it in XJ.

Figure B.52   Subroutine PSTEP3.

```
      SUBROUTINE PSTEP3(XJ1,XJ,A,B,P,DINV,
     1   BTEMP,S,TJ1,N,M,MTOT,ELL,K,JOUT,JJ)
      IMPLICIT DOUBLE PRECISION(A-H,O-Z)
      DIMENSION XJ1(N),XJ(N),A(MTOT,N),B(MTOT),
     1   P(MTOT),BTEMP(MTOT),S(N),JJ(N),
     2   DINV(N,N)
      INTEGER ELL
C   DETERMINATION OF NEW ACTIVE CONSTRAINT
      DO 100 I=1,MTOT
  100 BTEMP(I) = B(I) + TJ1*P(I)
      DO 200 I=1,N
  200 S(I) = DINV(I,K)
      CALL STEP2(A,BTEMP,XJ1,S,JJ,SIGJ,ELL,N,
     1   M,MTOT,JOUT)
      IF(JOUT.EQ.2) RETURN
      DO 300 I=1,N
  300 XJ(I) = XJ1(I) - SIGJ*S(I)
      RETURN
      END
```

SUBROUTINE PSTEP4 (Figure B.53) implements Step 4 of Algorithm 4. The DO 100 and DO 200 loops compute W (ω_j) and the associated index k. If $a_i'c_{ij} \leq 10^{-6}$ for $i = 1, \ldots, n$ then the problem has no feasible solution for $t > t_{j+1}$. This is communicated to ALG4 by setting JOUT to 3. Otherwise, PSTEP4 concludes by setting XJ to $x_j(t_{j+1})$.

```
      SUBROUTINE PSTEP4(A,XJ,XJ1,G1,G2,
     1  DINV,TJ1,ELL,K,N,MTOT,JOUT)
      IMPLICIT DOUBLE PRECISION(A-H,O-Z)
      DIMENSION A(MTOT,N),XJ(N),XJ1(N),
     1  G1(N),G2(N),DINV(N,N)
      INTEGER ELL
      TOL = 1.D-6
      JOUT = 0
      K = 0
      W = 1.D30
      DO 200 I=1,N
      BOT = 0.D0
      DO 100 J=1,N
  100 BOT = BOT + A(ELL,J)*DINV(J,I)
      IF(BOT.LE.TOL) GO TO 200
      RATIO = (G1(I) + TJ1*G2(I))/BOT
      IF(RATIO.GE.W) GO TO 200
      W = RATIO
      K = I
  200 CONTINUE
      IF(K.EQ.0) JOUT = 3
      IF(K.EQ.0) RETURN
      DO 300 I=1,N
  300 XJ(I) = XJ1(I)
      RETURN
      END
```

Figure B.53   Subroutine PSTEP4.

SUBROUTINE PSTEP5 (Figure B.54) implements Step 5 of Algorithm 4. The DO 100 loop copies a_l into WORK, and D_j^{-1} is replaced with D_{j+1}^{-1} by calling PHI (Figure B.8). The active set is updated, and t_j and j are replaced with t_{j+1} and $j + 1$, respectively.

```
      SUBROUTINE PSTEP5(DINV,A,JJ,WORK,N,MTOT,TJ,TJ1,
    1   K,ELL,ITER)
      IMPLICIT DOUBLE PRECISION(A-H,O-Z)
      DIMENSION DINV(N,N),WORK(N),JJ(N),A(MTOT,N)
      INTEGER ELL
C   UPDATE
      DO 100 I=1,N
  100 WORK(I) = A(ELL,I)
      CALL PHI(DINV,WORK,K,N)
      JJ(K) = ELL
      TJ = TJ1
      ITER = ITER + 1
      RETURN
      END
```

Figure B.54   Subroutine PSTEP5.

SUBROUTINE SUMM (Figure B.55) prints a summary of the results for each critical interval as well as the reason for final termination. The active set and optimal primal solution are printed first. The complete optimal dual solution is computed from G1 and G2 and the active set using the DO 100, 200, and 300 loops. Next, a summary of the changes in the active set from Steps 2, 3, and 4 is printed. There are two possibilities. If $t_{j+1} = \tilde{t}_{j+1}$ (KOUT = 1), k comes from Step 2.1 and l comes from Step 3. If $t_{j+1} = t_{j+1}^*$ (KOUT = 2), l comes from Step 2.2 and k comes from Step 4. The current critical interval is printed. Finally, if the parametric analysis is complete, the appropriate message is printed using JOUT.

Figure B.55   Subroutine SUMM.

```
      SUBROUTINE SUMM(H1,H2,G1,G2,JJ,K,ELL,KOUT,TJ,
    1   TJ1,N,M,ITER,JOUT)
      IMPLICIT DOUBLE PRECISION(A-H,O-Z)
      DIMENSION H1(N),H2(N),G1(N),G2(N),JJ(N)
      INTEGER ELL
      COMMON /UNITS/ IIN,IOUT
C
      ZERO = 0.D0
      WRITE(IOUT,8000) ITER
C   WRITE ACTIVE SET
      WRITE(IOUT,8010) (I,I=1,N)
      WRITE(IOUT,8020) (JJ(I),I=1,N)
```

```
C  WRITE OPTIMAL PRIMAL SOLUTION
       WRITE(IOUT,8030)
       WRITE(IOUT,8040) (I,H1(I),H2(I),I=1,N)
C  WRITE OPTIMAL DUAL SOLUTION
       WRITE(IOUT,8050)
       DO 300 I=1,M
       DO 100 J=1,N
       IF(JJ(J).NE.I) GO TO 100
       J1 = J
       GO TO 200
  100 CONTINUE
C  HERE, CONSTRAINT I MUST BE INACTIVE
       WRITE(IOUT,8040) I,ZERO,ZERO
       GO TO 300
  200 WRITE(IOUT,8040) I,G1(J1),G2(J1)
  300 CONTINUE
       IF(JOUT.EQ.1) GO TO 400
C
C  RESULTS FROM STEP 2
       IF(KOUT.EQ.1) WRITE(IOUT,8060) JJ(K),K
       IF(KOUT.EQ.2) WRITE(IOUT,8070) ELL,ELL
       IF(JOUT.GT.0) GO TO 400
C  RESULTS FROM STEPS 3 AND 4
       IF(KOUT.EQ.1) WRITE(IOUT,8080) ELL,ELL
       IF(KOUT.EQ.2) WRITE(IOUT,8090) JJ(K),K
C
  400 ITER1 = ITER + 1
       WRITE(IOUT,8100) ITER,TJ,ITER1,TJ1
       IF(JOUT.EQ.1) WRITE(IOUT,8110)
       IF(JOUT.EQ.2) WRITE(IOUT,8120) TJ1
       IF(JOUT.EQ.3) WRITE(IOUT,8130) TJ1
       RETURN
C
 8000 FORMAT(///,18X,'ITERATION',I4)
 8010 FORMAT(/,20X,10I3)
 8020 FORMAT(10X,'ACTIVE SET',10I3)
 8030 FORMAT(/,15X,'OPTIMAL PRIMAL SOLUTION',
      1  /,10X,'I',6X,'H1(I)',7X,'H2(I)')
 8040 FORMAT(8X,I3,2F12.4)
 8050 FORMAT(//,15X,'OPTIMAL DUAL SOLUTION',
      1  /,10X,'I',6X,'G1(I)',7X,'G2(I)')
 8060 FORMAT(//,' STEP',/,2X,'2.1',2X,'NEW ',
      1  'INACTIVE CONSTRAINT IS',I3,3X,'K =',I3)
```

Figure B.55 *(continued)*

```
8070 FORMAT(//,' STEP',/,2X,'2.2',2X,'NEW ',
   1     '  ACTIVE CONSTRAINT IS',I3,3X,'L =',I3)
8080 FORMAT(2X,'3',4X,'NEW   ACTIVE CONSTRAINT',
   1     ' IS',I3,3X,'L =',I3)
8090 FORMAT(2X,'4',4X,'NEW INACTIVE CONSTRAINT',
   1     ' IS',I3,3X,'K =',I3)
8100 FORMAT(//,15X,'CRITICAL INTERVAL:',/,6X,
   1     'T(',I2,') =',F9.4,3X,'T(',I2,') =',F9.4)
8110 FORMAT(/,6X,'END OF PARAMETRIC INTERVAL')
8120 FORMAT(/,6X,'PROBLEM IS UNBOUNDED FROM',
   1     ' BELOW FOR',/,10X,'  T GREATER THAN',
   2   F9.4)
8130 FORMAT(/,6X,'PROBLEM HAS NO FEASIBLE',
   1     ' SOLUTION FOR',/,10X,' T GREATER THAN',
   2   F9.4)
     END
```

Figure B.55   *(continued)*

We illustrate ALG4 by using it to solve the problem of Example 5.6. The complete program consists of the main program of Figure B.48 and subroutines ALG4, PSTEP1, PSTEP2, PSTEP3, PSTEP4, PSTEP5, STEP2, SUMM, and PHI. The data file for Example 5.6 is shown in Figure B.56. The output file is shown in Figure B.57.

Figure B.56   Example 5.6 data file for ALG4.

```
6  2
0. 10.
-4. -1.
1. -0.33333333333333
9.  25.  54.  30.  0.  0.
-5.  0.  -4.  0.  0.  -2.
-4.  3.
-1.  4.
2.  5.
2.  1.
-1.  0.
0.  -1.
15.  0.
0.5 0.5
0.  -1.
4  6
```

ALGORITHM 4

ITERATION   0

```
                    1   2
ACTIVE SET    4   6
```

OPTIMAL PRIMAL SOLUTION

I	H1(I)	H2(I)
1	15.0000	-1.0000
2	0.0000	2.0000

OPTIMAL DUAL SOLUTION

I	G1(I)	G2(I)
1	0.0000	0.0000
2	0.0000	0.0000
3	0.0000	0.0000
4	2.0000	-0.5000
5	0.0000	0.0000
6	1.0000	-0.8333

STEP
```
2.1  NEW  INACTIVE  CONSTRAINT  IS   6   K =  2
3    NEW    ACTIVE  CONSTRAINT  IS   3   L =  3
```

CRITICAL INTERVAL:
```
T( 0) =    0.0000    T( 1) =    1.2000
```


ITERATION   1

```
                    1   2
ACTIVE SET    4   3
```

OPTIMAL PRIMAL SOLUTION

I	H1(I)	H2(I)
1	12.0000	0.5000
2	6.0000	-1.0000

Figure B.57 Example 5.6 output from ALG4.

```
               OPTIMAL  DUAL  SOLUTION
        I        G1(I)           G2(I)
        1       0.0000          0.0000
        2       0.0000          0.0000
        3      -0.2500          0.2083
        4       2.2500         -0.7083
        5       0.0000          0.0000
        6       0.0000          0.0000
```

STEP
2.2 NEW   ACTIVE CONSTRAINT IS 6   L = 6
4   NEW INACTIVE CONSTRAINT IS 4   K = 1

```
              CRITICAL  INTERVAL:
    T( 1) =    1.2000   T( 2) =    2.0000
```


               ITERATION   2

```
                    1  2
        ACTIVE SET  6  3
```

```
              OPTIMAL  PRIMAL  SOLUTION
        I        H1(I)           H2(I)
        1      27.0000         -7.0000
        2       0.0000          2.0000
```

```
              OPTIMAL  DUAL  SOLUTION
        I        G1(I)           G2(I)
        1       0.0000          0.0000
        2       0.0000          0.0000
        3       2.0000         -0.5000
        4       0.0000          0.0000
        5       0.0000          0.0000
        6       9.0000         -2.8333
```

STEP
2.2 NEW   ACTIVE CONSTRAINT IS 1   L = 1

```
              CRITICAL  INTERVAL:
    T( 2) =    2.0000   T( 3) =    3.0000
```

PROBLEM HAS NO FEASIBLE SOLUTION FOR
        T GREATER THAN   3.0000

Figure B.57 *(continued)*

Algorithm 4 does not account for the possibility of degenerate critical values. If such intervals are encountered, one or more intervals of length zero may be computed. Theoretically, an active set could be repeated so that the cycle of zero length intervals would continue indefinitely. However, each update introduces small numerical errors so that cycling is unlikely. A better way to proceed is to modify ALG4 by using the Phase 1−Phase 2 procedure of Section 5.3.

B.4 DUAL METHODS

We give computer programs for the dual active set method Algorithm 5 (ALG5) and the dual simplex method (DSMPLX). The name and purpose of each subroutine is summarized in Table B.8. The FORTRAN identifiers for ALG5 are the same as those for the previous active set methods (see Table B.3). Those for the dual simplex are the same as those for the revised simplex method (see Table B.5).

TABLE B.8   Name and Purpose of FORTRAN Subroutines for Dual Methods

Subroutine Name	Figure	Purpose
	B.58	main program for Algorithm 5
ALG5	B.59	Algorithm 5
DSTEP1	B.60	Step 1 of Algorithm 5
DSTEP2	B.61	Step 2 of Algorithm 5
DSTEP3	B.62	Step 3 of Algorithm 5
PHI	B.8	Procedure Φ
PHIPRM	B.40	$[\Phi((B^{-1})', A_k, l)]'$; i.e., update of B^{-1}
	B.65	main program for the dual simplex method
DSMPLX	B.66	dual simplex method
DRSTP1	B.67	Step 1 of the dual simplex method
DRSTP2	B.68	Step 2 of the dual simplex method
DRSTP3	B.69	Step 3 of the dual simplex method
RSOLN	B.41	prints optimal primal and dual solutions from SIMPLX and DSMPLX
PDSOLN	B.32	prints optimal primal and dual solutions from ALG1, ALG2, ALG3, P1P2A, P1P2B, P1P2C, and ALG5

B.4.1 Algorithm 5

The main program for Algorithm 5 (Figure B.58) reads the data for m, n, c, b, A, x_0, J_0, and D_0^{-1} into M, N, C, B, A, X, JJ, and DINV, respectively. A call is then made to ALG5 which solves the problem using Algorithm 5. Upon return, X and OBJ contain the optimal solution and optimal objective function value, respectively, and DINV and JJ contain the final values of D_j^{-1} and J_j, respectively. The output condition code JOUT is also returned.   JOUT $= 1$ means that an optimal solution was obtained. JOUT $= 3$ means that the problem has no feasible solution. Note that we have previously used JOUT $= 2$ to indicate that the problem is unbounded from below and that this case cannot occur for Algorithm 5. A call is made next to PDSOLN (Figure B.32) which prints both primal and dual optimal solutions.

Figure B.58   Main program for Algorithm 5.

```
IMPLICIT DOUBLE PRECISION(A-H,O-Z)
DIMENSION A(7,2),B(7),C(2),X(2),DINV(2,2),
1    S(2),V(2),W(2),WORK(2),JJ(2)
INTEGER R
COMMON /UNITS/ IIN,IOUT
IIN = 37
IOUT = 38
READ(IIN,*) M,N
MTOT = M
R = 0
READ(IIN,*) (C(I),I=1,N)
READ(IIN,*) (B(I),I=1,M)
DO 100 I=1,M
100 READ(IIN,*) (A(I,J),J=1,N)
READ(IIN,*) (X(I),I=1,N)
READ(IIN,*) (JJ(I),I=1,N)
DO 200 I=1,N
200 READ(IIN,*) (DINV(I,J),J=1,N)
CALL ALG5(A,B,C,X,DINV,S,V,W,WORK,JJ,OBJ,N,
1    M,MTOT,JOUT)
CALL PDSOLN(JOUT,X,N,M,R,OBJ,C,DINV,JJ)
STOP
END
```

SUBROUTINE ALG5 (Figure B.59) organizes the calls to DSTEP1, DSTEP2, and DSTEP3 which implement Steps 1, 2, and 3, respectively, of Algorithm 5. ALG5 initializes by setting the iteration counter ITER (j) to zero and by using the DO 100 loop to compute $c'x_0$ and placing it in OBJ. The call to DSTEP1 computes the search direction S (s_j), the index ELL (l) of the most violated constraint, and the column index K (k) corresponding to the next inactive constraint (α_{kj}). If the current solution is optimal (JOUT = 1) or if it has been determined that the problem has no feasible solution (JOUT = 3), the final objective function value, ELL, and the final active set JJ (J_j) are printed, and control passes back to the calling program. Otherwise, DSTEP2 is called. SUBROUTINE DSTEP2 computes SIGJ (σ_j) using a_l, x_j, b_l, and s_j. Upon return, a summary of the results x_j, $c'x_j$, k, l, and J_j, are printed.

Figure B.59   Subroutine ALG5.

```
      SUBROUTINE ALG5(A,B,C,X,DINV,S,V,W,WORK,
     1   JJ,OBJ,N,M,MTOT,JOUT)
      IMPLICIT DOUBLE PRECISION(A-H,O-Z)
      DIMENSION A(MTOT,N),B(MTOT),C(N),X(N),
     1   DINV(N,N),S(N),V(N),W(N),WORK(N),JJ(N)
      INTEGER ELL
      COMMON /UNITS/ IIN,IOUT
C
C
C   INITIALIZE
      WRITE(IOUT,8000) (I,I=1,N)
      ITER = 0
      SUM = 0.D0
      DO 100 I=1,N
  100 SUM = SUM + C(I)*X(I)
      OBJ = SUM
  200 CONTINUE
C
C
      CALL DSTEP1(A,B,C,X,S,V,DINV,N,M,MTOT,K,JOUT,
     1   JJ,W,ELL)
      IF( (JOUT.EQ.1) .OR. (JOUT.EQ.3) )
     1   WRITE(IOUT,8010) ITER,OBJ,0,ELL,
     2                    (JJ(I),I=1,N)
      IF( (JOUT.EQ.1) .OR. (JOUT.EQ.3) ) RETURN
C
C
```

```
          CALL DSTEP2(A,B,X,S,N,M,MTOT,ELL,SIGJ)
          WRITE(IOUT,8010) ITER,OBJ,K,ELL,(JJ(I),I=1,N)
C
C

          CALL DSTEP3(X,C,S,DINV,A,WORK,SIGJ,OBJ,JJ,
        1 ELL,K,N,M,MTOT,ITER)
          GO TO 200
C
    8000 FORMAT(/,20X,'ALGORITHM 5',//,40X,'ACTIVE SET'
        1 ,/,7X,'ITER',6X,'OBJECTIVE',4X,'K',4X,'L',
        2 2X,(T39,10I3))
    8010 FORMAT(6X,I5,F15.5,2I5,2X,(T39,10I3))
          END
```

Figure B.59   *(continued)*

Control then passes to SUBROUTINE DSTEP3 which updates X, JJ, DINV, and ITER (x_j, J_j, D_j^{-1}, and j, respectively). This completes one iteration and the next one begins when control transfers back to statement 200.

SUBROUTINE DSTEP1 (Figure B.60) implements Step 1 of Algorithm 5 as follows. The DO 300 loop computes the smallest index l such that

$$a_l' x_j - b_l = \max\{a_i' x_j - b_i \mid \text{all } i \notin J_j \text{ with } i = 1, \ldots, m\}.$$

Each constraint I (i) is considered in turn. The DO 100 loop checks if $i \in J_j$; if so, constraint i is not considered further. Otherwise, the DO 200 loop computes $a_i' x_j - b_i$ and stores it in SUM. SUM is then compared with the current estimate of the largest of these values (BIG). If SUM is strictly greater than BIG, BIG and ELL are updated accordingly. Following the DO 300 loop, if $a_i' x_j - b_l \leq 10^{-6}$, primal feasibility has been satisfied and the current solution is optimal. This is communicated to ALG5 by returning with JOUT $= 1$. Otherwise, the computations proceed via the DO 400 and DO 500 loops to compute w_i and v_i, for $i = 1, \ldots, n$. The DO 600 loop computes the smallest index k such that

$$\frac{v_k}{w_k} = \min\left\{\frac{v_i}{w_i} \mid \text{all } i \text{ with } w_i > 10^{-6}\right\}.$$

If $k = 0$ at the end of the DO 600 loop, then $w_i \leq 10^{-6}$ for $i = 1, \ldots, n$. In this case, the problem has no feasible solution and this is communicated to ALG5 by setting JOUT $= 3$. If $k > 0$, the DO 700 loop computes $s_j = c_{kj}$ and control then returns to ALG5.

```
      SUBROUTINE DSTEP1(A,B,C,X,S,V,DINV,N,M,MTOT,K,
     1  JOUT,JJ,W,ELL)
      IMPLICIT DOUBLE PRECISION(A-H,O-Z)
      DIMENSION A(MTOT,N),B(MTOT),C(N),S(N),V(N),
     1  DINV(N,N),JJ(N),W(N),X(N)
      INTEGER ELL
C
      JOUT = 0
      TOL = 1.D-6
C
      ELL = 0
      BIG = -1.D30
      DO 300 I=1,M
      DO 100 J=1,N
      IF(JJ(J).EQ.I) GO TO 300
  100 CONTINUE
      SUM = - B(I)
      DO 200 J=1,N
  200 SUM = SUM + A(I,J)*X(J)
      IF(SUM.LE.BIG) GO TO 300
      BIG = SUM
      ELL = I
  300 CONTINUE
      IF(BIG.LE.TOL) JOUT = 1
      IF(BIG.LE.TOL) RETURN
C
      DO 500 I=1,N
      SUMW = 0.D0
      SUMV = 0.D0
      DO 400 J=1,N
      SUMW = SUMW + A(ELL,J)*DINV(J,I)
  400 SUMV = SUMV + C(J)*DINV(J,I)
      W(I) = SUMW
  500 V(I) = - SUMV
      K = 0
      SMALL = 1.D30
      DO 600 I=1,N
      IF(W(I).LE.TOL) GO TO 600
      RATIO = V(I)/W(I)
      IF(RATIO.GE.SMALL) GO TO 600
      SMALL = RATIO
      K = I
```

Figure B.60 Subroutine DSTEP1.

```
600  CONTINUE
     IF(K.EQ.0) JOUT = 3
     IF(K.EQ.0) RETURN
     DO 700 I=1,N
700  S(I) = DINV(I,K)
     RETURN
     END
```

Figure B.60 *(continued)*

SUBROUTINE DSTEP2 (Figure B.61) computes SIGJ (σ_j). The DO 100 loop computes $a_l' x_j$ and $a_l' s_j$.

```
SUBROUTINE DSTEP2(A,B,X,S,N,M,MTOT,ELL,SIGJ)
IMPLICIT DOUBLE PRECISION(A-H,O-Z)
DIMENSION A(MTOT,N),B(MTOT),X(N),S(N)
INTEGER ELL
SUM1 = 0.D0
SUM2 = 0.D0
DO 100 I=1,N
SUM1 = SUM1 + A(ELL,I)*X(I)
100 SUM2 = SUM2 + A(ELL,I)*S(I)
SIGJ = (SUM1 - B(ELL))/SUM2
RETURN
END
```

Figure B.61   Subroutine DSTEP2.

SUBROUTINE DSTEP3 (Figure B.62) updates x_j, $c' x_j$, D_j^{-1}, and J_j. The updating is identical to that of Algorithms 1, 2, and 3. Indeed, DSTEP3 *is* identical to STEP3 (Figure B.7).

Figure B.62   Subroutine DSTEP3.

```
SUBROUTINE DSTEP3(X,C,S,DINV,A,WORK,SIGJ,OBJ,
1    JJ,ELL,K,N,M,MTOT,ITER)
IMPLICIT DOUBLE PRECISION(A-H,O-Z)
DIMENSION X(N),C(N),S(N),WORK(N),DINV(N,N),
1    A(MTOT,N),JJ(N)
INTEGER ELL
DO 100 I=1,N
```

```
100  X(I) = X(I) - SIGJ*S(I)
     SUM = 0.D0
     DO 200 I=1,N
200  SUM = SUM + C(I)*X(I)
     OBJ = SUM
     DO 300 I=1,N
300  WORK(I) = A(ELL,I)
     CALL PHI(DINV,WORK,K,N)
     JJ(K) = ELL
     ITER = ITER + 1
     RETURN
     END
```

Figure B.62 *(continued)*

We illustrate **ALG5** by applying it to the problem of Example 6.3. The complete program consists of the main program of Figure B.58 and subroutines **ALG5**, **DSTEP1**, **DSTEP2**, **DSTEP3**, **PDSOLN**, and **PHI**. The data file for Example 6.3 is shown in Figure B.63. The output file is shown in Figure B.64.

Figure B.63 Example 6.3 data file for ALG5.

```
7 2
-16. -11.
-4. 0. 8. 17. 43. 8. 0.
-2. -1.
-1. 0.
0. 1.
1. 2.
5. 3.
1. 0.
0. -1.
8. 8.
6 3
1. 0.
0. 1.
```

ALGORITHM 5

ITER	OBJECTIVE	K	L	1	2
0	-216.00000	1	5	6	3
1	-148.80000	2	4	5	3
2	-146.00000	0	3	5	4

OPTIMAL OBJECTIVE FUNCTION VALUE IS   -146.00000

OPTIMAL PRIMAL SOLUTION
X(1) =   5.0000000
X(2) =   6.0000000

OPTIMAL DUAL SOLUTION
U(1) =   0.0000000
U(2) =   0.0000000
U(3) =   0.0000000
U(4) =   1.0000000
U(5) =   3.0000000
U(6) =   0.0000000
U(7) =   0.0000000

Figure B.64   Example 6.3 output from ALG5.

B.4.2   The Dual Simplex Method

The main program for the dual simplex method (Figure B.65) reads the data for m, n, c, b, A, x_B, u, I_B, and B^{-1} into M, N, C, B, A, XB, U, IB, and BINV, respectively. Note that, strictly speaking, neither x_B nor u need be part of the input data for DSMPLX since once b, c, I_B, and B^{-1} are available, $x_B = B^{-1}b$ and $u = -(B^{-1})'c_B$ can be computed. However, we feel that it is clearer to include them with the input data. A call is then made to DSMPLX which solves the problem using the dual simplex method. Upon return, XB and OBJ contain the optimal basic feasible solution and the optimal objective function value, respectively, and BINV and IB contain the final values of B^{-1} and I_B, respectively. The output condition is also returned. JOUT = 1 means that an optimal solution was obtained.   JOUT = 3 means that the problem has no feasible solution. Note that because DSMPLX begins with a dual feasible solution, the primal problem cannot be unbounded from below. Next, a call is made to RSOLN which prints both primal and dual optimal solutions.

```
      IMPLICIT DOUBLE PRECISION(A-H,O-Z)
      DIMENSION A(2,5),B(2),C(5),XB(2),BINV(2,2),
    1   SB(2),WORK(2),U(2),IB(2)
      COMMON /UNITS/ IIN,IOUT
      IIN = 37
      IOUT = 38
      READ(IIN,*) M,N
      READ(IIN,*) (C(I),I=1,N)
      READ(IIN,*) (B(I),I=1,M)
      DO 100 I=1,M
  100 READ(IIN,*) (A(I,J),J=1,N)
      READ(IIN,*) (XB(I),I=1,M)
      READ(IIN,*) (U(I),I=1,M)
      READ(IIN,*) (IB(I),I=1,M)
      DO 200 I=1,M
  200 READ(IIN,*) (BINV(I,J),J=1,M)
      CALL DSMPLX(A,B,C,XB,BINV,SB,U,WORK,IB,OBJ,N,
    1   M,JOUT)
      CALL RSOLN(JOUT,XB,U,C,A,IB,M,N,OBJ)
      STOP
      END
```

Figure B.65   Main program for the dual simplex method.

SUBROUTINE DSMPLX (Figure B.66) organizes the calls to DRSTP1, DRSTP2, and DRSTP3 which implement Steps 1, 2, and 3, respectively, of the dual simplex method.   DSMPLX initializes by setting the iteration counter ITER to zero and by using the DO 100 loop to compute $c_B' x_B$ and store it in OBJ. The call to DRSTP1 computes the search direction SB (s_B) and ELL (l) associated with the most violated nonnegativity constraint (β_l). If the current solution is feasible, and therefore optimal, (JOUT = 1), the final objective function value, ELL, and the final active set IB (I_B) are printed and control passes back to the calling program. Otherwise, DRSTP2 is called.

SUBROUTINE DRSTP2 determines the index k (K) of the new nonbasic variable. If it is determined that the problem has no feasible solution (JOUT = 3), the final objective function value, k, l, and I_B are printed and control passes back to the calling program. Otherwise, control passes to SUBROUTINE DRSTP3 which updates XB, IB, BINV, U, and ITER (x_B, I_B, B^{-1}, u, and j, respectively). This completes one iteration and the next begins when control transfers back to statement 200.

```
      SUBROUTINE DSMPLX(A,B,C,XB,BINV,SB,U,WORK,IB,
    1   OBJ,N,M,JOUT)
      IMPLICIT DOUBLE PRECISION(A-H,O-Z)
      DIMENSION A(M,N),B(M),C(N),XB(M),BINV(M,M),
    1   SB(M),U(M),WORK(M),IB(M)
      INTEGER ELL
      COMMON /UNITS/ IIN,IOUT
C
C
C  INITIALIZE
      WRITE(IOUT,8000) (I,I=1,M)
      ITER = 0
      SUM = 0.D0
      DO 100 I=1,M
  100 SUM = SUM + C(IB(I))*XB(I)
      OBJ = SUM
C
C
  200 CONTINUE
      CALL DRSTP1(XB,BINV,B,SB,ELL,M,JOUT)
      IF(JOUT.EQ.1) WRITE(IOUT,8010) ITER,OBJ,0,ELL,
    1   (IB(I),I=1,M)
      IF(JOUT.EQ.1) RETURN
C
C
      CALL DRSTP2(A,C,SB,U,IB,M,N,K,JOUT)
      WRITE(IOUT,8010) ITER,OBJ,K,ELL,(IB(I),I=1,M)
      IF(JOUT.EQ.3) RETURN
C
C
      CALL DRSTP3(BINV,A,B,C,XB,WORK,U,IB,K,ELL,N,
    1   M,ITER,OBJ)
      GO TO 200
C
 8000 FORMAT(/,17X,' DUAL SIMPLEX METHOD',//,40X,
    1   'BASIC SET',/,7X,'ITER',6X,'OBJECTIVE',4X,
    2   'K',4X,'L',2X,(T39,10I3))
 8010 FORMAT(6X,I5,F15.5,2I5,2X,(T39,10I3))
      END
```

Figure B.66   Subroutine DSMPLX.

SUBROUTINE DRSTP1 (Figure B.67) implements Step 1 of the dual simplex method. The DO 100 loop computes ELL (l) such that

$$(x_B)_l = \min\{(x_B)_i \mid i = 1, \ldots, m\}.$$

If $(x_B)_l \geq -10^{-6}$, the current solution is feasible and therefore optimal. In this case, DRSTP1 returns to DSMPLX with JOUT = 1. Otherwise, s_B' is assigned the value of the lth row of B^{-1} and control returns to DSMPLX.

```
      SUBROUTINE DRSTP1(XB,BINV,B,SB,ELL,M,JOUT)
      IMPLICIT DOUBLE PRECISION(A-H,O-Z)
      DIMENSION XB(M),BINV(M,M),B(M),SB(M)
      INTEGER ELL
      TOL = 1.D-6
      JOUT = 0
      SMALL = 1.D30
      ELL = 0
      DO 100 I=1,M
      IF(XB(I).GE.SMALL) GO TO 100
      SMALL = XB(I)
      ELL = I
  100 CONTINUE
      IF(SMALL.GE.-TOL) JOUT = 1
      IF(SMALL.GE.-TOL) RETURN
      DO 200 I=1,M
  200 SB(I) = - BINV(ELL,I)
      RETURN
      END
```

Figure B.67 Subroutine DRSTP1.

SUBROUTINE DRSTP2 (Figure B.68) implements Step 2 of the dual simplex method. The DO 300 loop computes K (k) such that

$$\frac{A_k'u + c_k}{A_k's_B} = \min\left\{\frac{A_i'u + c_i}{A_i's_B} \mid \text{all } i \notin I_B \text{ with } A_i's_B > 0\right\}.$$

If there were no candidates for the minimum (K = 0), then $A_i's_B \leq 10^{-6}$ for all $i = 1, \ldots, n$, $i \notin I_B$, and the problem has no feasible solution. This is communicated to DSMPLX by setting JOUT = 3. In either case, control returns to DSMPLX.

```
      SUBROUTINE DRSTP2(A,C,SB,U,IB,M,N,K,JOUT)
      IMPLICIT DOUBLE PRECISION(A-H,O-Z)
      DIMENSION A(M,N),C(N),SB(M),U(M),IB(M)
      TOL = 1.D-6
      JOUT = 0
      K = 0
      SMALL = 1.D30
      DO 300 I=1,N
      DO 100 J=1,M
      IF(IB(J).EQ.I) GO TO 300
  100 CONTINUE
      SUMU = 0.D0
      SUMSB = 0.D0
      DO 200 J=1,M
      SUMU = SUMU + A(J,I)*U(J)
  200 SUMSB = SUMSB + A(J,I)*SB(J)
      IF(SUMSB.LE.TOL) GO TO 300
      RATIO = (SUMU + C(I))/SUMSB
      IF(RATIO.GE.SMALL) GO TO 300
      SMALL = RATIO
      K = I
  300 CONTINUE
      IF(K.EQ.0) JOUT = 3
      RETURN
      END
```

Figure B.68   Subroutine DRSTP2.

SUBROUTINE DRSTP3 (Figure B.69) updates x_B, $c'_B x_B$, B^{-1}, u, and I_B. The DO 300 loop computes u. The remaining portion of DRSTP3 is quite similar to RSTEP3 (Step 3 of the revised simplex method, Figure B.39).

Figure B.69   Subroutine DRSTP3.

```
      SUBROUTINE DRSTP3(BINV,A,B,C,XB,WORK,U,IB,K,
     1    ELL,N,M,ITER,OBJ)
      IMPLICIT DOUBLE PRECISION(A-H,O-Z)
      DIMENSION BINV(M,M),A(M,N),B(M),WORK(M),U(M),
     1    IB(M),C(N),XB(M)
      INTEGER ELL
C
      DO 100 I=1,M
  100 WORK(I) = A(I,K)
      CALL PHIPRM(BINV,WORK,ELL,M)
      IB(ELL) = K
C
```

```
C   UPDATE U
         DO 300 I=1,M
         SUM = 0.D0
         DO 200 J=1,M
    200 SUM = SUM + BINV(J,I)*C(IB(J))
    300 U(I) = - SUM
C
C   UPDATE XB
         DO 500 I=1,M
         SUM = 0.D0
         DO 400 J=1,M
    400 SUM = SUM + BINV(I,J)*B(J)
    500 XB(I) = SUM
C
C   UPDATE OBJECTIVE FUNCTION
         SUM = 0.D0
         DO 600 I=1,M
    600 SUM = SUM + C(IB(I))*XB(I)
         OBJ = SUM
         ITER = ITER + 1
         RETURN
         END
```

Figure B.69   *(continued)*

We illustrate DSMPLX by applying it to the problem of Example 6.4. The complete program consists of the main program of Figure B.65 and subroutines DSMPLX, DRSTP1, DRSTP2, DRSTP3, RSOLN, and PHIPRM. The data file for Example 6.4 is shown in Figure B.70. The output file is shown in Figure B.71.

Figure B.70   Example 6.4 data file for DSMPLX.

```
2 5
2. 14. 36. 0. 0.
5. 2.
-2. 1. 4. -1. 0.
-1. -2. -3. 0. 1.
-2. -1.
0. 2.
1 4
0. -1.
-1. 2.
```

DUAL SIMPLEX METHOD

				BASIC SET	
ITER	OBJECTIVE	K	L	1	2
0	-4.00000	5	1	1	4
1	0.00000	3	2	5	4
2	45.00000	0	2	5	3

OPTIMAL OBJECTIVE FUNCTION VALUE IS   45.00000

OPTIMAL PRIMAL SOLUTION
X(1) =   0.0000000
X(2) =   0.0000000
X(3) =   1.2500000
X(4) =   0.0000000
X(5) =   5.7500000

OPTIMAL DUAL SOLUTION
EQUALITY CONSTRAINTS
U(1) =   -9.0000000
U(2) =   0.0000000

NONNEGATIVITY CONSTRAINTS
V(1) =   20.0000000
V(2) =   5.0000000
V(3) =   0.0000000
V(4) =   9.0000000
V(5) =   0.0000000

Figure B.71   Example 6.4 output from DSMPLX.

B.5 UPPER BOUNDING TECHNIQUES

In this section we give computer programs for the upper-bounded simplex method and the generalized upper-bounded simplex method as formulated in Sections 7.1 and 7.2, respectively. The name and purpose of the various subroutines are summarized in Table B.9.

**TABLE B.9   Name and Purpose of FORTRAN Subroutines
for Upper Bounding Methods**

Subroutine Name	Figure	Purpose
	B.72	main program for the upper-bounded simplex method
	B.80	main program for the generalized upper-bounded simplex method
USMPLX	B.73	upper-bounded simplex method
USTEP1	B.74	Step 1 of the upper-bounded simplex method
USTEP2	B.75	Step 2 of the upper-bounded simplex method
USTEP3	B.76	Step 3 of the upper-bounded simplex method
USOLN	B.77	prints optimal primal and dual solutions from USMPLX
PHIPRM	B.40	$[\Phi((B^{-1})', A_k, l)]'$; i.e., update of B^{-1}
GUB	B.81	generalized upper-bounded simplex method
GSTEP1	B.83	Step 1 of the generalized upper-bounded simplex method
GSTEP2	B.84	Step 2 of the generalized upper-bounded simplex method
GSTEP3	B.85	Step 3 of the generalized upper-bounded simplex method
GSOLN	B.86	prints optimal primal and dual solutions from GUB
ELT	B.82	given l, ELT finds ν such that $l \in S_\nu$

B.5.1 The Upper-Bounded Simplex Method

The most important FORTRAN identifiers used in coding the upper-bounded simplex method and their meanings are summarized in Table B.10. The main program (Figure B.72) reads the data for m, n, c, b, A, d, e, x_B, B^{-1}, and I_B into M, N, C, B, A, D, E, XB, BINV, and IB, respectively. The sets L and U are represented by LOGICAL vectors LOW and UP, respectively. For example, if LOW(I) has value .TRUE., then x_i is at its lower bound of d_i. Similarly, if UP(I) has value .TRUE. then x_i is at its upper bound of e_i. Initial values for LOW and UP are read in.

A call is made to USMPLX which is responsible for solving the problem using the upper-bounded simplex method. If an optimal solution is obtained, the optimal basic variables are returned in XB, OBJ contains the optimal objective function value, LOW and UP describe which components of the optimal solution are at their bounds, and U (u), V (v), and W (w) contain the optimal solution for the dual problem. The output condition indicator JOUT is set to 1 in this case.

Since the components of D (d) and E (e) are real numbers (specifically excluding $-\infty$ and $+\infty$), the problem presented to USMPLX has a bounded feasible region and therefore possesses an optimal solution. However, we

**TABLE B.10   Meaning of FORTRAN Identifiers
for the Upper-Bounded Simplex Method**

FORTRAN Identifier	Notation in Main Text
N	n
M	m
A	A
B	b
C	c
D	d
E	e
IB	I_B
BINV	B^{-1}
XB	x_B
U	u
V	v
W	w
SB	s_B
LOW	L
UP	U
K	k
K1	k_1
K2	k_2
ELL	l
ELL1	l_1
ELL2	l_2
ELL3	l_3
SIGB	σ_B
SIG1	σ_1
SIG2	σ_2
SIG3	σ_3
OBJ	$c_B' x_B + \sum_{i \in L} c_i d_i + \sum_{i \in U} c_i e_i$
JOUT	$= \begin{cases} 1, & \text{optimal solution} \\ 2, & \text{problem is unbounded from below} \\ 3, & \text{problem has no feasible solution} \end{cases}$

adopt the convention that $e_i \geq 10^{15}$ means that x_i has no upper bound. Similarly, $d_i \leq -10^{15}$ means that x_i has no lower bound. By taking these conventions into account, **USMPLX** may determine that the problem is unbounded from below. This possibility is noted with **JOUT** $= 2$.

```
      IMPLICIT DOUBLE PRECISION(A-H,O-Z)
      DIMENSION A(2,4),B(2),C(4),D(4),E(4),IB(2),
     1  BINV(2,2),XB(2),U(2),V(4),W(4),WORK(2),
     2  SB(2)
      LOGICAL LOW(4),UP(4)
      COMMON /UNITS/ IIN,IOUT
      IIN = 37
      IOUT = 38
      READ(IIN,*) M,N
      READ(IIN,*) (C(I),I=1,N)
      READ(IIN,*) (B(I),I=1,M)
      DO 100 I=1,M
  100 READ(IIN,*) (A(I,J),J=1,N)
      READ(IIN,*) (D(I),I=1,N)
      READ(IIN,*) (E(I),I=1,N)
      READ(IIN,*) (XB(I),I=1,M)
      DO 200 I=1,M
  200 READ(IIN,*) (BINV(I,J),J=1,M)
      READ(IIN,*) (IB(I),I=1,M)
      READ(IIN,*) (LOW(I),I=1,N)
      READ(IIN,*) (UP(I),I=1,N)
      CALL USMPLX(A,B,C,D,E,BINV,IB,XB,SB,U,V,W,
     1  WORK,LOW,UP,M,N,OBJ,JOUT)
      CALL USOLN(JOUT,XB,D,E,U,V,W,IB,LOW,UP,M,
     1  N,OBJ)
      STOP
      END
```

Figure B.72 Main program for the upper-bounded simplex method.

SUBROUTINE USMPLX (Figure B.73) organizes the calls to USTEP1, USTEP2, and USTEP3 which implement Steps 1, 2, and 3, respectively, of the upper-bounded simplex method. USMPLX initializes by setting the iteration counter ITER to zero and using the DO 100 and DO 200 loops to compute the initial objective function value and store it in OBJ. The call to USTEP1 computes the search direction SB (s_B) and the index K (k) of the new basic variable. If the optimality condition is satisfied, USTEP1 returns with JOUT $= 1$. In this case, the final objective function value, column index k, and index set IB (I_B) are printed, and control returns to the main program. Otherwise, USTEP2 is called.

Figure B.73 Subroutine USMPLX.

```
      SUBROUTINE USMPLX(A,B,C,D,E,BINV,IB,XB,SB,U,
     1  V,W,WORK,LOW,UP,M,N,OBJ,JOUT)
      IMPLICIT DOUBLE PRECISION(A-H,O-Z)
      LOGICAL LOW(N),UP(N),SWITCH
```

```
      INTEGER ELL,ELL1,ELL2,ELL3
      DIMENSION A(M,N),B(M),C(N),D(N),E(N),
     1  BINV(M,M),IB(M),XB(M),U(M),V(N),W(N),
     2  WORK(M),SB(M)
      COMMON /UNITS/ IIN,IOUT
C
C
C  INITIALIZE
      WRITE(IOUT,8000) (I,I=1,M)
      ITER = 0
      SUM = 0.D0
      DO 100 I=1,M
  100 SUM = SUM + C(IB(I))*XB(I)
      DO 200 I=1,N
      IF(LOW(I)) SUM = SUM + C(I)*D(I)
      IF(UP(I)) SUM = SUM + C(I)*E(I)
  200 CONTINUE
      OBJ = SUM
C
C
  300 CONTINUE
      CALL USTEP1(A,C,BINV,IB,SB,U,V,W,LOW,UP,M,
     1  N,K,K1,K2,JOUT)
      IF(JOUT.EQ.1) WRITE(IOUT,8010) ITER,OBJ,K,0,
     1  (IB(I),I=1,M)
      IF(JOUT.EQ.1) RETURN
C
C
      CALL USTEP2(XB,SB,IB,ELL,ELL1,ELL2,ELL3,K,D,
     1  E,M,N,JOUT,SWITCH)
      WRITE(IOUT,8010) ITER,OBJ,K,ELL,(IB(I),I=1,M)
      IF(JOUT.EQ.2) RETURN
C
C
      CALL USTEP3(XB,BINV,IB,A,B,C,D,E,WORK,LOW,
     1  UP,K,K1,K2,ELL,ELL1,ELL2,ELL3,M,N,OBJ,ITER,
     2  SWITCH)
      GO TO 300
C
 8000 FORMAT(/,23X,'UPPER-BOUNDED',/,
     1  23X,'SIMPLEX METHOD',//,40X,
     2  'BASIC SET',/,7X,'ITER',6X,'OBJECTIVE',4X,
     3  'K',4X,'L',2X,(T39,10I3))
 8010 FORMAT(6X,I5,F15.5,2I5,2X,(T39,10I3))
      END
```

Figure B.73 *(continued)*

USTEP2 computes the maximum feasible step size and the associated indices ELL, ELL1, ELL2, and ELL3 (l, l_1, l_2, and l_3, respectively). It also sets the logical variable SWITCH to .TRUE. if $\sigma_B = \sigma_3$ and .FALSE., otherwise. SWITCH is used in the updating in USTEP3. If USTEP2 determines that the problem is unbounded from below, it returns with JOUT = 2 and control returns to the main program.

Control now passes to USTEP3 which updates XB, IB, BINV, LOW, UP, and ITER (x_B, I_B, B^{-1}, L, U, and j, respectively). This completes one iteration and the next begins when control transfers back to statement 300.

SUBROUTINE USTEP1 (Figure B.74) implements Step 1 of the upper-bounded simplex method as follows. The DO 100 and DO 200 loops compute U ($u = -(B^{-1})'c_B$). The DO 400 and DO 500 loops set $v_i = A_i'u + c_i$ for all $i \in L$ and $w_i = -A_i'u - c_i$ for all $i \in U$. The smallest v_i and its associated index k_1 are determined by the DO 600 loop. The DO 700 loop finds the smallest such w_i and its index k_2. The next two statements set $k = k_1$ if $v_{k_1} \le w_{k_2}$ and $k = k_2$ if $v_{k_1} > w_{k_2}$. If $v_{k_1} > -10^{-6}$ and $w_{k_2} > -10^{-6}$, the optimality condition is satisfied and USTEP1 returns with JOUT = 1. Otherwise, the DO 900 and DO 1000 loops set $s_B = B^{-1}A_k$ if $k = k_1$ and $s_B = -B^{-1}A_k$ if $k = k_2$. Control then returns to USMPLX.

Figure B.74   Subroutine USTEP1.

```
      SUBROUTINE USTEP1(A,C,BINV,IB,SB,U,V,W,LOW,
     1   UP,M,N,K,K1,K2,JOUT)
      IMPLICIT DOUBLE PRECISION(A-H,O-Z)
      LOGICAL LOW(N),UP(N)
      DIMENSION A(M,N),C(N),BINV(M,M),IB(M),SB(M),
     1   U(M),V(N),W(N)
C
      TOLCON = 1.D-6
      JOUT = 0
C
C   COMPUTE U
      DO 200 J=1,M
      SUM = 0.D0
      DO 100 I=1,M
  100 SUM = SUM + BINV(I,J)*C(IB(I))
  200 U(J) = - SUM
C
      DO 300 I=1,N
      V(I) = 0.D0
```

```
  300 W(I) = 0.D0
      DO 500 I=1,N
      IF(.NOT.(LOW(I).OR.UP(I))) GO TO 500
      SUM = C(I)
      DO 400 J=1,M
  400 SUM = SUM + A(J,I)*U(J)
      IF(LOW(I)) V(I) = SUM
      IF(UP(I)) W(I) = - SUM
  500 CONTINUE
C
C  FIND K1 AND K2
      K1 = 0
      SMALLV = 1.D30
      DO 600 I=1,N
      IF(.NOT.LOW(I)) GO TO 600
      IF(V(I).GE.SMALLV) GO TO 600
      SMALLV = V(I)
      K1 = I
  600 CONTINUE
      K2 = 0
      SMALLW = 1.D30
      DO 700 I=1,N
      IF(.NOT.UP(I)) GO TO 700
      IF(W(I).GE.SMALLW) GO TO 700
      SMALLW = W(I)
      K2 = I
  700 CONTINUE
      IF(SMALLV.LE.SMALLW) K = K1
      IF(SMALLV.GT.SMALLW) K = K2
      IF( (SMALLV.LE.-TOLCON) .OR.
     1      (SMALLW.LE.-TOLCON) ) GO TO 800
      JOUT = 1
      RETURN
  800 CONTINUE
      DO 1000 I=1,M
      SUM = 0.D0
      DO 900 J=1,M
  900 SUM = SUM + BINV(I,J)*A(J,K)
      IF(K.EQ.K1) SB(I) = SUM
      IF(K.EQ.K2) SB(I) = - SUM
 1000 CONTINUE
      RETURN
      END
```

Figure B.74 *(continued)*

SUBROUTINE USTEP2 (Figure B.75) implements Step 2 of the upper-bounded simplex method. The DO 100 loop computes σ_1 and l_1. The DO 200 loop computes σ_2 and l_2. If $l_1 = 0$, $l_2 = 0$, $d_k \leq -10^{15}$, and $e_k \geq 10^{15}$, the problem is unbounded from below. In this case, USTEP2 sets JOUT = 2 and returns. Otherwise, the computations proceed following statement 300 to evaluate σ_3 and l_3 and then σ_B and l. As previously noted, logical variable SWITCH is set to .TRUE. if $\sigma_B = \sigma_3$ and .FALSE. otherwise.

Figure B.75   Subroutine USTEP2.

```
      SUBROUTINE USTEP2(XB,SB,IB,ELL,ELL1,ELL2,
     1 ELL3,K,D,E,M,N,JOUT,SWITCH)
      IMPLICIT DOUBLE PRECISION(A-H,O-Z)
      LOGICAL SWITCH
      INTEGER ELL,ELL1,ELL2,ELL3
      DIMENSION XB(M),SB(M),IB(M),D(N),E(N)
C
      EPS = 1.D-6
      SIG1 = 1.D30
      ELL1 = 0
      DO 100 I=1,M
      IF(SB(I).LE.EPS) GO TO 100
      IF(D(IB(I)).LE.-1.D15) GO TO 100
      TERM = (XB(I)-D(IB(I)))/SB(I)
      IF(TERM.GE.SIG1) GO TO 100
      SIG1 = TERM
      ELL1 = I
  100 CONTINUE
      SIG2 = 1.D30
      ELL2 = 0
      DO 200 I=1,M
      IF(SB(I).GE.-EPS) GO TO 200
      IF(E(IB(I)).GE.1.D15) GO TO 200
      TERM = (XB(I) - E(IB(I)))/SB(I)
      IF(TERM.GE.SIG2) GO TO 200
      SIG2 = TERM
      ELL2 = I
  200 CONTINUE
      IF((ELL1.GT.0).OR.(ELL2.GT.0)) GO TO 300
      IF(D(K).GT.-1.D15) GO TO 300
      IF(E(K).LT.1.D15) GO TO 300
      JOUT = 2
      RETURN
```

```
300 CONTINUE
    SIG3 = E(K) - D(K)
    ELL3 = K
    SWITCH = .FALSE.
    SIGB = DMIN1(SIG1,SIG2,SIG3)
    IF(SIGB.EQ.SIG1) ELL = ELL1
    IF(SIGB.EQ.SIG2) ELL = ELL2
    IF(SIGB.EQ.SIG3) ELL = ELL3
    IF(SIGB.EQ.SIG3) SWITCH = .TRUE.
    RETURN
    END
```

Figure B.75   *(continued)*

SUBROUTINE USTEP3 (Figure B.76) implements Step 3 of the upper-bounded simplex method.   LOW and UP are first updated. If $\sigma_B < \sigma_3$, the DO 100 loop copies A_k into WORK, PHIPRM (Figure B.40) updates B^{-1}, and I_B is updated by setting $\beta_l = k$. Next, the DO 300, 400, 600, and 700 loops compute

$$b - \sum_{i \in L} d_i A_i - \sum_{i \in U} e_i A_i$$

and store it in WORK. The DO 800 and DO 900 loops compute B^{-1} times WORK and store it in XB. Finally, the DO 1000 and DO 1100 loops compute the new objective function value and store it in OBJ.

Figure B.76   Subroutine USTEP3.

```
    SUBROUTINE USTEP3(XB,BINV,IB,A,B,C,D,E,WORK,
   1  LOW,UP,K,K1,K2,ELL,ELL1,ELL2,ELL3,M,N,OBJ,
   2  ITER,SWITCH)
    IMPLICIT DOUBLE PRECISION(A-H,O-Z)
    LOGICAL LOW(N),UP(N),SWITCH
    DIMENSION XB(M),BINV(M,M),IB(M),A(M,N),B(M),
   1  C(N),D(N),E(N),WORK(M)
    INTEGER ELL,ELL1,ELL2,ELL3
C
    IF(K.EQ.K1) LOW(K) = .FALSE.
    IF(K.EQ.K2) UP(K) = .FALSE.
C
```

```
      IF((.NOT.SWITCH).AND.(ELL.EQ.ELL1))
    1                           LOW(IB(ELL)) = .TRUE.
      IF((.NOT.SWITCH).AND.(ELL.EQ.ELL2))
    1                           UP(IB(ELL)) = .TRUE.
C
      IF(SWITCH.AND.(K.EQ.K1)) UP(K) = .TRUE.
      IF(SWITCH.AND.(K.EQ.K2)) LOW(K) = .TRUE.
C
      IF(SWITCH) GO TO 200
      DO 100 I=1,M
  100 WORK(I) = A(I,K)
      CALL PHIPRM(BINV,WORK,ELL,M)
      IB(ELL) = K
  200 CONTINUE
      DO 300 I=1,M
  300 WORK(I) = B(I)
      DO 700 I=1,N
      IF(.NOT.(LOW(I).OR.UP(I))) GO TO 700
      IF(UP(I)) GO TO 500
C  X(I) IS AT ITS LOWER BOUND
      DO 400 J=1,M
  400 WORK(J) = WORK(J) - D(I)*A(J,I)
      GO TO 700
  500 CONTINUE
      DO 600 J=1,M
  600 WORK(J) = WORK(J) - E(I)*A(J,I)
  700 CONTINUE
      DO 900 I=1,M
      SUM = 0.D0
      DO 800 J=1,M
  800 SUM = SUM + BINV(I,J)*WORK(J)
  900 XB(I) = SUM
      SUM = 0.D0
      DO 1000 I=1,M
 1000 SUM = SUM + C(IB(I))*XB(I)
      DO 1100 I=1,N
      IF(LOW(I)) SUM = SUM + C(I)*D(I)
      IF(UP(I))  SUM = SUM + C(I)*E(I)
 1100 CONTINUE
      OBJ = SUM
      ITER = ITER + 1
      RETURN
      END
```

Figure B.76   *(continued)*

SUBROUTINE USOLN (Figure B.77) prints the optimal primal and dual solutions determined by USMPLX. It first uses the output condition JOUT to see if an optimal solution has been found. If none has been found, control transfers to statement 500 where an appropriate message is printed. If an optimal solution has been found, the DO 300 loop prints the optimal primal solution. For each component i, the DO 100 loop decides if $i \in I_B$. If so, control transfers to statement 200 where i and the appropriate component of x_B are printed. If not, component i must be at a bound. Together with i, the appropriate bound is then printed.

Figure B.77   Subroutine USOLN.

```
      SUBROUTINE USOLN(JOUT,XB,D,E,U,V,W,IB,
     1  LOW,UP,M,N,OBJ)
      IMPLICIT DOUBLE PRECISION(A-H,O-Z)
      LOGICAL LOW(N),UP(N)
      DIMENSION XB(M),U(M),V(N),W(N),IB(M),D(N),
     1  E(N)
      COMMON /UNITS/ IIN,IOUT
      IF(JOUT.GE.2) GO TO 500
      WRITE(IOUT,8000) OBJ
      WRITE(IOUT,8010)
      DO 300 I=1,N
      DO 100 J=1,M
      INDEX = J
      IF(I.EQ.IB(J)) GO TO 200
  100 CONTINUE
      IF(LOW(I)) WRITE(IOUT,8020) I,D(I)
      IF(UP(I))  WRITE(IOUT,8020) I,E(I)
      GO TO 300
  200 WRITE(IOUT,8020) I,XB(INDEX)
  300 CONTINUE
      WRITE(IOUT,8030)
      WRITE(IOUT,8040) (I,U(I),I=1,M)
      WRITE(IOUT,8050)
      DO 400 I=1,N
      TEMP1 = 0.D0
      TEMP2 = 0.D0
      IF(LOW(I)) TEMP1 = V(I)
      IF(UP(I))  TEMP2 = W(I)
  400 WRITE(IOUT,8060) I,TEMP1,I,TEMP2
  500 CONTINUE
      IF(JOUT.EQ.2) WRITE(IOUT,8070)
      IF(JOUT.EQ.3) WRITE(IOUT,8080)
      RETURN
    C
```

```
8000 FORMAT(//,6X,'OPTIMAL OBJECTIVE FUNCTION ',
   1  'VALUE IS',F12.5)
8010 FORMAT(//,17X,'OPTIMAL PRIMAL SOLUTION')
8020 FORMAT(18X,'X(',I2,') =',F14.7)
8030 FORMAT(//,18X,'OPTIMAL DUAL SOLUTION',
   1  /,18X,'EQUALITY CONSTRAINTS')
8040 FORMAT(18X,'U(',I2,') =',F14.7)
8050 FORMAT(/,20X,'BOUND CONSTRAINTS',
   1  /,14X,'LOWER BOUND',14X,'UPPER BOUND')
8060 FORMAT(4X,'V(',I2,') =',F14.7,4X,'W(',I2,
   1  ') =',F14.7)
8070 FORMAT(//,6X,'PROBLEM IS UNBOUNDED FROM ',
   1  'BELOW')
8080 FORMAT(//,6X,'PROBLEM HAS NO FEASIBLE ',
   1  'SOLUTION')
     END
```

Figure B.77   *(continued)*

The optimal solution for the dual is then printed. All of U, V, and W were computed in USMPLX, then returned to the main program, and finally passed to USOLN. First, U is printed. The DO 400 loop prints V and W.

We illustrate USMPLX by using it to solve the problem of Example 7.1. The complete program consists of the main program of Figure B.72 and subroutines USMPLX, USTEP1, USTEP2, USTEP3, PHIPRM, and USOLN. The data file for Example 7.1 is shown in Figure B.78. The output file is shown in Figure B.79.

Figure B.78   Example 7.1 data file for USMPLX.

```
2 4
-4. -3. 0. 0.
0. 9.
1. 1. -1. 0.
2. 4. 0. -1.
1. 1. 3. 0.
4. 3. 6. 1.D20
2. 1.
1. 0.
4. -1.
2 4
T F T F
F F F F
```

UPPER - BOUNDED
SIMPLEX METHOD

				BASIC SET	
ITER	OBJECTIVE	K	L	1	2
0	-10.00000	3	1	2	4
1	-13.00000	1	1	3	4
2	-21.00000	2	1	1	4
3	-22.00000	1	0	2	4

OPTIMAL OBJECTIVE FUNCTION VALUE IS   -22.00000

OPTIMAL PRIMAL SOLUTION
X(1) =   4.0000000
X(2) =   2.0000000
X(3) =   6.0000000
X(4) =   7.0000000

OPTIMAL DUAL SOLUTION
EQUALITY CONSTRAINTS
U(1) =   3.0000000
U(2) =   0.0000000

BOUND CONSTRAINTS

	LOWER BOUND		UPPER BOUND
V(1) =	0.0000000	W(1) =	1.0000000
V(2) =	0.0000000	W(2) =	0.0000000
V(3) =	0.0000000	W(3) =	3.0000000
V(4) =	0.0000000	W(4) =	0.0000000

Figure B.79   Example 7.1 output from USMPLX.

B.5.2   The Generalized Upper-Bounded Simplex Method

The most important FORTRAN identifiers and their meanings are summarized in Table B.11. The main program (Figure B.80) reads the data for m, n, r, $(n_0, \ldots, n_r)$, c, b, A, I_B, and B^{-1} into M, N, R, (NJ(1), ..., NJ(R+1)), C, B, A, XB, IB, and BINV, respectively. A call is made to GUB which is responsible for solving the problem using the generalized upper-bounded simplex method. If an optimal solution is obtained, the optimal basic variables are returned in the first M (m) components of XB (($x'_B, x'_K)'$), and the optimal key variables are returned in the last R (r) components. The optimal objective function value is returned in OBJ. The output condition indicator JOUT is set to 1 in this case. If the problem is unbounded from below, JOUT is returned from GUB with value 2.

**TABLE B.11   Meaning of FORTRAN Identifiers
for the Generalized Upper-Bounded Simplex Method**

FORTRAN Identifier	Notation in Main Text
N	n
M	m
R	r
R1	$r + 1$
MR	$m + r$
A	A
B	b
C	c
IB	$I_{\hat{B}} = \{\beta_1, \ldots, \beta_m, \beta_{m+1}, \ldots, \beta_{m+r}\}$
BINV	B^{-1}
XB	$(x_B', x_K')'$
SB	$(s_B', s_K')'$
NJ(1),...,NJ(R+1)	$n_0, \ldots, n_r$
NU	ν
RHO	ρ
U1	u_1
V	v
K	k
ELL	l
OBJ	$c_B' x_B + c_K' x_K$

Figure B.80   Main program for the generalized upper-bounded simplex method.

```
IMPLICIT DOUBLE PRECISION(A-H,O-Z)
DIMENSION A(2,7),C(7),B(2),XB(4),SB(4),
1   DB(2),WORK(2),NJ(3),IB(4),U1(4),V(7),
2   BINV(2,2)
INTEGER R,R1
COMMON /UNITS/ IIN,IOUT
IIN = 37
IOUT = 38
READ(IIN,*) M,N,R
R1 = R + 1
MR = M + R
READ(IIN,*) (NJ(I),I=1,R1)
READ(IIN,*) (C(I),I=1,N)
READ(IIN,*) (B(I),I=1,M)
DO 100 I=1,M
```

```
100 READ(IIN,*) (A(I,J),J=1,N)
    READ(IIN,*) (XB(I),I=1,MR)
    READ(IIN,*) (IB(I),I=1,MR)
    DO 200 I=1,M
200 READ(IIN,*) (BINV(I,J),J=1,M)
    CALL GUB(A,B,C,BINV,XB,DB,U1,V,SB,WORK,NJ,
   1  IB,N,M,R,R1,MR,OBJ)
    CALL GSOLN(JOUT,XB,U1,V,C,A,IB,M,N,R,MR,
   1  OBJ)
    STOP
    END
```

Figure B.80   *(continued)*

SUBROUTINE GUB (Figure B.81) organizes the calls to GSTEP1, GSTEP2, and GSTEP3 which implement Steps 1, 2, and 3, respectively, of the generalized upper-bounded simplex method.   GUB initializes by setting the iteration counter ITER to zero and using the DO 100 loop to compute the initial objective function value and store it in OBJ. The call to GSTEP1 computes the index K (k) of the new basic variable and performs the optimality test. If the optimality test is satisfied, GSTEP1 returns with JOUT $= 1$. In this case, the final objective function value, column index k, and index set I_B are printed, and control transfers back to the main program. Otherwise, GSTEP2 is called.

Figure B.81   Subroutine GUB.

```
    SUBROUTINE GUB(A,B,C,BINV,XB,DB,U1,V,SB,
   1  WORK,NJ,IB,N,M,R,R1,MR,OBJ)
    IMPLICIT DOUBLE PRECISION(A-H,O-Z)
    INTEGER ELL,R,R1
    DIMENSION A(M,N),B(M),C(N),XB(MR),DB(M),
   1  NJ(R1),U1(M),IB(MR),SB(MR),BINV(M,M),
   2  WORK(M),V(N)
    COMMON /UNITS/ IIN,IOUT
C
C   INITIALIZE
    WRITE(IOUT,8000) (I,I=1,MR)
    ITER = 0
    SUM = 0.D0
    DO 100 I=1,MR
100 SUM = SUM + C(IB(I))*XB(I)
    OBJ = SUM
C
```

```
    200 CONTINUE
        CALL GSTEP1(A,C,BINV,DB,IB,U1,V,NJ,K,N,M,
      1    R,R1,MR,JOUT)
        IF(JOUT.EQ.1) WRITE(IOUT,8010) ITER,OBJ,K,
      1    0,(IB(I),I=1,MR)
        IF(JOUT.EQ.1) RETURN
  C
  C

        CALL GSTEP2(SB,XB,BINV,A,NJ,IB,K,ELL,M,N,
      1    R,R1,MR,JOUT)
        WRITE(IOUT,8010) ITER,OBJ,K,ELL,
      1    (IB(I),I=1,MR)
        IF(JOUT.EQ.2) RETURN
  C
  C

        CALL GSTEP3(A,B,C,NJ,IB,XB,BINV,ELL,K,OBJ,
      1    WORK,M,N,R,R1,MR,ITER)
        GO TO 200
  C
   8000 FORMAT(/,17X,'GENERALIZED UPPER-BOUNDED',/,
      1    23X,'SIMPLEX METHOD',//,40X,
      2    'BASIC SET',/,7X,'ITER',6X,'OBJECTIVE',4X,
      3    'K',4X,'L',2X,(T39,10I3))
   8010 FORMAT(6X,I5,F15.5,2I5,2X,(T39,10I3))
        END
```

Figure B.81 *(continued)*

GSTEP2 computes the search direction s_B (only the defining index k was determined in GSTEP1), the maximum feasible step size σ_B, and the associated index l. Note that SB $((s'_B, s'_K)')$ has dimension $m + r$. The first m components of SB are those of s_B, and the last r are those of s_K. Similarly, XB contains first the m basic variables and then the r key variables. Finally, IB is also $(m + r)$-dimensional and contains $\beta_1, \ldots, \beta_m, \beta_{m+1}, \ldots, \beta_{m+r}$. If GSTEP2 determines that the problem is unbounded from below, it sets JOUT = 2 and returns to GUB.

Control now passes to GSTEP3 which updates XB, IB, BINV, ITER, and OBJ (x_B, I_B, B^{-1}, j, and $c'_B x_B + c'_K x_K$, respectively). This completes one iteration and the next begins when control transfers back to statement 200.

Before discussing SUBROUTINE GSTEP1 it is convenient to introduce SUBROUTINE ELT (Figure B.82). The following situation arises throughout the statement of the generalized upper-bounded simplex method. Given an index i with $1 \leq i \leq n$, determine ν such that $i \in S_\nu$. ELT determines the ν by using the definition of $S_0, S_1, \ldots, S_r$. It first checks if $i \in S_0$. If not, it checks if $i \in S_1$, and so on.

```
      SUBROUTINE ELT(INDEX,NU,NJ,R1)
      INTEGER R,R1
      DIMENSION NJ(R1)
      R = R1 - 1
      IF(INDEX.GT.NJ(1)) GO TO 100
      NU = 0
      RETURN
  100 CONTINUE
      DO 200 I=1,R
      IF(INDEX.GT.NJ(I+1)) GO TO 200
      NU = I
      GO TO 300
  200 CONTINUE
  300 CONTINUE
      RETURN
      END
```

Figure B.82 Subroutine ELT.

SUBROUTINE GSTEP1 (Figure B.83) implements Step 1 of the generalized upper-bounded simplex method. The DO 100 loop computes d_i for $i = 1, \ldots, n$ and stores them in DB (d_B). The DO 200 and DO 300 loops construct (U1 =) $u_1 = -(B^{-1})'d_B$. The DO 800 and included loops construct v_i for all $i \notin I_B$. Each constraint i (I) is considered in turn. The DO 400 loop determines if $i \in I_{\hat{B}}$. If it is, constraint i is not considered further. Otherwise, ELT is used to determine NU (ν) such that $i \in S_\nu$. Next, v_i is determined by the DO 500 loop (if $\nu = 0$) or the DO 700 loop (if $\nu > 0$). The DO 500 loop determines the smallest relevant v_i, v_k. If $v_k \geq -10^{-5}$, the optimality criterion is satisfied and GSTEP1 returns with JOUT $= 1$. Otherwise, a return is made with JOUT $= 0$.

```
      SUBROUTINE  GSTEP1(A,C,BINV,DB,IB,U1,V,NJ,
     1   K,N,M,R,R1,MR,JOUT)
      IMPLICIT DOUBLE PRECISION(A-H,O-Z)
      INTEGER R,R1
      DIMENSION C(N),BINV(M,M),IB(MR),U1(M),
     1   NJ(R1),DB(M),V(N),A(M,N)
C
      JOUT = 0
      TOL = 1.D-5
      DO 10 I=1,N
   10 V(I) = 0.D0
C
C   COMPUTE D
      DO 100 I=1,M
      INDEX = IB(I)
      DB(I) = C(INDEX)
      CALL ELT(INDEX,NU,NJ,R1)
      IF(NU.GT.0) DB(I) = DB(I) - C(IB(M+NU))
  100 CONTINUE
C   COMPUTE U1
      DO 300 I=1,M
      SUM = 0.D0
      DO 200 J=1,M
  200 SUM = SUM + BINV(J,I)*DB(J)
      U1(I) = - SUM
  300 CONTINUE
C   COMPUTE V
      DO 800 I=1,N
      DO 400 J=1,MR
      IF(I.EQ.IB(J)) GO TO 800
  400 CONTINUE
      INDEX = I
      CALL ELT(INDEX,NU,NJ,R1)
      IF(NU.GT.0) GO TO 600
      SUM = C(I)
      DO 500 J=1,M
  500 SUM = SUM + A(J,I)*U1(J)
      V(I) = SUM
      GO TO 800
  600 CONTINUE
      INDEX = IB(M+NU)
      SUM = C(I) - C(INDEX)
      DO 700 J=1,M
```

Figure B.83   Subroutine GSTEP1.

```
700 SUM = SUM + (A(J,I)-A(J,INDEX))*U1(J)
    V(I) = SUM
800 CONTINUE
    K = 0
    SMALL = 1.D30
    DO 1000 I=1,N
    DO 900 J=1,MR
    IF(I.EQ.IB(J)) GO TO 1000
900 CONTINUE
    IF(V(I).GE.SMALL) GO TO 1000
    SMALL = V(I)
    K = I
1000 CONTINUE
    IF(SMALL.GE.-TOL) JOUT = 1
    RETURN
    END
```

Figure B.83   *(continued)*

SUBROUTINE GSTEP2 (Figure B.84) implements STEP2 of the generalized upper-bounded simplex method. It begins by using ELT to determine RHO (ρ) such that $k \in S_\rho$. As previously discussed, the composite vectors $(s_B', s_K')'$ and $(x_B', x_K')'$ are maintained in SB and XB, respectively. If $\rho = 0$, the components of SB are computed using the DO 200 and DO 400 loops as follows. The DO 200, and included DO 100 loop, sets $s_B = B^{-1}A_k$. The DO 300 loop computes s_K. If $\rho > 0$, the DO 700 and DO 900 loops perform the analogous computations for s_B and s_K, respectively.

Figure B.84   Subroutine GSTEP2.

```
    SUBROUTINE GSTEP2(SB,XB,BINV,A,NJ,IB,K,
   1   ELL,M,N,R,R1,MR,JOUT)
    IMPLICIT DOUBLE PRECISION(A-H,O-Z)
    INTEGER ELL,R,R1,RHO
    DIMENSION SB(MR),BINV(M,M),A(M,N),NJ(R1),
   1   XB(MR),IB(MR)
C
    TOL = 1.D-5
    CALL ELT(K,RHO,NJ,R1)
    IF(RHO.GT.0) GO TO 500
    DO 200 I=1,M
    SUM = 0.D0
    DO 100 J=1,M
100 SUM = SUM + BINV(I,J)*A(J,K)
```

```
 200  SB(I) = SUM
      DO 400 I=1,R
      SB(M+I) = 0.D0
      DO 300 NU=1,M
      INDEX = IB(NU)
      IF(((NJ(I)+1).LE.INDEX).AND.
     1  (INDEX.LE.NJ(I+1)))  SB(M+I) = SB(M+I)
     2                                     - SB(NU)
 300  CONTINUE
 400  CONTINUE
      GO TO 1000
 500  CONTINUE
      ICOL = IB(M+RHO)
      DO 700 I=1,M
      SUM = 0.D0
      DO 600 J=1,M
 600  SUM = SUM + BINV(I,J)*(A(J,K)-A(J,ICOL))
 700  SB(I) = SUM
      DO 900 I=1,R
      SB(M+I) = 0.D0
      IF(I.EQ.RHO)  SB(M+I) = 1.D0
      DO 800 NU=1,M
      INDEX = IB(NU)
      IF(((NJ(I)+1).LE.INDEX).AND.
     1  (INDEX.LE.NJ(I+1)))  SB(M+I) = SB(M+I)
     2                                     - SB(NU)
 800  CONTINUE
 900  CONTINUE
1000  CONTINUE
C
      ELL = 0
      SMALL = 1.D30
      DO 1100 I=1,MR
      IF(SB(I).LE.TOL) GO TO 1100
      RATIO = XB(I)/SB(I)
      IF(RATIO.GE.SMALL) GO TO 1100
      SMALL = RATIO
      ELL  = I
1100  CONTINUE
      IF(ELL.EQ.0) JOUT = 2
      RETURN
      END
```

Figure B.84 *(continued)*

The maximum feasible step size SIGB (σ_B) and the associated index ELL (l) are computed by the DO 1100 loop. If $(s_{\hat{B}})_i \leq 10^{-5}$ for $i = 1, \ldots, m + r$, the problem is unbounded from below. In this case, JOUT is set to 2 and control returns to GUB. Otherwise, a return to GUB is made with the critical index l.

SUBROUTINE GSTEP3 (Figure B.85) implements Step 3 of the generalized upper-bounded simplex method. It first decides which substep to use as follows. If $l \leq m$, control transfers to statement 200 (Step 3.1). Otherwise ELT is used to determine ν such that $\beta_l \in S_\nu$. The DO 100 loop determines if there is a ρ with $\beta_\rho \in S_\nu$. If there is, control transfers to statement 500 (Step 3.3). If there is no such ρ, control transfers to statement 400 (Step 3.2).

Figure B.85   Subroutine GSTEP3.

```
      SUBROUTINE GSTEP3(A,B,C,NJ,IB,XB,BINV,ELL,
     1   K,OBJ,WORK,M,N,R,R1,MR,ITER)
      IMPLICIT DOUBLE PRECISION(A-H,O-Z)
      INTEGER ELL,R,R1,RHO,BETAL
      DIMENSION A(M,N),B(M),NJ(R1),IB(MR),
     1   XB(MR),BINV(M,M),WORK(M),C(N)
C
      IF(ELL.LE.M) GO TO 200
      BETAL = IB(ELL)
      CALL ELT(BETAL,NU,NJ,R1)
      DO 100 I=1,M
      RHO = I
      IF(((NJ(NU)+1).LE.IB(I)).AND.
     1     (IB(I).LE.NJ(NU+1))) GO TO 500
  100 CONTINUE
      GO TO 400
C
C   STEP 3.1
  200 CONTINUE
      CALL ELT(K,NU,NJ,R)
      DO 300 I=1,M
      TEMP = A(I,K)
      IF(NU.GT.0) TEMP = TEMP - A(I,IB(M+NU))
  300 WORK(I) = TEMP
      CALL PHIPRM(BINV,WORK,K,M)
      IB(ELL) = K
      GO TO 1100
C
C   STEP 3.2
```

```
    400 CONTINUE
        IB(ELL) = K
        GO TO 1100
C
C   STEP 3.3
    500 CONTINUE
        DO 600 I=1,M
    600 WORK(I) = 0.D0
        DO 800 I=1,M
        INDEX = IB(I)
        CALL ELT(INDEX,ISET,NJ,R1)
        IF(ISET.NE.NU) GO TO 800
        DO 700 J=1,M
    700 WORK(J) = WORK(J) - BINV(I,J)
    800 CONTINUE
        DO 900 I=1,M
    900 BINV(RHO,I) = WORK(I)
        CALL ELT(K,NU,NJ,R)
        DO 1000 I=1,M
        TEMP = A(I,K)
        IF(NU.GT.0) TEMP = TEMP - A(I,IB(M+NU))
   1000 WORK(I) = TEMP
        CALL PHIPRM(BINV,WORK,RHO,M)
        IB(ELL) = IB(RHO)
        IB(RHO) = K
C
C   STEP 3.4
   1100 CONTINUE
        DO 1200 I=1,M
   1200 WORK(I) = B(I)
        DO 1400 I=1,R
        INDEX = IB(M+I)
        DO 1300 J=1,M
   1300 WORK(J) = WORK(J) - A(J,INDEX)
   1400 CONTINUE
        DO 1600 I=1,M
        SUM = 0.D0
        DO 1500 J=1,M
   1500 SUM = SUM + BINV(I,J)*WORK(J)
   1600 XB(I) = SUM
        DO 1800 I=1,R
        SUM = 1.D0
```

Figure B.85   *(continued)*

```
      DO 1700 NU=1,M
      INDEX = IB(NU)
      IF(((NJ(I)+1).LE.INDEX).AND.
    1      (INDEX.LE.NJ(I+1))) SUM = SUM -XB(NU)
 1700 CONTINUE
 1800 XB(M+I) = SUM
      SUM = 0.D0
      DO 1900 I=1,MR
 1900 SUM = SUM + C(IB(I))*XB(I)
      OBJ = SUM
      ITER = ITER + 1
      RETURN
      END
```

Figure B.85   *(continued)*

Statement 200 begins Step 3.1. ELT is used to determine ν such that $k \in S_\nu$. $\tilde{B}_\rho$ is constructed according to $\nu = 0$ or $\nu > 0$ and the result is stored in WORK. B^{-1} is then updated using PHIPRM (Figure B.40).

Statement 400 begins Step 3.2 which simply replaces β_l with k.

Statement 500 begins Step 3.3. The DO 800 loop computes the new column ρ of $(B^{*-1})'$. The DO 1000 loop constructs $\tilde{B}_\rho$ and places it in WORK. The new B^{-1} is then obtained from PHIPRM. The indices β_l and β_ρ are updated in order.

On completion, each of Steps 3.1, 3.2, and 3.3 transfer to Step 3.4, which begins at statement 1100. The DO 1200, 1300, and 1400 loops construct

$$b - \sum_{i=1}^{r} A_{\beta_{m+i}}$$

and store it in WORK. The DO 1500 and DO 1600 loops compute B^{-1} times the above quantity and store it in x_B. The DO 1500 through DO 1800 loops construct x_K. Finally, the DO 1900 loop computes the new objective function value and stores it in OBJ. ITER is increased by one and control returns to GUB.

SUBROUTINE GSOLN (Figure B.86) prints the optimal primal and dual solutions found by GUB. On entry to GSOLN, JOUT is checked to see if an optimal solution has been obtained. If not, an appropriate message is issued. Otherwise, the DO 300 loop prints the primal solution as follows. Each component I (i) is considered in turn. The DO 100 loop checks if $i \in I_{\hat{B}}$. If so, the appropriate component of XB is printed, annotated with "key variable" if that is the case. If $i \notin I_{\hat{B}}$, then variable i is nonbasic and has value zero.

```
      SUBROUTINE GSOLN(JOUT,XB,U1,V,C,A,IB,M,N,
     1   R,MR,OBJ)
      IMPLICIT DOUBLE PRECISION(A-H,O-Z)
      INTEGER R
      DIMENSION XB(MR),U1(M),C(N),A(M,N),IB(MR),
     1   V(N)
      COMMON /UNITS/ IIN,IOUT
C
      ZERO = 0.D0
      IF(JOUT.GE.2) GO TO 800
      WRITE(IOUT,8000) OBJ
      WRITE(IOUT,8010)
      DO 300 I=1,N
C IS X(I) BASIC?
      DO 100 J=1,MR
      INDEX = J
      IF(IB(J).EQ.I) GO TO 200
  100 CONTINUE
      WRITE(IOUT,8020) I,ZERO
      GO TO 300
  200 CONTINUE
      IF(INDEX.LE.M) WRITE(IOUT,8020) I,XB(INDEX)
      IF(INDEX.GT.M) WRITE(IOUT,8025) I,XB(INDEX)
  300 CONTINUE
C
C
      WRITE(IOUT,8030)
      WRITE(IOUT,8040) (I,U1(I),I=1,M)
C  DUAL VARIABLES FOR SUM EQUALITY CONSTRAINTS
      DO 500 I=1,R
      INDEX = IB(M+I)
      SUM = - C(INDEX)
      DO 400 J=1,M
  400 SUM = SUM - U1(J)*A(J,INDEX)
      JDEX = M + I
  500 WRITE(IOUT,8040) JDEX,SUM
C  DUAL VARIABLES FOR NONNEGATIVITY CONSTRAINTS
      WRITE(IOUT,8050)
      WRITE(IOUT,8060) (I,V(I),I=1,N)
      RETURN
  800 CONTINUE
      IF(JOUT.EQ.2) WRITE(IOUT,8070)
      IF(JOUT.EQ.3) WRITE(IOUT,8080)
      RETURN
C
```

Figure B.86 Subroutine GSOLN.

```
8000 FORMAT(//,6X,'OPTIMAL OBJECTIVE FUNCTION ',
    1   'VALUE IS',F12.5)
8010 FORMAT(//,17X,'OPTIMAL PRIMAL SOLUTION')
8020 FORMAT(18X,'X(',I2,') =',F14.7)
8025 FORMAT(4X,'KEY VARIABLE',2X,'X(',I2,') =',
    1   F14.7)
8030 FORMAT(//,18X,'OPTIMAL DUAL SOLUTION',
    1   /,18X,'EQUALITY CONSTRAINTS')
8040 FORMAT(18X,'U(',I2,') =',F14.7)
8050 FORMAT(/,15X,'NONNEGATIVITY CONSTRAINTS')
8060 FORMAT(18X,'V(',I2,') =',F14.7)
8070 FORMAT(//,6X,'PROBLEM IS UNBOUNDED FROM ',
    1   'BELOW')
8080 FORMAT(//,6X,'PROBLEM HAS NO FEASIBLE ',
    1   'SOLUTION')
     END
```

Figure B.86 *(continued)*

The dual variables for the first m equality constraints are explicitly available from GUB in U1. These are printed. The dual variables for the remaining r equality constraints, u_2, were derived in Section 7.2 following (7.20) as

$$(u_2)_i = -c_{\beta_{m+i}} - A'_{\beta_{m+i}}u_1, \quad i = 1, \ldots, r.$$

These are computed using the DO 400 and DO 500 loops.

The dual variables for the nonnegativity constraints are explicitly available from GUB in V (v). These are printed and control returns to the calling program.

We illustrate GUB by using it to solve the problem of Example 7.2. The complete program consists of the main program of Figure B.80 and subroutines GUB, GSTEP1, GSTEP2, GSTEP3, ELT, GSOLN, and PHIPRM. The data file for Example 7.2 is shown in Figure B.87. The output file is shown in Figure B.88.

Figure B.87   Example 7.2 data file for GUB.

```
2 7 2
2 5 7
-7. -4. 5. -3. -2. -2. -4.
16. 13.
3. 2. 18. 14. 8. 2. -4.
2. 1. 12. 8. 4. 2. 2.
0.25 0.5 0.75 0.5
4 6 3 7
0. -0.25
0.1666666666666 -0.1666666666666
```

GENERALIZED UPPER-BOUNDED
SIMPLEX METHOD

				BASIC SET			
ITER	OBJECTIVE	K	L	1	2	3	4
0	0.00000	1	3	4	6	3	7
1	-17.00000	5	3	1	6	4	7
2	-30.00000	2	2	1	6	5	7
3	-32.00000	6	0	1	2	5	7

OPTIMAL OBJECTIVE FUNCTION VALUE IS   -32.00000

OPTIMAL PRIMAL SOLUTION

	X(1) =	2.0000000
	X(2) =	3.0000000
	X(3) =	0.0000000
	X(4) =	0.0000000
KEY VARIABLE	X(5) =	1.0000000
	X(6) =	0.0000000
KEY VARIABLE	X(7) =	1.0000000

OPTIMAL DUAL SOLUTION
EQUALITY CONSTRAINTS

U(1) =   1.0000000
U(2) =   2.0000000
U(3) =   -14.0000000
U(4) =   4.0000000

NONNEGATIVITY CONSTRAINTS

V(1) =   0.0000000
V(2) =   0.0000000
V(3) =   33.0000000
V(4) =   13.0000000
V(5) =   0.0000000
V(6) =   8.0000000
V(7) =   0.0000000

Figure B.88   Example 7.2 output from GUB.

REFERENCES

American National Standards Institute, *American National Standard Programming Language FORTRAN, ANSI X3.9-1978*, American National Standards Institute, New York, 1978.

Bartels, R. H., and Golub, G. H., "The simplex method of linear programming using LU decomposition," *Communications of the Association for Computing Machinery*, 12 (1969), 266-268.

Benenson, P., and Glassey, C. R., "A linear economic model of fuel and energy use in the United States," Electric Power Research Institute, ES-115 (Volumes 1 and 2), Palo Alto, Calif., 1975A.

Benenson, P., and Glassey, C. R., "A quadratic programming problem analysis of energy in the United States economy," Electric Power Research Institute, ES-116, Palo Alto, Calif., 1975B.

Best, M. J., "A compact formulation of an elastoplastic analysis problem," *Journal of Optimization Theory and Applications*, 37 (1982), 343-353.

Bland, R. G. "New finite pivoting rules for the simplex method," *Mathematics of Operations Research*, 2 (1977), 103-107.

Chvátal, V., *Linear Programming*, W. H. Freeman, New York, 1983.

Cohn, M. Z., and Maier G. (editors), *Engineering Plasticity by Mathematical Programming*, Pergamon Press, New York, 1979.

Dantzig, G. B., *Linear Programming and Extensions*, Princeton University Press, Princeton, N.J., 1963.

Dantzig, G. B., Orden, A., and Wolfe, P., "The generalized simplex method for minimizing a linear form under linear inequality constraints," *Pacific Journal of Mathematics*, 5 (1955), 183-195.

Gass, S. I., *Linear Programming Methods and Applications*, McGraw-Hill, New York, 1975.

Maier, G., Grierson, D. E., and Best, M. J., "Mathematical programming methods for deformation analysis at plastic collapse," *Computers and Structures*, 7 (1977), 599-612.

Mangasarian, O. L., *Nonlinear Programming*, McGraw-Hill, New York, 1969.

Miller, M. D., *Elements of Graduation*, The Actuarial Society of America and the American Institute of Actuaries, 1946.

Sherman, J., and Morrison, W. J., "Adjustment of an inverse matrix corresponding to changes in the elements of a given column or a given row of the original matrix," *The Annals of Mathematical Statistics*, 20 (1949), 621.

Woodbury, M., "Inverting modified matrices," Memorandum Report 42, Statistical Research Group, Princeton University, Princeton, N.J., 1950.

INDEX